Practical Finite Element Modeling in Earth Science Using Matlab

Practical Finite Element Modeling in Earth Science Using Matlab

Guy Simpson
Department of Earth Science
University of Geneva
Switzerland

Registered Office
John Wiley & Sons Ltd, The Atrium, Southern Gate, Chichester, West Sussex, PO19 8SQ, UK

Editorial Offices
111 River Street, Hoboken, NJ 07030, USA
9600 Garsington Road, Oxford, OX4 2DQ, UK
The Atrium, Southern Gate, Chichester, West Sussex, PO19 8SQ, UK

For details of our global editorial offices, customer services, and more information about Wiley products visit us at www.wiley.com.

Wiley also publishes its books in a variety of electronic formats and by print-on-demand. Some content that appears in standard print versions of this book may not be available in other formats.

Library of Congress Cataloging-in-Publication data applied for

ISBN: 9781119248620

Cover design: Wiley
Cover image: ©Nongkran_ch/iStockphoto

Set in 10/12pt Warnock by SPi Global, Pondicherry, India
Printed in Singapore by C.O.S. Printers Pte Ltd

10 9 8 7 6 5 4 3 2 1

All difficult things have their origin in that which is easy, and great things in that which is small.

—Lao Tzu

Brief Contents

Contents

Preface

Over the past few decades, mathematical models have become an increasingly important tool for Earth scientists to understand and make predictions about how our planet functions and evolves through time and space. These models often consist of partial differential equations (PDEs) that are discretized with a numerical method and solved on a computer. The most commonly used discretization methods are the finite difference method (FDM), the finite volume method, the finite element method (FEM), the discrete element method, the boundary element method, and various spectral methods. In theory, each method provides the same solution to the original PDEs. However, in practice, certain methods are better suited to certain problems than others. Often, one method dominates within any given discipline and in the Earth sciences, the FDM is the most prevalent, due to its simplicity. Although the FEM is arguably better suited to many Earth science problems—especially those with complicated geometry and/or material behavior—Earth scientists have been hesitant to wholeheartedly embrace this technique because it is regarded as being complicated to implement compared to other schemes. However, this perceived difficulty largely reflects the fact that most textbooks on this method are written by engineers or mathematicians for engineers who have a different educational background as compared to Earth scientists and who are interested in different applications. This is unfortunate because the FEM is a remarkably flexible and powerful tool with enormous potential in the Earth sciences that is no more difficult (or even easier) to implement than other numerical schemes.

The text is intended for students and researchers in Earth science, attempting their first steps with the FEM. It provides a practical guide on how the FEM can easily be used to solve various Earth science problems using Matlab. For the most part, I assume that the equations governing the processes of interest are known. Emphasis is on how one actually computes the solution using the FEM. The text does not deal in detail with benchmarking and interpretation of model results or with application of the model results to specific case studies. To guide readers, many sample finite element Matlab scripts are presented. These scripts are written with an emphasis on simplicity and clarity, not on modularity and efficiency. However, once the underlying concepts are clear, these standalone codes could easily be modularized, optimized, and transported to other more efficient languages such as Fortran or C/C++. It is assumed that the reader is familiar with linear algebra and PDEs and has basic programming experience. Some of these aspects are covered briefly in Chapter 1 and Appendix B. The text is directed toward graduate students, advanced undergraduates, and Earth science researchers. While the text is intended to show how finite element programs can be written from scratch, it should also be of interest to researchers who use existing FEM software (e.g., ABAQUS and COMSOL) but who want to know more about what goes on within the "black box". Because the level of the material presented is quite basic compared to other finite element texts, readers are strongly advised to consult other more advanced books once they understand the basics and see how programs are constructed in practice. The following titles are good starting points: *The Finite Element Method* (three volumes, by Zienkiewicz and Taylor, 2000a, b, c) and *The Finite Element Method* (by Hughes, 2000).

This book is structured in two parts. Part I begins with a general introduction to numerical modeling before passing to a series of chapters that show how an archetypical mathematical model (i.e., the diffusion equation) is discretized with the FEM, programmed in Matlab, and solved on a computer. Each chapter builds on the previous one and introduces one (or more) key aspect of the FEM. Chapter 7 generalizes the concepts introduced in Chapters 1–6 by showing how the FEM can extended from single parabolic equations to systems of equations and also to elliptic and hyperbolic equations. By the end of Part I, the reader should understand the essentials of the FEM and be able to write their own Matlab scripts from scratch to solve the most commonly encountered PDEs in one dimension (1D), two dimension (2Ds), and three dimension (3Ds). This material can be taught as a one semester course on numerical modeling in Earth science for master and PhD students. Part II comprises a series of independent chapters, each of which focuses on how the FEM can be applied in different contexts in Earth science. The problems investigated are heat transfer in the crust, landscape evolution modeling, fluid flow in porous media, flexure, and deformation of Earth's crust. Although readers can choose to read only the chapter(s) that fall closest to their topic of interest, each chapter introduces a different aspect of the FEM, and so every chapter should be studied by readers interested in eventually mastering the technique.

I am very much indebted to Yuri Podladchikov who initially inspired my interest in the subject presented and who contributed in a major way to my understanding of the FEM. During the same period, I also benefited enormously from discussions and interaction with many other colleagues from the ETH in Zurich, including, in particular, Alan Thompson, Jamie Connolly, Neil Mancktelow, Jean-Pierre Burg, Steve Miller, Luigi Burlini, Katja Petrini, Taras Gerya, Stefan Schmalholz, Daniel Schmid, Boris Kaus, and Dave May. Line Probst from the University of Geneva is thanked for many comments and corrections on the text. I gratefully acknowledge financial support for my research from the Swiss National Science Foundation, the Department of Earth Science at the University of Geneva, and the Schmidheiny Foundation. Finally, I thank the encouragement and support from my parents, my wife Katja, and my children Luca, Lara, and Fabio.

Guy Simpson
Geneva

Symbols

Matlab variable	Dimensions	Description
b	[sdof, 1]	Global right-hand-side vector
bee	[nst, ntot]	Kinematic strain—displacement matrix
bcdof	[1 ndn]	Array containing list of equations where Dirichlet boundary conditions are imposed
ndn	scalar	Number of equations to which Dirichlet conditions are applied
bcval	[1 ndn]	Array containing fixed boundary values
bx0	[1 ny] in 2D	Arrays containing list of nodes on $x = 0$ boundary
bxn	[1 ny] in 2D	Arrays containing list of nodes on $x = lx$ boundary
coord	[nod, ndim]	Node coordinates for one element
dee	[nst, nst]	Material deformation matrix
der	[ndim nod]	Derivatives of shape functions in local coordinates, evaluated at an integration point
der_s	[ndim nod nip]	Derivatives of shape functions saved for all integration points
deriv	[ndim nod]	Derivatives of shape functions in global coordinates, evaluated at an integration point
detjac	scalar	Determinant of the Jacobian matrix
displ	[sdof, 1]	Global solution vector
displ0	[sdof, 1]	Global solution vector from previous time step
dt	scalar	Time increment
dx	scalar	Element dimension in the x-direction
F	[ntot, 1]	Element load vector
ff	[sdof, 1]	Global right-hand-side load vector
fun	[1 nod]	Shape functions, evaluated at an integration point
fun_s	[nip nod]	Shape functions saved for all integration points
g	[ntot 1]	Equation list for one element
g_coord	[ndim nn]	Node coordinates for entire mesh
g_g	[ntot nels]	Equation numbers of each element for entire mesh
g_num	[nod nels]	Node numbers of each element for entire mesh
invjac	[ndim, ndim]	Inverse of the Jacobian matrix
jac	[ndim, ndim]	Jacobian matrix
KM	[ntot, ntot]	Element stiffness matrix

Matlab variable	Dimensions	Description
`lhs`	`[sdof, sdof]`	Global stiffness matrix (sparse)
`lx`	scalar	Total domain length in the x-direction
`MM`	`[ntot, ntot]`	Element mass matrix
`ndim`	scalar	Number of spatial dimensions
`ndof`	scalar	Number of degrees of freedom (unknowns) per node
`nels`	scalar	Total number of elements in mesh
`nf`	`[ndof nn]`	Equation numbers on each node for entire mesh
`nip`	scalar	Number of Gauss–Legendre integration points for an element
`nod`	scalar	Number of nodes in one element
`nn`	scalar	Total number of nodes in mesh
`nst`	scalar	Number of stress (strain) components
`ntot`	scalar	Number of degrees of freedom per element (=`ndof*nod`)
`nx`	scalar	Number of mesh nodes in the x-direction
`num`	`[nod 1]`	Node list for one element
`phase`	`[1 nels]`	Phase index for each element
`points`	`[nip ndim]`	Positions of Gauss–Legendre integration points (in local coordinates)
`rhs`	`[sdof, sdof]`	Global right-hand-side matrix (sparse)
`sdof`	scalar	Total number of unknowns (equations) in the global system
`stress`	`[nst, 1]`	Stress components (e.g., σ_{xx}, σ_{zz}, and σ_{xz}) for a single integration point
`tensor`	`[nst nip nels]`	Stresses at each integration point for all elements
`wts`	`[1 nip]`	Weights of Gauss–Legendre integration points

About the Companion Website

This book is accompanied by a companion website:

www.wiley.com/go/simpson

This website includes:

- (.mat) files

Part I

The Finite Element Method with Matlab

Part I has two main purposes. The first purpose is to introduce readers to the Galerkin form of the finite element method (FEM), which is a numerical technique for discretizing partial differential equations (PDEs). The second purpose is to show practically how the resulting equations are programmed and solved on a computer using Matlab. Each chapter builds on the previous one and introduces one or more key concept. We will consider how the FEM is applied in one dimension (1D), two dimension (2D), and three dimension (3D), using a parabolic (diffusion) equation as an example. Chapter 7 generalizes the concepts and extends application of the FEM to systems of equations and to elliptic and hyperbolic problems.

Practical Finite Element Modeling in Earth Science Using Matlab, First Edition. Guy Simpson.

Companion website: www.wiley.com/go/simpson

1

Preliminaries

This chapter provides a short introduction to mathematical models consisting of systems of partial differential equations (PDEs) along with auxiliary (boundary and initial) conditions. We discuss how these equations can be solved, either exactly or using numerical methods. We also briefly consider the important issues of precision and stability of a numerical solution. A Matlab script is provided at the end of the chapter to enable readers to compare an analytical solution with its corresponding numerical approximation.

1.1 Mathematical Models

The application of the principles of conservation of mass, momentum, and energy combined with experimentally derived laws produces sets of PDEs that describe variations in velocity (or displacement), pressure, and temperature in space and time. When combined with boundary and initial conditions, these equations constitute mathematical models that can be solved and studied in a way somewhat similar to performing experiments in a laboratory. Whether a model is mathematical or analogue, both are simplified abstractions of reality. However, such models are useful because they can help isolate the influence of certain parameters or scenarios, study complex system interactions, and make predictions.

An example of a mathematical model that has important application in Earth science is the heat conduction equation, often more generally referred to as the diffusion equation. A complete derivation of the heat conduction equation is given in Appendix A. In one dimension (1D), the heat conduction equation can be written as follows:

$$\rho c \frac{\partial T}{\partial t} = k \frac{\partial^2 T}{\partial x^2} + A \tag{1.1}$$

Here, T is the temperature (K), x is the distance (m), t is the time (s), ρ is the rock density (kg m^{-3}), c is the specific heat capacity (J kg^{-1} K^{-1}), k is the thermal conductivity (W m^{-1} K^{-1}), and A is the rate of internal heat production per unit volume (J s^{-1} m^{-3}). In Equation 1.1, the temperature (the unknown) is referred to as the dependent variable, while t and x are known as independent variables. This type of equation is called a "partial differential equation" since the dependent variable depends on more than one independent variable. The physical parameters ρ, c, k, and A are assumed to be known. Obtaining a solution to the equation means finding the function $T(x, t)$ (i.e., T as a function of x and t) that satisfies the PDE.

Practical Finite Element Modeling in Earth Science Using Matlab, First Edition. Guy Simpson.

Companion website: www.wiley.com/go/simpson

More generally, the heat equation just introduced is also referred to mathematically as a parabolic (initial value) problem, which are typically of the form

$$\frac{\partial u}{\partial t} = \frac{\partial^2 u}{\partial x^2} + \frac{\partial^2 u}{\partial y^2} \tag{1.2}$$

Parabolic equations involve time-dependent behavior (term 1) and dissipation (terms 2 and 3), together which tend to smooth the solution with increasing time (at least for linear problems). Note that the signs in front of the second-order spatial derivatives on the right-hand side of 1.2 are necessarily positive; otherwise, the solutions grow rather than decay in time. Note also that the solution to parabolic equations depends on the initial value of the solution at $t = 0$ (hence the name initial value problems). The other two major classes of PDEs are elliptic (boundary value) problems and hyperbolic. Elliptic equations are typically associated with steady-state problems. Examples of elliptic equations are Poisson's equation,

$$\frac{\partial^2 u}{\partial x^2} + \frac{\partial^2 u}{\partial y^2} = f \tag{1.3}$$

and Laplace's equation

$$\frac{\partial^2 u}{\partial x^2} + \frac{\partial^2 u}{\partial y^2} = 0 \tag{1.4}$$

which govern incompressible potential flow and steady heat transfer. Note that these equations don't involve any time derivatives and so their solutions depend only on the boundary conditions (hence the name boundary value problems) and any source (if present). Hyperbolic (initial value) PDEs involve time-dependent wave-like solutions. An example of a hyperbolic equation is the first-order wave equation

$$\frac{\partial u}{\partial t} = \frac{\partial u}{\partial x} + \frac{\partial u}{\partial y} \tag{1.5}$$

Here, the first term accounts for time-dependent behavior, while the second and third terms translate the solution laterally without any dissipation. Hyperbolic equations are common in problems involving flowing fluids.

1.2 Boundary and Initial Conditions

The solution to a PDE is not unique until boundary conditions are imposed. Boundary conditions essentially "ground" the solution to some specific physical scenario. There are four types of boundary conditions commonly encountered in the solution of PDEs:

1) Dirichlet, where the value of the solution is imposed on the boundary
2) Neumann boundary conditions, where the derivative of the solution is imposed on the boundary
3) Robin boundary conditions, where one specifies some linear combination of the solution and its derivative
4) Periodic (or repeating) boundary conditions, where one assumes that the solution at one end of the model domain is equal to the solution at the other end

The number of boundary conditions necessary to determine a solution to a differential equation matches the order of the highest spatial derivative in the differential equation. For example, Equation 1.1 contains a second-order spatial derivative and so two boundary conditions must be specified, one at each end of the domain. The equation also contains a first-order time derivative, so we must also provide an initial condition. This means we must define the value of T everywhere (over the entire domain) at $t = 0$. Equation 1.5 has only first-order spatial derivatives and so requires only one boundary condition in each direction. In this case, the boundary condition should be imposed at the end of the domain from where flow arrives, whereas the downstream end should be left unconstrained so that the flow can exit uninhibited.

1.3 Analytical Solutions

For relatively simple PDEs and for certain boundary conditions and initial conditions, it may be possible to find an exact (also known as a closed-form or analytical) solution. As an example, consider 1D heat transfer about a steadily creeping, narrow, planar, vertical fault. In this case, Equation 1.1 needs to be solved with A given by (e.g., see McKenzie and Brune, 1972)

$$A = \delta(x_0)\tau v \tag{1.6}$$

where τ is the (constant) shear stress (Pa) resolved on the fault plane, v is the fault slip rate (m s^{-1}), and $\delta(x_0)$ is the Dirac function, that is, ∞ when $x_0 = 0$, 0 when $x \neq 0$, and $\int_{-\infty}^{\infty} \delta(x_0)dx_0 = 1$. The initial temperature at $t = 0$ is assumed to be 0°C everywhere. The spatial domain extends horizontally from $-\infty$ to $+\infty$ on either side of the fault located at $x = 0$. The boundary conditions are that the first derivative of the temperature vanishes at $\pm\infty$. The exact solution to Equation 1.1 combined with 1.6 can be written down directly using the Green's function for this equation (Morse and Feshbach ,1953, p. 981). The solution is

$$T(x,t) = \frac{\tau v}{\kappa\sqrt{\pi}\rho c}\left(|x|\sqrt{\pi}\,\mathrm{erf}\left(\frac{|x|}{2t\sqrt{\kappa t}}\right) + 2t\exp\left(\frac{-|x|^2}{4t\kappa}\right)\sqrt{\frac{\kappa}{t}} - |x|\sqrt{\pi}\right) \tag{1.7}$$

where κ $(= k/(\rho c))$ is the thermal diffusivity (m^2 s^{-1}) and erf is the error function ($\mathrm{erf}(x) = 2/\pi \int_0^x \exp(-t^2)dt$). This solution can easily be evaluated exactly at any desired x and t once the values for the various physical parameters are specified (as done in the following).

1.4 Numerical Solutions

Although it is normally always desirable to obtain exact solutions to the PDE(s) being investigated, in practice this is often not possible. A closed-form solution may either not exist, or it may be too complicated to be of practical use. This may be due a number of factors, including nonlinearities in the governing equation, variable material properties, complicated geometries or boundary conditions, and so on. In such cases, one must resort to numerical methods that provide an approximate solution to the governing differential equation(s). Today, with powerful computers, many complicated problems can be solved quickly using numerical techniques.

The process of obtaining a computational solution consists of two stages shown schematically in Figure 1.1. The first stage converts the continuous PDE and auxiliary conditions (boundary and initial conditions) into a discrete system of algebraic equations. This first stage is called "discretization" and may be performed using various methods (one of which is the finite element method or FEM). The second stage involves solving the system of algebraic equations (normally performed on a computer,

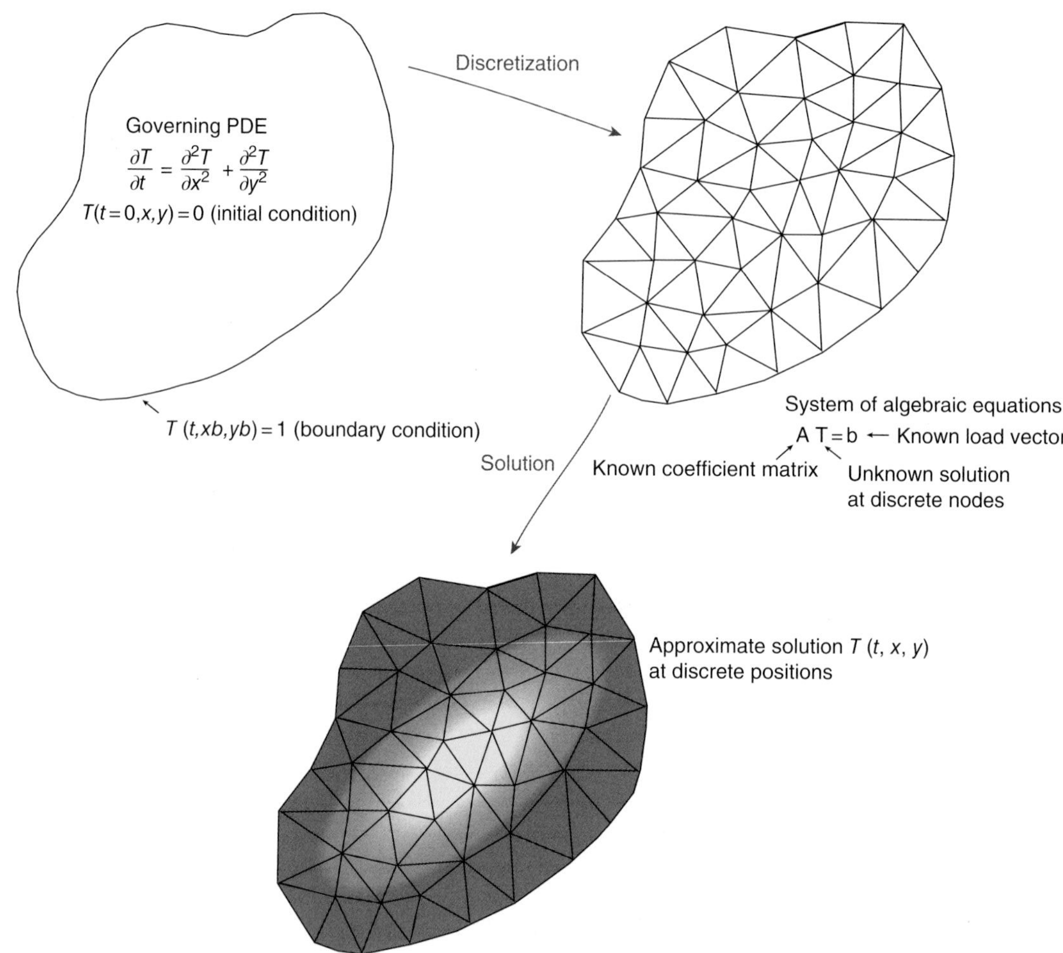

Figure 1.1 Major steps involved in obtaining a numerical solution to a PDE.

see Appendix B) to obtain an approximate solution to the original PDE. This second stage typically will involve some standard mathematical method such as Gaussian elimination.

Two important issues that must be considered when obtaining a numerical solution to PDEs are *error* and *stability*. All numerical methods introduce discretization errors, which in principle can be reduced by increasing the spatial and temporal resolution. This can be achieved by increasing the number of nodes (in time or space) where the solution is computed, or equivalently, by decreasing the spacing between nodes. It both cases, this should be performed without changing the total spatial or temporal extent of the model domain. Ideally, a numerical solution will converge to the exact solution as the resolution is increased. Even if an exact solution doesn't exist, one should always check that the numerical solution doesn't change significantly as the numerical resolution is changed, indicating that convergence has been achieved. Other errors may also arise (e.g., round-off errors produced during the solution of systems of linear algebraic equations), though these are usually small in comparison to discretization errors.

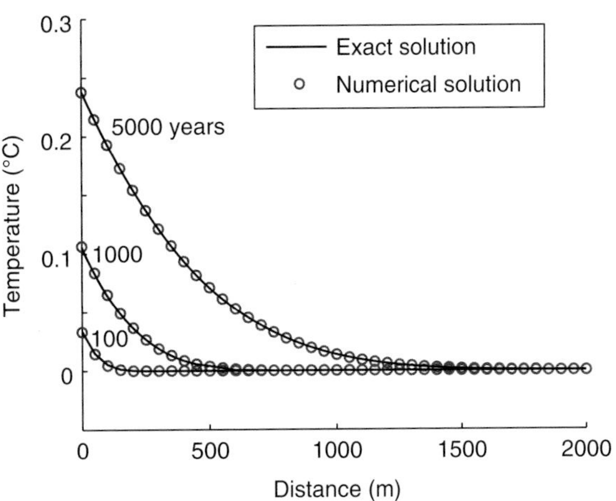

Figure 1.2 Comparison between the numerical (circles) and analytical solution (line, see Equation 1.7) for the temperature around a creeping fault after 100, 1000, and 5000 years (see Equations 1.1 and 1.6). The fault (located at $x = 0$) creep generates frictional heat that conducts outward into the surrounding rocks. Only the domain to the right of the fault is shown (the temperature is symmetrical about $x = 0$). The numerical solution is computed using the FEM. The Matlab script used to compute these results is provided at the end of the chapter.

The issue of stability concerns whether numerical errors, which are always present, decay or grow with time. A stable solution is one where the errors decay with time. An unstable solution is one where the errors grow with time, something that will eventually lead to large oscillations that have no physical meaning (i.e., they are simply numerical errors). Numerical methods are typically referred to as being either stable, unstable, or conditionally stable (meaning it can exhibit both behaviors depending on certain conditions). A stable method is an essential property of any numerical scheme. However, it is important to emphasize that a stable method can still be inaccurate. Thus, it is also important to assess the precision of a numerical solution. The best way this can be achieved is by directly comparing the numerical solution with an exact solution (as done in Figure 1.2). This approach is desirable because a numerical solution may look correct and may display the expected behavior but may be completely wrong (e.g., due to a simple erroneous factor in the numerical code). When an exact solution is not available, one should attempt to compare the numerical solution with other published numerical results.

Figure 1.2 shows a comparison between a numerical solution (computed using the FEM) to Equations 1.1 and 1.6, along with the analytical solution to the same equations (i.e., Equation 1.7). The Matlab code used to generate the figure is reproduced in Section 1.6. In this example, one sees that the agreement between the approximate and exact solutions is very good, indicating that the numerical solution is indeed a faithful representation of the original governing PDE. This comparison illustrates the importance of exact analytical solutions, since they provide a means of verifying the accuracy of a numerical solution.

1.5 Numerical Solution Methods

There are many different numerical methods available for solving PDEs, including the FEM, finite different method, finite volume method, boundary element method, discrete element method, and spectral methods. A comparison between three of these methods for a simple problem is given in Appendix C. In theory, each numerical method should provide the same (correct) solution to the original differential equation. However, in practice, some methods are better suited to certain types of equations and model geometries than others. Often the best approach is to choose the method that best suits the problem being investigated. This approach, however, requires considerable experience.

The text focuses entirely on the FEM that is widely regarded as being one of the most powerful, flexible, and robust techniques, while also being mathematically sound (Hughes, 2000; Zienkiewicz and Taylor, 2000a). The technique is slightly harder to learn than, for example, the finite difference technique. However, as you will see later, the effort invested in initially learning the FEM pays off later in the wide range of problems that the method is capable of solving. The FEM is especially well suited (though not restricted) to solving mechanical problems and problems that involve complex shapes. Another advantage of programming with the FEM is that the main structure of the code remains the same even for very different physical problems. Thus, once you learn this basic structure, you can easily modify it to solve various problems with minimal effort.

1.6 Matlab Script

The following is a Matlab script used to compute the numerical and analytical solution for Equations 1.1 and 1.6 presented in Figure 1.2. The reader is advised to reinspect the script after reading Chapters 2 and 3. Details of the time-stepping scheme are presented in Appendix E.

```
%-----------------------------------------------
% Program diffn1d.m
% 1-D FEM solution of diffusion equation
% and comparison with analytical solution
%-----------------------------------------------

clear % clear memory from current workspace
seconds_per_yr = 60*60*24*365; % number of seconds in 1 year

% physical parameters
lx    = 2000 ;       % length of spatial domain
Cp    = 1e3  ;       % rock heat capacity J/kg/K
rho   = 2700 ;       % rock density
K     = 3.3  ;       % bulk thermal conductivity W/m/K
kappa = K/(Cp*rho); % thermal diffusivity
tau   = 10e6 ;       % shear stress resolved on fault (Pa)
udot  = 10e-3/seconds_per_yr ;                  % fault slip rate (m/s)
dTdx  = (1/2)*tau*udot/(rho*Cp*kappa) ; % T gradient at fault

% numerical parameters
dt      = seconds_per_yr ;  % time step (s)
theta   = 1                 % time stepping parameter [0,1]
ntime   = 5000 ;            % number of time steps
nels    = 200 ;             % total number of elements
nod     = 2 ;               % number of nodes per element
nn      = nels+1            % total number of nodes
dx      = lx/nels ;         % element size
g_coord = [0:dx:lx] ;       % spatial domain (1-D mesh)

% explicit time stepping options
lumped_explicit = 'N';
if theta==0 % if fully explicit
    lumped_explicit = input('Would you like to lump the mass matrix? Y/N [N]:','s');
    if isempty(lumped_explicit)
      lumped_explicit = 'N';
    end
end
```

```
%  define boundary conditions
 bcdof = [   nn ]   ; % boundary nodes
 bcval = [   0 ]    ; % boundary values

% define connectivity and equation numbering
g_num       = zeros(nod,nels) ;
g_num(1,:) = [1:nn-1]   ;
g_num(2,:) = [2:nn]     ;

% initialise matrices and vectors
b       = zeros(nn,1);            % system rhs vector
lhs     = sparse(nn,nn);          % system lhs matrix
rhs     = sparse(nn,nn);          % system rhs matrix
displ   = zeros(nn,1);            % initial temperature (°C)
lumped_diag = zeros(nn,1) ;       % storage for lumped diagonal
%----------------------------------------------
% matrix assembly
%----------------------------------------------
 for iel=1:nels % loop over all elements
   num = g_num(:,iel)   ;              % retrieve equation number
   dx  = abs(diff(g_coord(num))) ; % length of element
   MM  = dx*[1/3  1/6 ; 1/6 1/3 ] ;% mass matrix
   KM  = [kappa/dx -kappa/dx ; -kappa/dx kappa/dx ];% diffn matrix
   if lumped_explicit=='N'
     lhs(num,num) = lhs(num,num) + MM/dt + theta*KM   ; % assemble lhs
     rhs(num,num) = rhs(num,num) + MM/dt - (1-theta)*KM   ; % assemble rhs
   else
     lumped_diag(num) = lumped_diag(num) + sum(MM)'/dt ; % lumped diagonal
     rhs(num,num) = rhs(num,num) + diag(sum(MM))/dt - (1-theta)*KM   ; % assemble rhs
    end
 end    % end of element loop
%----------------------------------------------

% time loop
t   = 0 ; % time
k   = 1 ; % counter
ii = [100 1000 5000]; % array used for plotting
for n=1:ntime
   n
   t   = t + dt ;                      % compute time
   b   = rhs*displ   ;                 % form rhs load vector

   % impose boundary conditions
   lhs(bcdof,:) = 0 ;                  % zero the relevent equations
   tmp = spdiags(lhs,0) ;              % store diagonal
   tmp(bcdof)=1 ;                      % place 1 on stored-diagonal
   lhs=spdiags(tmp,0,lhs);             % reinsert diagonal
   b(bcdof) = bcval ;                  % set rhs

   b(1) = b(1) + dTdx*kappa ;          % add heat flux at left boundary

   if lumped_explicit=='N'
       displ = lhs \ b ; % solve system of equations
   else
       displ = b./lumped_diag ; % fully explicit, diagonalised solution
   end

   % evaluate analytical solution
```

```
    x      = g_coord ;
    term1  = abs(x).*sqrt(pi).*erf((1/2)*abs(x)./(sqrt(kappa./t).*t));
    term2  = 2*t.*exp(-(1/4)*abs(x).^2./(kappa*t)).*sqrt(kappa./t);
    term3  = abs(x)*sqrt(pi) ;
    term4  = (tau*udot)/(2*kappa*sqrt(pi)*rho*Cp);
    Texact = term4*(term1+term2-term3);

  % plotting
  if mod(n,ii(k))==0
    k = k+1;
    hold on
    figure(1)
    plot(g_coord,displ,'o-',g_coord,Texact,'r')
    title(['Time (kyr) = ', num2str(t/seconds_per_yr/1e3)])
    xlabel('Distance away from fault (m)')
    ylabel('Temperature (°C)')
    drawnow
    t/seconds_per_yr
    pause
  end
  hold off
end % end of time loop

%-----------------------------------------------
```

1.7 Exercises

1) Depending on the values of the constant coefficients A, B, C, D, E, F, and G, the following PDE

$$A\frac{\partial^2 u}{\partial x^2} + B\frac{\partial^2 u}{\partial x \partial y} + C\frac{\partial^2 u}{\partial y^2} + D\frac{\partial u}{\partial x} + E\frac{\partial u}{\partial y} + Fu + G = 0$$

can be classified according to the following scheme (Garabedian, 1964):

Elliptic:	$B^2 - 4AC < 0$
Parabolic:	$B^2 - 4AC = 0$
Hyperbolic:	$B^2 - 4AC > 0$

Using this scheme, classify the following PDEs:

$$\frac{\partial T}{\partial t} = \kappa\frac{\partial^2 T}{\partial x^2}$$

$$\frac{1}{c^2}\frac{\partial^2 T}{\partial t^2} = \frac{\partial^2 T}{\partial x^2}$$

$$\frac{\partial^2 T}{\partial x^2} + \frac{\partial^2 T}{\partial y^2} = 0$$

$$\frac{\partial T}{\partial t} + u\frac{\partial T}{\partial x} = 0$$

$$\frac{\partial T}{\partial t} + u\frac{\partial T}{\partial x} = \kappa\frac{\partial^2 T}{\partial x^2}$$

2) Match the following exact solutions

$$\phi = \sin(2\pi x - t)$$
$$\phi = \sin \pi x \cos \pi t$$
$$\phi = \sin(\pi x)e^{-\pi y}$$
$$\phi = \sin(\pi x)e^{-\pi^2 t}$$

with their corresponding PDEs:

$$\frac{\partial^2 \phi}{\partial t^2} = \frac{\partial^2 \phi}{\partial x^2}$$
$$\frac{\partial \phi}{\partial t} = \frac{\partial^2 \phi}{\partial x^2}$$
$$\frac{\partial \phi}{\partial t} + \frac{\partial \phi}{\partial x} = 0$$
$$\frac{\partial^2 \phi}{\partial x^2} + \frac{\partial^2 \phi}{\partial y^2} = 0$$

In each case, explain your reasoning. Write a matlab script to evaluate and compare the analyical solutions. In each case, assume the independent variables extend from 0 to 1. Make a list of the most important characteristics of each.

3) In 1D, the equation governing diffusion-advection of a passive scalar $T(x, t)$ is

$$\frac{\partial T}{\partial t} + u\frac{\partial T}{\partial x} = \alpha\frac{\partial^2 T}{\partial x^2}$$

where u is the flow velocity and α is the diffusivity. For the conditions

$$T(x < 0, t = 0) = 1$$
$$T(x > 0, t = 0) = 0$$
$$T(x = -2, t) = 1$$
$$T(x = 2, t) = 0$$

the exact solution (valid before the influence of the step reaches the boundary) is

$$T(x,t) = \frac{1}{2} - \frac{2}{\pi}\sum_{k=1}^{N} \sin\left((2k-1)\frac{\pi(x-ut)}{L}\right)\frac{\exp\left(-\alpha(2k-1)^2\pi^2 t/L^2\right)}{2k-1}$$

Write a Matlab script to evaluate the exact solution for $L = 4$, $u = 1$, $\alpha = 0.1$, and t between 0 and 1. Study the effect of changing α and u.

4) Use the script listed in Section 1.6 to study the accuracy and stability of the numerical solution (with respect to the exact solution). Try a range of different values for `dx` (modified by varying the number of finite elements, `nels`) and `dt`. The parameter θ can be used to study different time-stepping schemes, ranging from fully implicit ($\theta = 1$) to fully explicit ($\theta = 0$) (see Appendix E). In addition, the explicit case can be solved more efficiently by diagonalizing the system stiffness matrix. Warning: For the boundary conditions considered, the finite element solution can only match the analytical solution until the temperature perturbation influences the lateral boundary. To study longer times, one would need to move the lateral boundary further away (i.e., increase `lx`) to avoid boundary effects.

Suggested Reading

E. A. Bender, *An Introduction to Mathematical Modeling,* Dover Publications Inc., New York, 2000.

C. A. J. Fletcher, *Computational Techniques for Fluid Dynamics,* Springer, Berlin, 2000.

A. Ismail-Zadeh and P. Tackley, *Computational Methods for Geodynamics,* Cambridge University Press, Cambridge, 2000.

R. Slinderland and L. Kump, *Mathematical Modeling of Earth's Dynamical Systems,* Princeton University Press, Princeton, NJ, 2011.

G. Strang, *Introduction to Applied Mathematics,* Wellesley-Cambridge Press, Cambridge, MA, 1986.

2

Beginning with the Finite Element Method

This chapter provides a detailed, step-by-step account of how the finite element method (FEM) is applied to solve the diffusion equation in one dimension 1D. The chapter derives and evaluates the element matrices and local load vector, shows how they are assembled into their global equivalents for a small mesh, and illustrates how Dirichlet boundary conditions are imposed in the final system of algebraic equations. Solution of the system is performed in Chapter 3 using Matlab.

The FEM is a technique for solving partial differential equations (PDEs). As with other common numerical techniques for solving PDEs, this involves converting the continuous PDE(s) and auxiliary (boundary and initial) conditions into a discrete system of algebraic equations (Figure 1.1). Usually, only the spatial derivatives are discretized with the FEM, whereas the finite difference method (FDM) is used to discretize time derivatives. Spatial discretization is carried out locally over small regions of simple but arbitrary-shaped (finite) elements. This discretization process results in matrix equations relating the loads (input) at specified points in the element (called nodes) to the solution (output) at these same points. In order to solve the equations over larger regions, one sums node by node the matrix equations for the smaller subregions (elements) resulting in a global matrix equation. This system of equations can then be solved simultaneously by standard linear algebra techniques to yield the solution at nodes. This last step completes the numerical solution of the governing PDE.

2.1 The Governing PDE

The various steps involved in performing the FEM are best illustrated with a simple example. Consider the parabolic PDE that governs transient heat conduction in 1D

$$\frac{\partial T}{\partial t} = \kappa \frac{\partial^2 T}{\partial x^2} + H \tag{2.1}$$

with the initial conditions

$$T(x, t = 0) = 0 \quad \forall\, x \in [0; lx] \tag{2.2}$$

and the boundary conditions

$$T(x = 0, t) = 0 \text{ and } T(x = lx, t) = 0 \tag{2.3}$$

Here $T(x, t)$ is the unknown temperature, κ is the thermal diffusivity ($m^2\ s^{-1}$), H is a constant heat source ($K\ s^{-1}$), and the spatial domain extends from 0 to lx (Figure 2.1). Equation 2.1 is generally known as a diffusion equation with a source term (derived in Appendix A). Variants of this equation appear in numerous different contexts in Earth science, including conductive heat transfer, chemical

Practical Finite Element Modeling in Earth Science Using Matlab, First Edition. Guy Simpson.

Companion website: www.wiley.com/go/simpson

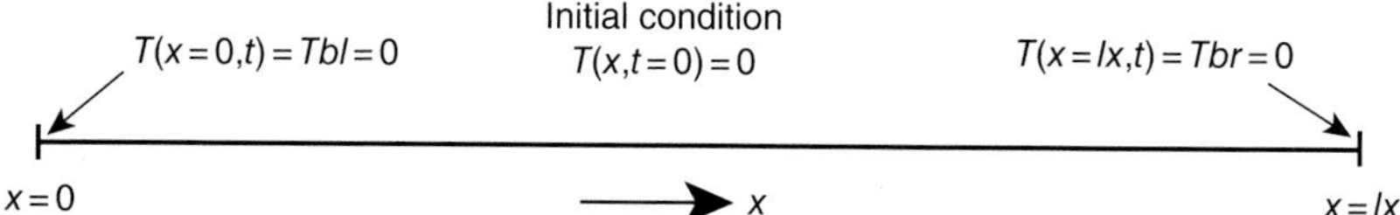

Figure 2.1 Illustration of the domain, boundary and initial conditions for the model of transient heat conduction in 1D.

diffusion, fluid flow in porous media, and erosion. We are interested in computing the temperature function $T(x, t)$ that satisfies Equation 2.1 along with the auxiliary conditions using the FEM.

2.2 Approximating the Continuous Variable

The starting point for the FEM is to approximate (or interpolate) the continuous variable $T(x, t)$ in the original governing equation in terms of nodal variables T_i using simple functions called shape or trial (or basis) functions, denoted by N_i. This procedure is also generally known as discretization. To achieve this, we begin by defining a single, 1D finite element (i.e., a line) that contains two nodes, one at each end (Figure 2.2a). For the moment, we don't make any explicit link between the size and location of this element and the model domain illustrated in Figure 2.1. Using the shape functions, the approximation of the continuous variable can be written as

$$T \simeq \sum_{i=1}^{n} N_i(x) T_i \tag{2.4}$$

where T_i are the unknown nodal temperatures, N_i are the shape functions, and the summation is carried out over all nodes of the element (in this case 2). To make this more explicit, we can also rewrite the equation as

$$T \simeq N_1 T_1 + N_2 T_2 \tag{2.5}$$

or using matrix notation

$$T \simeq [N_1\, N_2] \begin{Bmatrix} T_1 \\ T_2 \end{Bmatrix} = [N]\{T\} \tag{2.6}$$

Note that **N** is defined as a row vector (denoted using square brackets), whereas **T** is a column vector (using curly braces). In the following, square brackets are also used for matrices.

The shape functions are typically chosen to be low order, piecewise polynomials. For the two-node element chosen here, the shape functions are two-parameter polynomials (i.e., linear functions),

$$N_1 = 1 - \frac{x}{L}, \quad N_2 = \frac{x}{L} \tag{2.7}$$

where L is the length of the element and x is the spatial variable that varies from 0 at node 1 to L at node 2 (Figure 2.2b). The shape functions have the following important properties:

- $N_1 = 1$ at node 1, while $N_1 = 0$ at node 2.
- $N_2 = 0$ at node 1, while $N_2 = 1$ at node 2.
- $N_1 + N_2 = 1$ for all x in an element.
- The functions are only defined locally (i.e., they connect adjacent nodes within a single element).

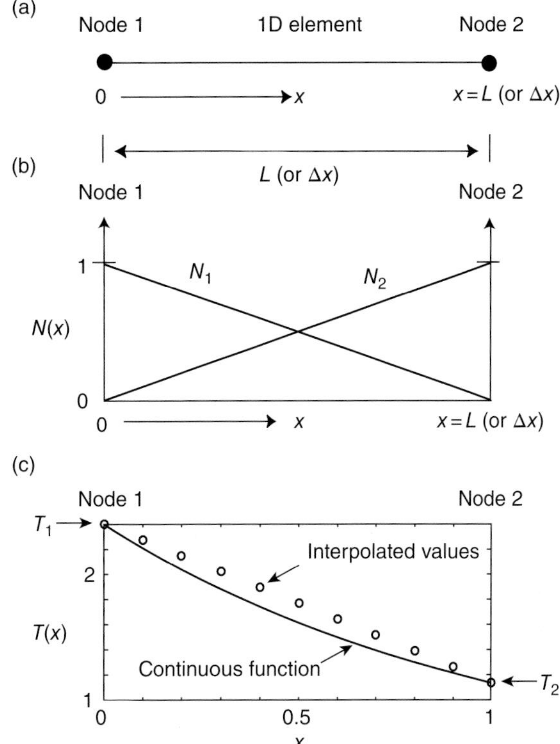

Figure 2.2 A single 1D finite element of length L (later denoted Δx) containing two nodes (a). Plot of the two linear shape functions (N_1 and N_2) for the two-node element (b). Approximation (or interpolation) of a continuous nonlinear function (solid line) using the two linear shape functions and the nodal values (T_1 and T_2), evaluated at different positions across the element (c).

Figure 2.2c shows an example of approximating a continuous variable T using two linear shape functions and the nodal values T_1 and T_2. Note that, in this case, a linear approximation does not provide an accurate representation of the true continuous function, which is nonlinear. However, in practice, far better accuracy can be achieved by decreasing the element size.

Now that the approximate solution is completely defined (Equation 2.6), it can be substituted into the governing PDE (Equation 2.1), resulting in

$$\frac{\partial}{\partial t}[N_1\,N_2]\left\{\begin{array}{c}T_1\\T_2\end{array}\right\} - \kappa\frac{\partial^2}{\partial x^2}[N_1\,N_2]\left\{\begin{array}{c}T_1\\T_2\end{array}\right\} - H = R \tag{2.8}$$

where R (the residual) is a measure of the error introduced during discretization (since Equation 2.6 is an approximate relation). If the trial function were the exact solution, the residual would be zero. The problem now reduces to finding values for T_1 and T_2 (the nodal temperatures) such that the residual is minimized. However, this is a single equation with two unknowns, which cannot be solved exactly. To find a solution, we have to generate from 2.8 a system of equations where the number of equations equals the number of unknowns.

2.3 Minimizing the Residual

In the FEM, the unknown nodal variables (i.e., T_i) are determined by minimizing the residual in a weighted average (integral) sense (Finlayson, 1972). To achieve this step practically, one must multiply (or "weight") the residual in Equation 2.8 by a set of weighting functions (each in turn), integrate over the element, and equate to zero, that is,

$$\int_0^L R(x)\, W_i(x)\, dx = 0 \qquad i = 1, 2 \tag{2.9}$$

where $R(x)$ is the residual and the number of weighting functions $W_i(x)$ is exactly equal to the number of unknown coefficients (T_i) in Equation 2.4 (i.e., 2). This is known as the method of weighted residuals. Several different submethods (e.g., collocation, subdomain, least squares, and Galerkin) can be used to achieve this process, the difference between which depends on the choice of the weighting functions (see Chapter 13). In the text, we consider almost always the Galerkin FEM, which assumes that the weighting functions **W** are identical to the shape functions **N** (see Equation 2.6). In what follows, we use the same symbols for both shape and weighting functions, without distinguishing the two. By performing the steps just described in Equation 2.8, one obtains

$$\int_0^L \begin{Bmatrix} N_1 \\ N_2 \end{Bmatrix} \frac{\partial}{\partial t} [N_1\, N_2] \begin{Bmatrix} T_1 \\ T_2 \end{Bmatrix} dx - \int_0^L \begin{Bmatrix} N_1 \\ N_2 \end{Bmatrix} \kappa \frac{\partial^2}{\partial x^2} [N_1\, N_2] \begin{Bmatrix} T_1 \\ T_2 \end{Bmatrix} dx$$
$$- \int_0^L \begin{Bmatrix} N_1 \\ N_2 \end{Bmatrix} H\, dx = \begin{Bmatrix} 0 \\ 0 \end{Bmatrix} \tag{2.10}$$

which can be expanded to give

$$\int_0^L N_1 N_1 \frac{\partial T_1}{\partial t} dx + \int_0^L N_1 N_2 \frac{\partial T_2}{\partial t} dx - \int_0^L \kappa N_1 \frac{\partial^2 N_1}{\partial x^2} T_1 dx$$
$$- \int_0^L \kappa N_1 \frac{\partial^2 N_2}{\partial x^2} T_2 dx - \int_0^L N_1 H dx = 0 \tag{2.11}$$

$$\int_0^L N_2 N_1 \frac{\partial T_1}{\partial t} dx + \int_0^L N_2 N_2 \frac{\partial T_2}{\partial t} dx - \int_0^L \kappa N_2 \frac{\partial^2 N_1}{\partial x^2} T_1 dx$$
$$- \int_0^L \kappa N_2 \frac{\partial^2 N_2}{\partial x^2} T_2 dx - \int_0^L N_2 H dx = 0 \tag{2.12}$$

Referring to Equation 2.10 (or equivalently to Equations 2.11 and 2.12), several points are noteworthy. First, there are now two equations with two unknowns, T_1 and T_2, as desired. Second, the three terms on the left-hand side of 2.10 come directly from the three respective terms in the governing PDE (see Equation 2.1). Third, because the shape functions **N** don't depend on t, the first term can be simplified by moving the shape functions outside the time differential. Fourth, the nodal unknowns (i.e., T_1 and T_2) don't depend on x and so can be moved outside the integrals appearing in terms 1 and 2. Finally, note that in this example where the shape functions are assumed to be linear, double differentiation of these functions in term 2 would cause them to vanish. This would result in the physically unrealistic situation that the conduction term completely disappears. This possibility can be avoided by applying *integration by parts* to term 2. The application of integration by parts to a shorthand version of term 2 results in (Appendix D) the following:

$$\int_0^L N_i \frac{\partial^2 N_j}{\partial x^2} dx = \left[\frac{\partial N_j}{\partial x} N_i \right]_0^L - \int_0^L \frac{\partial N_i}{\partial x} \frac{\partial N_j}{\partial x} dx \tag{2.13}$$

Thus, the original integral involving twice differentiable shape functions is converted into a boundary term (the first term on the right of 2.13) and an integral involving only singly differentiable shape functions (the second term on the right-hand side of the equation). Normally, only the second of these terms is retained, while the first term can be neglected since it cancels at the boundaries between adjacent finite elements. This term doesn't cancel at real boundaries, but in this case it is normally overridden by applying Dirichlet conditions. Later in the text, we will show how this neglected boundary term can actually be used to implement flux boundary conditions (see Section 7.1).

Taking into account these various points, one can rewrite Equation 2.10 as

$$\int_0^L \begin{bmatrix} N_1N_1 & N_1N_2 \\ N_2N_1 & N_2N_2 \end{bmatrix} dx \frac{\partial}{\partial t} \begin{Bmatrix} T_1 \\ T_2 \end{Bmatrix} + \int_0^L \kappa \begin{bmatrix} \frac{\partial N_1}{\partial x}\frac{\partial N_1}{\partial x} & \frac{\partial N_1}{\partial x}\frac{\partial N_2}{\partial x} \\ \frac{\partial N_2}{\partial x}\frac{\partial N_1}{\partial x} & \frac{\partial N_2}{\partial x}\frac{\partial N_2}{\partial x} \end{bmatrix} dx \begin{Bmatrix} T_1 \\ T_2 \end{Bmatrix} - \int_0^L H \begin{Bmatrix} N_1 \\ N_2 \end{Bmatrix} dx = \begin{Bmatrix} 0 \\ 0 \end{Bmatrix} \tag{2.14}$$

or using matrix notation

$$[\mathbf{MM}]\frac{\partial}{\partial t}\{\mathbf{T}\} + [\mathbf{KM}]\,\{\mathbf{T}\} = \{\mathbf{F}\} \tag{2.15}$$

where

$$\mathbf{MM} = \int_0^L \mathbf{N}^T\mathbf{N}\,dx = \int_0^L \begin{bmatrix} N_1N_1 & N_1N_2 \\ N_2N_1 & N_2N_2 \end{bmatrix} dx \tag{2.16}$$

$$\mathbf{KM} = \int_0^L \kappa \frac{\partial \mathbf{N}^T}{\partial x}\frac{\partial \mathbf{N}}{\partial x} dx = \int_0^L \kappa \begin{bmatrix} \frac{\partial N_1}{\partial x}\frac{\partial N_1}{\partial x} & \frac{\partial N_1}{\partial x}\frac{\partial N_2}{\partial x} \\ \frac{\partial N_2}{\partial x}\frac{\partial N_1}{\partial x} & \frac{\partial N_2}{\partial x}\frac{\partial N_2}{\partial x} \end{bmatrix} dx \tag{2.17}$$

$$\mathbf{F} = \int_0^L H\,\mathbf{N}^T dx = \int_0^L H \begin{Bmatrix} N_1 \\ N_2 \end{Bmatrix} dx \tag{2.18}$$

and

$$\mathbf{T} = \begin{Bmatrix} T_1 \\ T_2 \end{Bmatrix} \tag{2.19}$$

At this stage, the governing PDE has been converted to its so-called weak (or integral) form. Equation 2.15 shows that the FEM has resulted in a series of element matrices (i.e., **MM** and **KM**) that connect the unknown variable (i.e., the temperature) defined at the element nodes (i.e., **T**) to loads (i.e., **F**) on the same nodes.

2.4 Evaluating the Element Matrices

For simple shape functions and constant material properties, the integrals appearing in **MM**, **KM**, and **F** can be evaluated exactly. For example, referring to the shape functions defined in 2.7, the second term on first line of **MM** (Equation 2.16) becomes

$$\mathbf{MM}_{12} = \int_0^L N_1N_2\,dx = \int_0^L \left[\left(1-\frac{x}{L}\right)\frac{x}{L}\right] dx = \frac{L}{6} \tag{2.20}$$

The integrals in **KM** (Equation 2.17) involve the derivatives of the shape functions, which for the linear shape functions defined in Equation 2.7 are

$$\frac{\partial N_1}{\partial x} = -\frac{1}{L}, \quad \frac{\partial N_2}{\partial x} = \frac{1}{L} \tag{2.21}$$

Therefore, for example, the second term on the second line of **KM** becomes

$$\mathbf{KM}_{22} = \kappa \int_0^L \frac{\partial N_2}{\partial x}\frac{\partial N_2}{\partial x}\,dx = \kappa \int_0^L \frac{1}{L}\frac{1}{L}\,dx = \frac{\kappa}{L} \tag{2.22}$$

Evaluating the other integrals in a similar manner and using Δx instead of L for the element length hereafter results in

$$\mathbf{MM} = \begin{bmatrix} \frac{\Delta x}{3} & \frac{\Delta x}{6} \\ \frac{\Delta x}{6} & \frac{\Delta x}{3} \end{bmatrix} \tag{2.23}$$

$$\mathbf{KM} = \kappa \begin{bmatrix} \frac{1}{\Delta x} & -\frac{1}{\Delta x} \\ -\frac{1}{\Delta x} & \frac{1}{\Delta x} \end{bmatrix} \tag{2.24}$$

and

$$\mathbf{F} = H \begin{Bmatrix} \frac{\Delta x}{2} \\ \frac{\Delta x}{2} \end{Bmatrix} \tag{2.25}$$

The matrix **MM** is typically referring to as an element mass matrix, **KM** is the element stiffness matrix, and **F** is the element load vector.

2.5 Time Discretization

Equation 2.15 still contains a continuous time derivative that must be discretized before it can be solved numerically. Although it is also possible to perform finite element discretization on time derivatives (e.g., see Zienkiewicz and Taylor (2000a)), the standard approach (and the one followed here) is to replace time derivatives using the FDM. Assuming a fully implicit (finite difference) time discretization (see Appendix E), one can rewrite 2.15 as

$$\mathbf{MM}\frac{\mathbf{T}^{n+1} - \mathbf{T}^{n}}{\Delta t} + \mathbf{KM}\ \mathbf{T}^{n+1} = \mathbf{F} \tag{2.26}$$

where Δt is the time interval between n and $n+1$, $\mathbf{T}^{n+1}$ is the future temperature at the nodes (i.e., the unknowns), and $\mathbf{T}^{n}$ is the vector of old (i.e., known) temperatures. Note that the superscripts involving n refer to the time level, not to powers of n. Rearranging, one can write this as

$$\left(\frac{\mathbf{MM}}{\Delta t} + \mathbf{KM}\right)\mathbf{T}^{n+1} = \frac{\mathbf{MM}}{\Delta t}\,\mathbf{T}^{n} + \mathbf{F} \tag{2.27}$$

or more compactly as

$$\mathbf{L}\ \mathbf{T}^{n+1} = \mathbf{R}\ \mathbf{T}^{n} + \mathbf{F} \tag{2.28}$$

where

$$\mathbf{L} = \begin{bmatrix} \frac{\Delta x}{3\Delta t} + \frac{\kappa}{\Delta x} & \frac{\Delta x}{6\Delta t} - \frac{\kappa}{\Delta x} \\ \frac{\Delta x}{6\Delta t} - \frac{\kappa}{\Delta x} & \frac{\Delta x}{3\Delta t} + \frac{\kappa}{\Delta x} \end{bmatrix} \tag{2.29}$$

$$\mathbf{R} = \begin{bmatrix} \frac{\Delta x}{3\Delta t} & \frac{\Delta x}{6\Delta t} \\ \frac{\Delta x}{6\Delta t} & \frac{\Delta x}{3\Delta t} \end{bmatrix} \tag{2.30}$$

$$\mathbf{F} = H \begin{Bmatrix} \frac{\Delta x}{2} \\ \frac{\Delta x}{2} \end{Bmatrix} \tag{2.31}$$

Here, the letters "**L**" and "**R**" are used to remind readers that they refer to element stiffness matrices appearing on the left- and right-hand sides, respectively, of Equation 2.28. In Equation 2.28, everything appearing on the right-hand side is known (and it combines to form a single vector). The matrix **L** is referred to as the total element stiffness matrix, whereas $\mathbf{T}^{n+1}$ is the element unknown vector.

2.6 Assembly

In the preceding sections, we have been concerned with discretization of the governing equation over a single finite element. Thus, so far, our discrete system of equations consists of two equations and two unknowns, one at each node (Equation 2.28). Although we could actually solve this system of equations (once initial and boundary conditions were defined), the resulting solution would generally be inaccurate, unless the true variation in temperature fortuitously matched our initial approximation (Equation 2.4). To obtain a precise solution, the normal approach is to subdivide the spatial domain (shown in Figure 2.1) into numerous finite elements, collectively referred to as a mesh. In doing so, we will see that we can compute very accurate solutions to cases where T varies strongly in time and space, even though, at the scale of a single element, the solution is assumed to be described by low-order polynomials (see Figure 2.2).

The process of passing from a one-element model, to a multielement mesh is very simple, which is one of the great strengths of the FEM. To illustrate the procedure, we consider a small 1D mesh

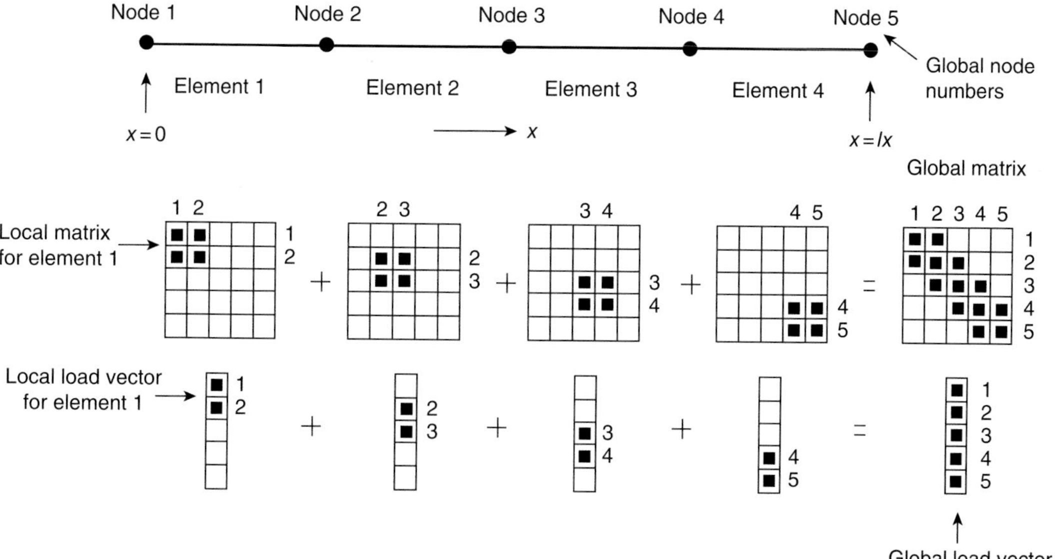

Figure 2.3 One-dimensional mesh consisting of four-, two-node elements, each of length Δx. Each local element matrix (e.g., **L** or **R**) contributes to a global counterpart (e.g., $\tilde{\mathbf{L}}$ or $\tilde{\mathbf{R}}$) in a position depending on the global nodes of that particular element. The same applies to the element load vectors **F**, which contribute to a global load vector $\tilde{\mathbf{F}}$.

consisting of four identical elements (Figure 2.3). Now, instead of having just two unknowns, we have five, related to the five nodes in the mesh. We introduce the following notation:

$$\mathbf{L}^1 = \begin{bmatrix} L_{11}^1 & L_{12}^1 \\ L_{21}^1 & L_{22}^1 \end{bmatrix}$$

Here, the superscripts refer to the element number and subscripts refer to the location within each element matrix. Thus, for example, by referring to Equation 2.29, it can be seen that the term L_{11}^1 has the value $\frac{\Delta x}{3\Delta t} + \frac{\kappa}{\Delta x}$. Similar notation is assumed for $\mathbf{R}$.

The assembly of the global system of equations is achieved by summing node by node the matrix equations derived for a single element (i.e., Equation 2.28). Thus, for example, the element matrix for the first element contributes to positions 1 and 2 in the global matrix because it shares these two global nodes, while the element matrix for the third element contributes to positions 3 and 4 in the global matrix because it shares these two global nodes (Figure 2.3). Performing this procedure for all elements (using the notation introduced above) leads to

$$\begin{bmatrix} L_{11}^1 & L_{12}^1 & 0 & 0 & 0 \\ L_{21}^1 & L_{22}^1 + L_{11}^2 & L_{12}^2 & 0 & 0 \\ 0 & L_{21}^2 & L_{22}^2 + L_{11}^3 & L_{12}^3 & 0 \\ 0 & 0 & L_{21}^3 & L_{22}^3 + L_{11}^4 & L_{12}^4 \\ 0 & 0 & 0 & L_{21}^4 & L_{22}^4 \end{bmatrix} \begin{Bmatrix} T_1 \\ T_2 \\ T_3 \\ T_4 \\ T_5 \end{Bmatrix}^{n+1}$$

$$= \begin{bmatrix} R_{11}^1 & R_{12}^1 & 0 & 0 & 0 \\ R_{21}^1 & R_{22}^1 + R_{11}^2 & R_{12}^2 & 0 & 0 \\ 0 & R_{21}^2 & R_{22}^2 + R_{11}^3 & R_{12}^3 & 0 \\ 0 & 0 & R_{21}^3 & R_{22}^3 + R_{11}^4 & R_{12}^4 \\ 0 & 0 & 0 & R_{21}^4 & R_{22}^4 \end{bmatrix} \begin{Bmatrix} T_1 \\ T_2 \\ T_3 \\ T_4 \\ T_5 \end{Bmatrix}^{n} + H\Delta x \begin{Bmatrix} \frac{1}{2} \\ 1 \\ 1 \\ 1 \\ \frac{1}{2} \end{Bmatrix} \tag{2.32}$$

Using matrix notion, this system of equations can be written compactly as

$$\tilde{\mathbf{L}}\,\tilde{\mathbf{T}}^{n+1} = \tilde{\mathbf{R}}\,\tilde{\mathbf{T}}^{n} + \tilde{\mathbf{F}} \tag{2.33}$$

where the tilde symbol ($\tilde{}$) indicates that the matrices and the vectors refer to the entire "global" problem and not to a single element. Note that the matrices $\tilde{\mathbf{L}}$ and $\tilde{\mathbf{R}}$ are symmetrical, which is an important (though not a necessary) property when it comes to solving the system of equations. Finally, note that most nonzero terms are clustered near the main diagonal (called "diagonal dominance"), a property that helps when the equations are solved. Equation 2.33 can also be written in the form

$$\tilde{\mathbf{L}}\,\tilde{\mathbf{T}}^{n+1} = \tilde{\mathbf{b}}^{n} \tag{2.34}$$

which is the classic form for a system of linear equations (often written as $A\,x = b$). In this form, it's more clear that the right-hand side of Equation 2.33 combines to form a single vector ($\tilde{\mathbf{b}}$). In this last expression, $\tilde{\mathbf{L}}$ is referred to as the stiffness (or coefficient) matrix (and is known), $\tilde{\mathbf{F}}$ is referred to as the "right-hand side vector" or "load vector," and $\tilde{\mathbf{T}}^{n+1}$ is the unknown "solution vector" or "reaction vector."

2.7 Boundary and Initial Conditions

Inspection of the governing PDE (2.1) shows that it contains a first-order time derivative, indicating that we must provide an initial temperature distribution across the entire model space. This is easily imposed by initially setting

$$\tilde{\mathbf{T}}^n = \mathbf{Ti} \tag{2.35}$$

on the right-hand side of Equation 2.33, where **Ti** is a vector containing the initial temperatures for all nodes of the mesh (in our example, $\mathbf{Ti} = 0$; see Figure 2.1). The original equation also contains a second-order spatial derivative, indicating that two spatial boundary conditions must be supplied, one at each end of the model domain. In this chapter, we consider Dirichlet boundary conditions where the temperatures at $x = 0$ and $x = lx$ are fixed to known values (Tbl and Tbr) and remain constant in time (note that both Tbl and Tbr are zero in this example; see Figure 2.1). The approach taken here is (1) form the global system of equations without boundary conditions (as done in the derivation of 2.33), and (2) modify these equations to get the correct boundary values.

Recall that for the four-element mesh considered previously, nodes 1 and 5 are located at the ends of the model domain (see Figure 2.3). Thus, the temperatures of T_1 and T_5 in the global solution vector need to be fixed (to Tbl and Tbr). An easy means of achieving this is by performing the following steps in the given order:

1) Zero-out entirely lines 1 and 5 in the global stiffness matrix $\tilde{\mathbf{L}}$.
2) Place "1" on the first and fifth diagonal entries of the global stiffness matrix $\tilde{\mathbf{L}}$.
3) Set the first and fifth entries of the global load vector $\tilde{\mathbf{b}}$ to the desired temperature (i.e., Tbl and Tbr).

Performing these steps on system 2.34 results in the following:

$$\begin{bmatrix} 1 & 0 & 0 & 0 & 0 \\ L_{21}^1 & L_{22}^1 + L_{11}^2 & L_{12}^2 & 0 & 0 \\ 0 & L_{21}^2 & L_{22}^2 + L_{11}^3 & L_{12}^3 & 0 \\ 0 & 0 & L_{21}^3 & L_{22}^3 + L_{11}^4 & L_{12}^4 \\ 0 & 0 & 0 & 0 & 1 \end{bmatrix} \begin{Bmatrix} T_1 \\ T_2 \\ T_3 \\ T_4 \\ T_5 \end{Bmatrix}^{n+1} = \begin{Bmatrix} Tbl \\ b_2 \\ b_3 \\ b_4 \\ Tbr \end{Bmatrix} \tag{2.36}$$

where the boundary conditions are seen to be exactly satisfied (i.e, $T_1^{n+1} = Tbl$ and $T_5^{n+1} = Tbr$).

2.8 Solution of the Algebraic Equations

At this stage, the governing PDE along with the boundary and initial conditions have been entirely converted to a system of linear algebraic equations. All that remains to be solved is this system of simultaneous equations (at each step). Recall that for our simple mesh, there are a total of five equations (one for each global node), for the five unknown temperatures. Actually, only the temperatures on the central three nodes are truly unknown because we fixed the temperatures at either end of the mesh (see Figure 2.3). Nevertheless, we have chosen to retain the boundary equations (1 and 5) in the global system and to solve for the temperatures at these nodes as if they are unknown.

Various methods can be used to solve the system of equations generated by the FEM, the details of which are well beyond the scope of the text. Relatively small systems are easily and rapidly solved using direct methods (e.g., Gaussian elimination), which, in the absence of rounding errors, provide an exact solution (to the algebraic system, not the original differential equation) in a finite number of operations. For larger systems, memory requirements and computation times can be drastically reduced by exploiting the sparsity of the global stiffness matrix (as is done in the following chapters). Sparsity here refers to the fact that most terms in the global stiffness matrices resulting from finite element discretization are zero. Nevertheless, for very large systems (e.g., involving more than 10^6 unknowns), it may be necessary to resort to iterative methods (e.g., Jacobi, Gauss–Seidel, Conjugate gradient, successive relaxation) that seek an approximate solution after a (hopefully) small number of iterations. The downside of iterative methods is that they may fail to converge. In Chapter 3, we show how Matlab can easily be used to obtain a direct solution to the system of equations derived with the FEM.

2.9 Exercises

1) Derive the three shape functions for a 1D quadratic (three node) element on the domain $[0\ L]$. Verify that $N_1 + N_2 + N_3 = 1$. Show a graph of the three shape functions on the interval $[0\ L]$ for $L = 1$. Hint: A quadradic polynomial can be written in the form

$$f = a + bx + cx^2$$

where a, b, and c are coefficients to be determined. Recall that by definition,

$$N_1(x=0)=1 \quad N_1\left(x=\frac{L}{2}\right)=0 \quad N_1(x=L)=0$$
$$N_2(x=0)=0 \quad N_2\left(x=\frac{L}{2}\right)=1 \quad N_2(x=L)=0$$

and

$$N_3(x=0)=0 \quad N_3\left(x=\frac{L}{2}\right)=0 \quad N_3(x=L)=1$$

Each of these conditions can be written as a linear system and solved for a, b, and c. For example, for N_1, the three equations (obtained by substituting the values for x into the given polynomial) are

$$\begin{aligned} a &= 1 \\ \frac{1}{4}cL^2 + \frac{1}{2}bL + a &= 0 \\ L^2c + Lb + a &= 0 \end{aligned}$$

which can be solved to give

$$a = 1 \quad b = -\frac{3}{L} \quad c = \frac{2}{L^2}$$

Therefore,

$$N_1 = 2\frac{x^2}{L^2} - 3\frac{x}{L} + 1$$

Doing the same for N_2 and N_3 leads to

$$N_2 = 4\frac{x}{L} - 4\frac{x^2}{L^2}$$

and

$$N_3 = 2\frac{x^2}{L^2} - \frac{x}{L}$$

2) Use the quadratic shape functions just derived to interpolate the function

$$T = (1 + \sin(\pi x))e^{-2x}$$

on the interval $x = [0\ 1]$ using 15 equally spaced points. Compare the result with that obtained by halving the domain (i.e., using $L = 0.5$).
3) Compute the first spatial derivatives for the quadratic shape functions derived before.
4) Later in the text, we will see that finite element discretization of the advection term $u\partial T/\partial x$ results in the element matrix

$$\mathbf{CM} = \int_0^L u \begin{bmatrix} N_1\frac{\partial N_1}{\partial x} & N_1\frac{\partial N_2}{\partial x} \\ N_2\frac{\partial N_1}{\partial x} & N_2\frac{\partial N_2}{\partial x} \end{bmatrix} dx$$

Assuming linear shape functions (given by 2.7) and a constant u, integrate this element matrix exactly. Note that the resulting matrix is asymmetrical.
5) Write out the system of equations resulting from discretization by the FEM for the ordinary differential equation

$$-\frac{\partial^2 u(x)}{\partial x^2} = 1$$

with

$$u(0) = 0$$

and

$$u(1) = 0$$

Assume linear shape functions, linear weighting functions and four equally spaced finite elements.

Suggested Reading

C. A. J. Fletcher, *Computational Techniques for Fluid Dynamics,* Springer, Berlin, 2000.

Y. W. Kwong, and H. C. Bang, *The Finite Element Method using Matlab,* CRC Press, New York, 2000.

I. M. Smith, and D. V. Griffith, *Programming the Finite Element Method,* John Wiley & Sons, Ltd, Chichester, 1998.

O. C. Zienkiewicz, and R. L. Taylor, *The Finite Element Method, Volume 1, The Basis,* Butterworth-Heinemann, Oxford/Boston, MA, 2000.

3

Programming the Finite Element Method in Matlab

The purpose of this chapter is to learn how to program the finite element method (FEM) in Matlab. To illustrate this we solve the diffusion (transient heat conduction) equation, discretized in Chapter 2. A complete stand-alone Matlab script to compute the numerical solution is provided at the end of the chapter. Readers relatively new to Matlab are advised to refer to Appendix B for a refresher on basics of linear algebra using Matlab.

3.1 Program Structure and Philosophy

The basic structure of a general finite element program normally consists of three main parts as follows:

1) *Preprocessor*, involving parameter definition and initialization
2) *Solution*, involving element integration, assembly, and solution
3) *Postprocessor*, possibly involving additional calculation (based on the solution) and visualization of results

Often, these parts are performed in separate independent modules, which themselves call other more specific modules to perform more specialized tasks. This so-called *modular structure* makes the programs relatively easy to modify and reuse to solve very different problems with minimal reprogramming. The downside is that these programs are quite difficult to understand for beginners, especially when a program call many large general-purpose routines of which only small portions are actually used. Because of this, here we construct simple, nonmodular (or stand-alone) programs. Once the principles are clear, the programs could readily be reprogrammed in modular form and possibly translated to other more efficient programming languages (i.e., Fortran or C/C++).

3.2 Summary of the Problem

Our task here is to compute a numerical solution to the parabolic partial differential equation

$$\frac{\partial T}{\partial t} = \kappa \frac{\partial^2 T}{\partial x^2} + H \tag{3.1}$$

with the initial condition

$$T(x, t = 0) = 0 \quad \forall\, x \in [0; lx] \tag{3.2}$$

Practical Finite Element Modeling in Earth Science Using Matlab, First Edition. Guy Simpson.

Companion website: www.wiley.com/go/simpson

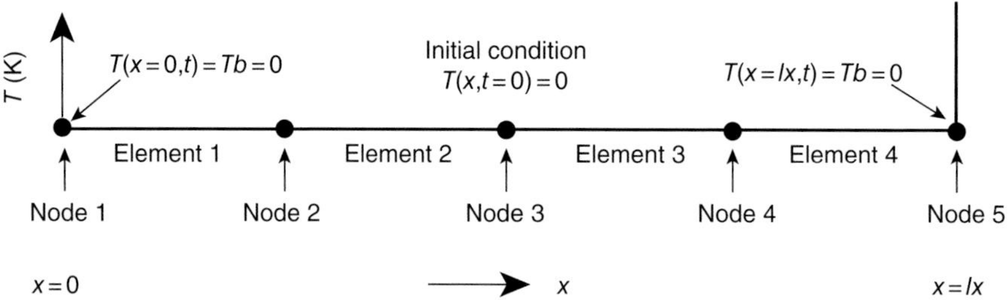

Figure 3.1 Summary of finite element mesh, boundary conditions, and initial condition for numerical solution of Equation 3.1. A constant heat source *H* extends across the model domain.

and the boundary conditions

$$T(x = 0, t) = 0 \quad \text{and} \quad T(x = lx, t) = 0 \tag{3.3}$$

where $T(x, t)$ is the unknown temperature, κ is the thermal diffusivity ($= 10^{-6}$ m^2 s^{-1}), H is a constant heat source ($= 10^{-6}$ K s^{-1}), and the spatial domain extends from 0 to lx ($= 10\,000$ m) (Figure 3.1).

3.3 Discretized Equations

In Chapter 2, we showed that applying the FEM to Equation 3.1 results in the following discrete system of equations for a single two-node element:

$$\mathbf{L}\,\mathbf{T}^{n+1} = \mathbf{R}\,\mathbf{T}^{n} + \mathbf{F} \tag{3.4}$$

Here, the various element vectors and matrices are given by

$$\mathbf{L} = \frac{\mathbf{MM}}{\Delta t} + \mathbf{KM} \tag{3.5}$$

$$\mathbf{R} = \frac{\mathbf{MM}}{\Delta t} \tag{3.6}$$

$$\mathbf{MM} = \begin{bmatrix} \frac{\Delta x}{3} & \frac{\Delta x}{6} \\ \frac{\Delta x}{6} & \frac{\Delta x}{3} \end{bmatrix} \tag{3.7}$$

$$\mathbf{KM} = \kappa \begin{bmatrix} \frac{1}{\Delta x} & -\frac{1}{\Delta x} \\ -\frac{1}{\Delta x} & \frac{1}{\Delta x} \end{bmatrix} \tag{3.8}$$

$$\mathbf{F} = H \begin{Bmatrix} \frac{\Delta x}{2} \\ \frac{\Delta x}{2} \end{Bmatrix} \tag{3.9}$$

$$\mathbf{T} = \begin{Bmatrix} T_1 \\ T_2 \end{Bmatrix} \tag{3.10}$$

The variable Δt is the time step and Δx[1] is the element length. The superscripts n and $n + 1$ on $\mathbf{T}$ refer to the current and future time levels, respectively. Thus, $\mathbf{T}^{n+1}$ is unknown. Recall also that when

1 Note the different notation for Δx compared to Chapter 2 (where it was denoted by L).

a multielement mesh is considered, the various element matrices and vectors need to be assembled into global equivalents. The global system of equations to be solved (derived in Chapter 2) is

$$\tilde{\mathbf{L}}\,\tilde{\mathbf{T}}^{n+1} = \tilde{\mathbf{R}}\,\tilde{\mathbf{T}}^{n} + \tilde{\mathbf{F}} = \tilde{\mathbf{b}} \tag{3.11}$$

where $\tilde{\mathbf{L}}$ is the global (left-hand side) stiffness matrix, $\tilde{\mathbf{R}}$ is the global right-hand side matrix, $\tilde{\mathbf{F}}$ is the global load vector, and $\tilde{\mathbf{T}}$ is the array of global nodal temperatures at the n and $n+1$ time levels.

3.4 The Program

A complete listing of a Matlab script to solve Equation 3.1 with the FEM is described in this section and is listed at the end of the chapter (Section 3.5). The basic steps that are performed within the computer program can be summarized as follows:

1) Define all physical parameters (e.g., diffusivity, source term, length of spatial domain) and numerical parameters (e.g., number of elements and nodes).
2) Define the finite element mesh where the solution will be obtained.
3) Define the relationship between the element node numbers and the global node numbers.
4) Define boundary node indices and the values of the solution at the boundary nodes.
5) Initialize the global matrices $\tilde{\mathbf{L}}$, $\tilde{\mathbf{R}}$, and $\tilde{\mathbf{F}}$ so that their dimensions are defined and that the matrices are filled with zeros.
6) Within an element loop, compute the element matrices **MM** and **KM** and the element load vector **F** (see Equations 3.7, 3.8, and 3.9). Use these to compute **L** and **R** (Equations 3.5 and 3.6). Assemble **L**, **R**, and **F** into their global equivalents (see Equation 3.11). If the element properties do not depend on time, these global matrices only need to be calculated once and can be saved for later use.
7) Within a time loop, perform the operations on the right-hand side of Equation 3.11 (i.e., first multiply $\tilde{\mathbf{R}}$ with the old temperature vector $\tilde{\mathbf{T}}^{n}$ and then add the resulting vector to $\tilde{\mathbf{F}}$ to form a single right-hand-side vector $\tilde{\mathbf{b}}$).
8) Apply boundary conditions (see Section 2.7).
9) Solve the linear system of equations (3.11) for the new temperature $\tilde{\mathbf{T}}^{n+1}$.
10) Plot or save the solution to a file, if desired.
11) Go to the next time step (step 7).

We now describe each of these steps in more detail. Steps 1–5 are considered preprocessing, steps 6–9 form part of the solution stage, and step 10 is postprocessing. In the following text, Matlab script snippets are shown as follows:

```
kappa = 1e-6 ; % This is some Matlab code
```

3.4.1 Preprocessor Stage

In the first part of the program, all parameters need to be defined. When doing this, it's advisable to clearly distinguish physical parameters (e.g., thermal diffusivity, length of model domain, and heat source term) from numerical parameters related to the discretization procedure (e.g., number of elements and time increment (Δt)). It's also a good idea to use meaningful names for variables and to make use of comments to help yourself (and possibly other users) to understand the meaning of the different variables.

Once the domain length and number of finite elements have been defined, we are ready to define the global node coordinates of the finite element mesh. For the one-dimensional (1D) problem treated here, these coordinates are saved in a line vector that can be constructed as follows:

```
g_coord = [0 : dx : lx] ; % global nodal coordinates
```

Here, `lx` is the physical length of the model domain and `dx` is the length of an individual element (equivalently the node spacing). Note that the "g" in `g_coord` refers to that fact that the coordinates are global (relating to the mesh, not an individual element). Note also that `dx=lx/nels` where `nels` is the number of elements. Finally, note that for two-node elements, the number of nodes in the entire finite element mesh is `nn = nels+1`. Thus, the dimensions of `g_coord` are `[1,nn]` (meaning that it has 1 line and `nn` columns).

An extremely important component of every finite element program is the array specifying nodal element connectivity, here saved in the array `g_num`. It provides a mapping from local to global node numbers for each element. For our problem that has 4×2-node elements and a total of five global nodes (Figure 3.1), `g_num` can be constructed with the lines:

```
g_num(1,:) = [1 2 3 4 ] ; %  connectivity for 4-element mesh
g_num(2,:) = [2 3 4 5 ] ;
```

Each column of `g_num` lists the two global nodes present in each element (4 in total). Thus, for example, the two *local* node numbers 1 and 2 in element 3 (column 3) correspond to the *global* node numbers 3 and 4, respectively. More generally (i.e., for any number of elements), `g_num` can be created with the lines

```
g_num(1,:) = [1 : nn-1 ] ; % connectivity for general 2-node mesh
g_num(2,:) = [2 : nn   ] ;
```

The matrix `g_num` has the dimensions `[nod nels]` where `nod` is the number of nodes per element (here 2) and `nels` is the total number of elements in the mesh (here 4).

Toward the end of the preprocessor, one needs to define to which global nodes boundary conditions are to be applied. Recall that in this problem we will apply Dirichlet boundary conditions, so we will also need to define the value of the temperature at the boundary nodes. All these can be done using the following Matlab commands:

```
bcdof =  [ 1   nn ] ; % specify boundary nodes
bcval =  [ Tb  Tb ] ; % specify boundary values
```

Here, `Tb` are the temperatures at the left- and right-hand boundaries (here zero) and must be defined in the parameter section. The first of these commands specifies that the first and last nodes in the mesh (i.e., 1 and `nn`) are to be constrained, whereas the second defines the value of the temperature at each respective boundary node. Note that these lines do not actually impose the boundary conditions, which is done after assembly of the global matrices (see the following text).

In the final part of the preprocessor, one initializes the global matrices and vectors (appearing in Equation 3.11) that will be filled later in the program. The matrices $\tilde{\mathbf{L}}$, $\tilde{\mathbf{R}}$ and the vector $\tilde{\mathbf{F}}$ can be initialized using the commands

```
lhs = sparse(nn,nn) ;  % Global left hand side matrix (L)
rhs = sparse(nn,nn) ;  % Global right hand side matrix (R)
ff  = zeros(nn,1)   ;  % Global rhs load vector (F)
```

where `nn` is the number of equations in the finite element mesh (which also equals the total number of nodes in this problem). Note that the matrices are both initialized as sparse matrices, which is not essential for small problems but requires far less memory than using the command `zeros(nn,nn)`. Finally, the initial temperature at $t = 0$ needs to be defined. For our specific problem where the initial temperature is zero, this can be achieved with the line

```
displ   = zeros(nn,1)    ;  % Initial temperature
```

Here, and throughout the text, `displ` refers to the global solution vector (i.e., `displ` $= \mathbf{T}$, see Equation 3.4). Also note that in what follows, `displ` will normally be repeatedly overwritten at each time step. Thus, we will not normally distinguish $\mathbf{T}^{n+1}$ from $\mathbf{T}^n$.

3.4.2 Solution Stage

The solution stage begins with an element loop. Within this loop one performs the following tasks:

1) Retrieve the global nodes for each element. For example, the global nodes for element `iel` can be obtained using the command

   ```
   num = g_num(:,iel) ; % get list of element nodes
   ```

2) Compute the element length as

   ```
   dx = abs(diff(g_coord(num))) ; % element length
   ```

3) Construct the element matrices **MM** and **KM** and the element load vector **F** (Equations 3.7, 3.8, and 3.9). For example, **KM** can be formed using the line

   ```
   KM = kappa*[1/dx  -1/dx  ;  -1/dx  1/dx] ; % elem. stiff. matrix
   ```

4) Compute **L** and **R** (Equations 3.5 and 3.6).
5) Sum the element matrices and vectors node by node to form their global counterparts. This is achieved with the following script:

   ```
   lhs(num,num)  = lhs(num,num)  + L   ; % assemble lhs global matrix
   rhs(num,num)  = rhs(num,num)  + R   ; % assemble rhs global matrix
   ff(num)       = ff(num)       + F   ; % assemble global load vector
   ```

After the given element loop, the global matrices have been completely assembled. The next major part of the program comprises a time loop within which one does the following:

1) Compute the right-hand side of Equation 3.11 with the line

   ```
   b = rhs*displ + ff ; % form rhs vector
   ```

   where `displ` is the temperature at the n time level (i.e., $\mathbf{T}^n$). The dimensions of `b` are `[nn,1]` (i.e., it's a column vector with `nn` lines).
2) Modify $\tilde{\mathbf{L}}$ and $\tilde{\mathbf{b}}$ such that the fixed boundary conditions are satisfied. Referring to Equation 2.36, this can be achieved with the lines

   ```
   lhs(bcdof,:)  = 0 ;   % zero rows of boundary equations
   tmp           = spdiags(lhs,0) ; % store diagonal
   tmp(bcdof)    = 1   ; % place 1 on entries of boundary equations
   lhs           = spdiags(tmp,0,lhs) ; % reinsert  diagonal
   b(bcdof)      = bcval  ;                % modify rhs vector
   ```

   The first of these lines places zeros in $\tilde{\mathbf{L}}$ on the rows of the equations where boundary conditions should be imposed, the second stores the diagonal of the stiffness matrix (in the vector `tmp`), the third places "ones" in the vector in positions (equations) where boundary conditions should be applied, the fourth reinserts the modified diagonal vector into the stiffness matrix $\tilde{\mathbf{L}}$, and the fifth puts the fixed temperature boundary values in the global right-hand side vector.

3) Solve the global system of equations simultaneously for the new temperature at the $n+1$ time level (i.e., $\mathbf{T}^{n+1}$).

```
displ = lhs \ b ;   % solve the linear system of equations
```

Note here the use of the Matlab operator\(backslash), which is a blackbox Matlab function that solves a linear system of equations (Appendix B). This function dispatches the most appropriate solver depending on the properties and structure of the stiffness matrix (`lhs`). Note also that directly after this line, `displ` has now overwritten the old solution vector. So `displ` becomes the old vector of nodal temperatures for the next time step. This now completes the solution at a single instant in time. One now returns to point 1 to compute the solution at the next time step.

3.4.3 Postprocessor Stage

The purpose of the postprocessor is to perform additional tasks that are not directly related to calculation of the solution. The most common tasks performed at this stage are saving results and/or visualization. One of the advantages of Matlab compared to languages such as Fortran and C/C++ is that visualization can be performed within the same environment in which calculations are performed. Thus, in order to see the results in "real" time, one can simply plot them directly, for example, using

```
plot(g_coord,displ)   % plotting x vs the solution
drawnow % ensure the plot is drawn at each time step
```

This line should be included within the given time loop, just after the solution is computed. Of course, often one may also want to save the results for later analysis and visualization. This could, for example, be performed by adding the line

```
% saving a file at each time step (n)
 dlmwrite(['results_',num2str(n)],[g_coord' displ],' \t')
```

to the time loop after the solution is computed. At each time step, a file named *results_n* (where n is the current time step) is created that contains the nodal coordinates and nodal temperatures in the first and second columns, respectively. Other tasks that are sometimes performed within the postprocessing stage are calculations based on the solution. For example, based on the calculated temperature, one may be interested in computing the heat flux q, defined as

$$q = -k\frac{\partial T}{\partial x}$$

where k is the thermal conductivity. This requires calculation of the gradient of the temperature, which is itself now known.

3.5 Matlab Script

The following Matlab script computes the numerical solution to Equation 3.1 using the FEM. An example of results produced by the program is illustrated in Figure 3.2. Shown for comparison is the exact analytical solution published by Carslaw and Jaeger (1959, page 130). For the four-element mesh discussed in this chapter, the numerical results are, not surprisingly, inaccurate. However, one

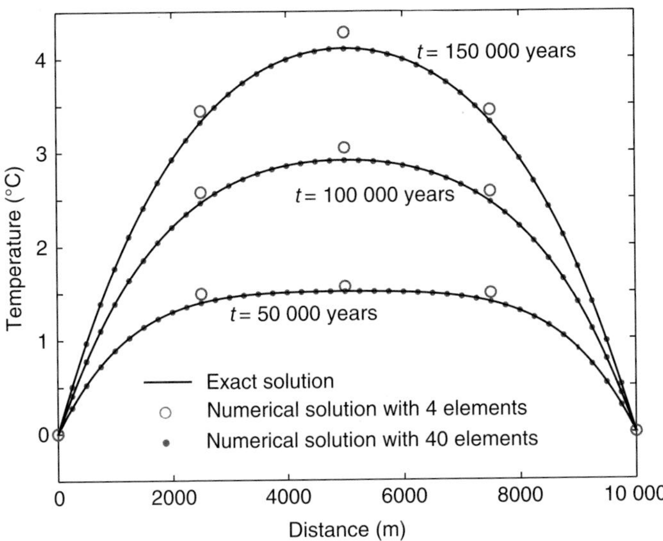

Figure 3.2 Comparison between the analytical and numerical solutions to Equation 3.1 using two different mesh resolutions and a numerical time step of 1000 years. Results were computed with the FEM program presented in Section 3.5.

observes a good agreement between the numerical and analytical solutions as the number of elements is increased.

```
%--------------------------------------------------------
% Program diffusion1d.m
% 1-D FEM solution of diffusion equation
% For the analytical solution see equation 7, page 130
% Carslaw and Jaeger (1959)
%--------------------------------------------------------

clear % clear memory from current workspace

% physical parameters
seconds_per_yr = 60*60*24*365; % number of seconds in one year
lx     = 10000 ;          % length of spatial domain (m)
Cp     = 1e3   ;          % rock heat capacity (J/kg/K)
rho    = 2700 ;           % rock density (kg/m^3)
K      = 3.3   ;          % bulk thermal conductivity (W/m/K)
kappa  = K/(Cp*rho);      % thermal diffusivity (m^2/s)
Tb     = 0 ;              % temperatures at boundaries (°C)
A      = 2.6e-6 ;         % heat production (W/m^3)
H      = A/(rho*Cp);      % heat source term (°K/s)

% numerical parameters
dt       = 1000*seconds_per_yr ; % time step (s)
ntime    = 5000 ;          % number of time steps
nels     = 40 ;            % total number of elements
nod      = 2 ;             % number of nodes per element
nn       = nels+1          % total number of nodes
dx       = lx/nels ;       % element size
```

```
g_coord = [0:dx:lx] ; % spatial domain (1-D mesh)

%  define boundary conditions
 bcdof = [  1     nn ]    ; % boundary nodes
 bcval = [  Tb    Tb ]    ; % boudary values

% define connectivity and equation numbering
g_num        = zeros(nod,nels) ;
g_num(1,:) = [1:nn-1]   ;
g_num(2,:) = [2:nn]     ;

% initialise matrices and vectors
ff      = zeros(nn,1);          % system load vector
b       = zeros(nn,1);          % system rhs vector
lhs     = sparse(nn,nn);        % system lhs matrix
rhs     = sparse(nn,nn);        % system rhs matrix
displ   = zeros(nn,1);          % initial temperature (°C)

%-------------------------------------------------
% matrix assembly
%-------------------------------------------------

   for iel=1:nels % loop over all elements
      num = g_num(:,iel)   ;            % retrieve equation number
      dx  = abs(diff(g_coord(num))) ; % length of element
      MM  = dx*[1/3  1/6 ; 1/6 1/3 ] ;% mass matrix
      KM  = [kappa/dx -kappa/dx ; -kappa/dx kappa/dx ];%diffn matrix
      F   = dx*H*[1/2 ; 1/2] ;          % load vector
      lhs(num,num) = lhs(num,num) + MM/dt + KM ; % assemble lhs
      rhs(num,num) = rhs(num,num) + MM/dt      ; % assemble rhs
      ff(num)      = ff(num)      + F          ; % assemble load
   end    % end of element loop

%-------------------------------------------------

% time loop
t   = 0 ; % time
k   = 1 ; % counter
ii = [100 1000 5000]; % array used for plotting
for n=1:ntime
   n
   t   = t + dt ;                     % compute time
   b   = rhs*displ + ff ;             % form rhs vector

   % impose boundary conditions
   lhs(bcdof,:) = 0 ;                 % zero the relevent equations
   tmp = spdiags(lhs,0) ;             % store diagonal
   tmp(bcdof)=1 ;                     % place 1 on stored-diagonal
   lhs=spdiags(tmp,0,lhs);            % reinsert diagonal
   b(bcdof) = bcval ;                 % set rhs vector

   displ = lhs \ b ;                  % solve system of equations

%-------------------------------------------------
   % evaluate analytical solution
   nterms = 100 ;
   L     = lx/2 ;
   x     = linspace(-L,L,100) ;
```

```
    sumv = 0 ;
    for ni=0:nterms
        et   = exp(-kappa*(2*ni+1)^2*pi^2*t/4/L^2);
        sumv = sumv + (-1)^ni/(2*ni+1)^3*cos((2*ni+1)/2/L*pi*x)*et ;
    end
    Texact = A*L^2/(2*K)*(1-x.^2/L^2-32/pi^3.*sumv) ;
%-------------------------------------------------------------

  % plotting
    figure(1)
    plot(g_coord,displ,'o')
    hold on
    plot(x+L,Texact,'r')
    title('Comparison between numerical and analytical solutions')
    xlabel('Distance (m)')
    ylabel('Temperature (°C)')
    hold off
    drawnow
    pause

end % end of time loop
%-------------------------------------------------------------
```

3.6 Exercises

1) Within the postprocessing portion of the script listed in Section 3.5, compute the heat flux ($q = -k\,\partial T/\partial x$) for each element. Make a plot of the heat flux versus x for several different times and superimpose a curve with the exact solution given by

$$q(x,t) = -\frac{AL^2}{2}\left(\frac{32}{\pi^3}S - \frac{2x}{L^2}\right)$$

where

$$S = \sum_{k=0}^{N} \frac{(-1)^k}{(2k+1)^3}\frac{(2k+1)\pi}{2L}\sin\frac{(2k+1)\pi x}{2L}\exp\frac{-\kappa(2k+1)^2\pi^2 t}{4L^2}$$

with $L = lx/2$ and noting that x goes from $-L$ to L. Hint: To compute the flux, you need to compute the temperature gradient in each element. Because the shape functions are linear, the gradient is constant for each element and can be computed by multiplying the shape function derivatives in global coordinates (`deriv`, see Equation 2.21) with the nodal temperature values for a given element, i.e., `dTdx = deriv*displ(num)`). The flux can then be obtained from Fourier's law, that is, $q = -k\,\partial T/\partial x$.

2) Modify the script listed in Section 3.5 to account for a variable element spacing by computing the mesh coordinates (`g_coord`) from

$$x = \frac{lx}{2}\left(1 - \cos\left(\pi\frac{j}{N}\right)\right) \quad \text{with } j = 0, 1, \ldots, N$$

where lx is the length of the model domain and N is the number of elements in the mesh. Check that the numerical results still agree with the exact solution.

3) Modify the script listed in Section 3.5 to solve the ordinary differential equation boundary value problem

$$\frac{\partial^2 u}{\partial x^2} = u^{4x}$$

with

$$-1 < x < 1, \quad u(\pm 1) = 0$$

Check your numerical results with the exact solution

$$u(x) = \frac{e^{4x} - x\sinh(4) - \cosh(4)}{16}$$

Hint: To solve this problem, you will need to derive the load vector for the case when the source term varies in space. Assume that H_1 and H_2 are the source values evaluated at nodes 1 and 2 of a particular element. They can be interpolated over the element using the shape functions, that is, `H = fun * [H_1 ; H_2]`). Recall that the element load vector is defined as (see 2.18)

$$\mathbf{F} = \int_0^L H\mathbf{N}^T dx = \int_0^L H \begin{Bmatrix} N_1 \\ N_2 \end{Bmatrix} dx$$

Evaluating the integral with the variable H yields

$$\mathbf{F} = \begin{Bmatrix} \frac{1}{6}\Delta x(H_2 + 2H_1) \\ \frac{1}{6}\Delta x(2H_2 + H_1) \end{Bmatrix}$$

which reduces to 3.9 when $H_1 = H_2$.

4) Resolve Equation 3.1 (along with 3.2 and 3.3) using the quadratic shape functions derived as an exercise in Chapter 2. Check your results with the exact solution and compare the accuracy of the numerical solution against the results computed with linear shape functions.

Suggested Reading

A. J. M. Ferreira, *MATLAB Codes for Finite Element Analysis: Solids and Structures (Solid Mechanics and Its Applications)*, Springer, Berlin, 2009.

Y. W. Kwong, and H. C. Bang, *The Finite Element Method using Matlab*, CRC Press, New York, 2000.

I. M. Smith, and D. V. Griffith, *Programming the Finite Element Method*, John Wiley & Sons, Ltd, Chichester, 1998.

4

Numerical Integration and Local Coordinates

In Chapter 3, the element matrices resulting from finite element discretization were integrated analytically. In this chapter, we consider how to evaluate the same integrals numerically, as is more usual. To do so, the shape functions and their derivatives will normally be defined in the local (normalized), rather than the physical coordinate system. As we will see later in the book, these important steps make the finite element method (FEM) very flexible in its ability to treat complex material behaviour and geometries. A full Matlab program is provided at the end of the chapter to show how numerical integration is implemented within the FEM in practice.

In Chapter 2, we saw that finite element discretization of the one-dimensional (1D) heat conduction equation led to the element matrices

$$\mathbf{MM} = \int_0^L \mathbf{N}^T\mathbf{N}\,dx = \int_0^L \begin{bmatrix} N_1N_1 & N_1N_2 \\ N_2N_1 & N_2N_2 \end{bmatrix} dx \tag{4.1}$$

$$\mathbf{KM} = \int_0^L \kappa\,\frac{\partial \mathbf{N}^T}{\partial x}\frac{\partial \mathbf{N}}{\partial x}dx = \int_0^L \kappa \begin{bmatrix} \frac{\partial N_1}{\partial x}\frac{\partial N_1}{\partial x} & \frac{\partial N_1}{\partial x}\frac{\partial N_2}{\partial x} \\ \frac{\partial N_2}{\partial x}\frac{\partial N_1}{\partial x} & \frac{\partial N_2}{\partial x}\frac{\partial N_2}{\partial x} \end{bmatrix} dx \tag{4.2}$$

$$\mathbf{F} = \int_0^L H\,\mathbf{N}^T dx = \int_0^L H \begin{Bmatrix} N_1 \\ N_2 \end{Bmatrix} dx \tag{4.3}$$

Here, κ is the thermal diffusivity, H is a heat source, L is the element length, and N_i are shape functions defined as

$$N_1 = 1 - \frac{x}{L}, \quad N_2 = \frac{x}{L} \tag{4.4}$$

Using matrix notation, the shape functions can be written as

$$\mathbf{N} = [N_1\ N_2] \tag{4.5}$$

while the shape function derivatives are denoted

$$\frac{\partial \mathbf{N}}{\partial x} = \begin{bmatrix} \frac{\partial N_1}{\partial x} & \frac{\partial N_2}{\partial x} \end{bmatrix} \tag{4.6}$$

As we have already seen (Section 2.4), it is sometimes possible to evaluate the given integrals exactly, especially if one is prepared to use a mathematical software such as Maple or Mathematica. This has the advantage in that it avoids having to compute the integrals within the FEM programs and is thus very efficient and accurate. However, in many circumstances (e.g., for variable coefficient problems or when elements are complicated), the exact evaluation of the integrals in Equations 4.1 and 4.2 leads

Practical Finite Element Modeling in Earth Science Using Matlab, First Edition. Guy Simpson.

Companion website: www.wiley.com/go/simpson

to very long and complex expressions. It is therefore often more convenient to evaluate the element integrals numerically within the finite element program. We will take this approach throughout the remainder of the book.

4.1 Gauss–Legendre Quadrature

The method we will use to perform numerical integration is called Gauss–Legendre quadrature. Although other methods could be used, Gauss–Legendre integration is accurate and fast and therefore widely used. To illustrate the method, consider the integral

$$\int_{-1}^{1} f(\xi)\, d\xi \tag{4.7}$$

that we wish to evaluate numerically (Figure 4.1). The rule for computing this integral with Gauss–Legendre quadrature is

$$\int_{-1}^{1} f(\xi)\, d\xi \simeq \sum_{k=1}^{n} f(\xi_k)\, w_k \tag{4.8}$$

where n is the number of integration points, ξ_k is the spatial coordinate of the kth integration point, and w_k is the weight of the kth integration point. Thus, to compute the integral, one must evaluate the known function f at certain predefined positions ξ_k, multiply the result with certain predefined weights w_k, and sum. The positions of the integration points and the weights for different numbers of integration points are provided in Table 4.1. These values have been carefully chosen so that the given approximation is exact for a polynomial of degree $p = 2n - 1$ or less. Because we normally use low-order shape functions (often first- or second-order polynomials), this condition indicates that very few integration points are required for good accuracy. For example, using only two integration points ($n = 2$), Gauss–Legendre quadrature produces exact results for first-, second-, and third-order polynomial functions.

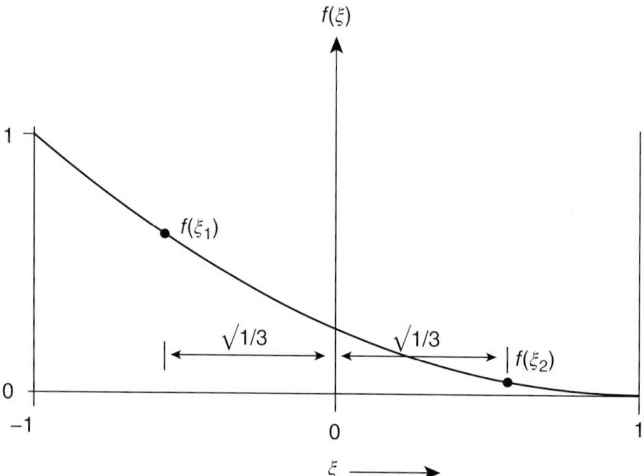

Figure 4.1 Gauss–Legendre integration of the function $f(\xi) = N_1(\xi)N_1(\xi)$ (see Equation 4.10) on the interval $[-1,1]$ using two integration points. The positions of the two points (i.e., $\pm\sqrt{1/3}$) are obtained from Table 4.1. The integral of this function evaluated using Gauss–Legendre quadrature is exact (=2/3).

Table 4.1 Gauss quadrature points (ξ_k) and weights (w_k).

n	ξ_k	w_k	Degree
2	$\pm\sqrt{\frac{1}{3}}$	1.000 000 000 000	3
3	0.000 000 000 000	0.888 888 888 888	
	±0.774 596 669 241	±0.555 555 555 555	5
4	±0.339 981 043 584	±0.652 145 154 862	
	±0.861 136 311 594	±0.347 854 845 137	7
5	0.000 000 000 000	0.568 888 888 888	
	±0.538 469 310 105	±0.478 628 670 499	
	±0.906 179 845 938	±0.236 926 885 056	9
6	±0.238 619 186 083	±0.467 913 934 572	
	±0.661 209 386 466	±0.360 761 573 048	
	±0.932 469 514 203	±0.171 324 492 379	11
7	0.000 000 000 000	0.417 959 183 673	
	±0.405 845 151 377	±0.381 830 050 505	
	±0.741 531 185 599	±0.279 705 391 489	
	±0.949 107 912 342	±0.129 484 966 168	13
8	±0.183 434 642 495	±0.362 683 783 378	
	± 0.525 532 409 916	±0.313 706 645 877	
	±0.796 666 477 413	±0.222 381 034 453	
	±0.960 289 856 597	±0.101 228 536 290	15
9	0.000 000 000 000	0.330 239 355 001	
	±0.324 253 423 403	±0.312 347 077 040	
	±0.613 371 432 700	±0.260 610 696 402	
	±0.836 031 107 326	±0.180 648 160 694	
	±0.968 160 239 507	±0.081 274 388 361	17

After Abramowitz and Stegun (1983).

4.2 Local Coordinates

Generally, we will be integrating functions $f(x)$ over $[0, L]$ and not $f(\xi)$ over $[-1, 1]$, as required by Gauss–Legendre quadrature. Thus, some means of mapping between the physical x-space and the local ξ-space must first be established (Figure 4.2). This concept was first introduced by Taig (1961) and is now routine in finite element analysis. The shape functions defined in 4.4 can be transformed to the local coordinate system [−1,1] with the following linear relation:

$$x = \frac{L}{2}(\xi + 1) \tag{4.9}$$

Substituting this into 4.4 gives the shape functions defined in terms of local coordinates as follows:

$$N_1(\xi) = \frac{(1-\xi)}{2}, \quad N_2(\xi) = \frac{(1+\xi)}{2} \tag{4.10}$$

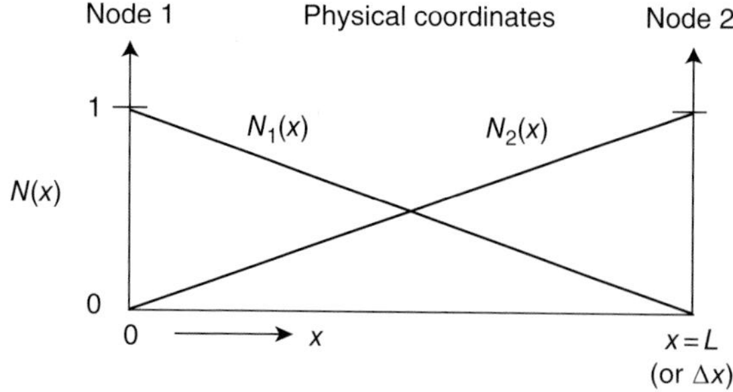

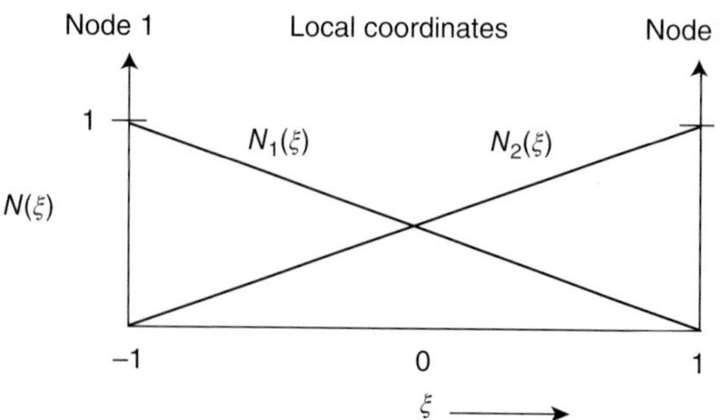

Figure 4.2 Linear shape functions defined in the physical (upper plot) and local (lower plot) coordinate systems. Note that these shape functions are identical in both coordinate systems.

Note that these shape functions in local coordinates can be used to map coordinates from the local coordinate system to the physical coordinate system. For example, if x_1 and x_2 are the nodal coordinates in physical space and N_1 and N_2 are both evaluated at a certain position ξ on the domain $[-1,1]$, then taking

$$x = N_1(\xi)x_1 + N_2(\xi)x_2 = \mathbf{N}\,\mathbf{x} \tag{4.11}$$

gives the coordinate x in the physical coordinate system that corresponds to ξ. Equation 4.11 is of identical form to the relation used to approximate the solution in the first step of the FEM (i.e., $T = N_1T_1 + N_2T_2$, see Equation 2.5). The elements that have this special property, where the shape functions defining the geometry and approximation of the solution are the same are called "isoparametric." For practical purposes, it means that we could have avoided this step altogether and simply defined the shape functions in the local coordinate system from the outset, which is the approach followed throughout the remainder of the text.

The derivatives of the shape functions can also be defined in the local coordinate system; but unlike the shape functions, they do need to be transformed from one coordinate system to another. Essentially, they need to be rescaled. This can be performed using the chain rule:

$$\frac{\partial N_i}{\partial x} = \frac{\partial N_i}{\partial \xi}\frac{\partial \xi}{\partial x} \tag{4.12}$$

The last term here can be obtained by taking the derivative of 4.9 with respect to ξ

$$\frac{\partial x}{\partial \xi} = \frac{L}{2} \tag{4.13}$$

and inverting the result, that is

$$\frac{\partial \xi}{\partial x} = \frac{2}{L} \tag{4.14}$$

Using this, the transformation rule for derivatives (4.12) becomes

$$\frac{\partial N_i}{\partial x} = \frac{\partial N_i}{\partial \xi}\frac{2}{L} \tag{4.15}$$

The derivatives of the shape functions in local coordinates, obtained by differentiating 4.10, are

$$\frac{\partial N_1}{\partial \xi} = -\frac{1}{2}, \quad \frac{\partial N_2}{\partial \xi} = \frac{1}{2} \tag{4.16}$$

which are constant in each element.

4.3 Evaluating the Integrals

Let us now consider how one goes about computing integrals such as those appearing in Equation 4.1 or 4.2 using Gauss–Legendre quadrature. As an example, say we want to calculate the following integral appearing in Equation 4.1:

$$\int_0^L N_1(x)N_1(x)\,dx \tag{4.17}$$

Our first task is to transform the integral to the domain $[-1, 1]$ so that it can be evaluated with Gauss–Legendre quadrature. This is achieved by taking 4.17, replacing the shape functions $N_1(x)$ with their local equivalents $N_1(\xi)$, and substituting $dx = (L/2)\,d\xi$ (from 4.13), leading to

$$\int_0^L N_1(x)N_1(x)\,dx = \frac{L}{2}\int_{-1}^{1} N_1(\xi)N_1(\xi)\,d\xi \tag{4.18}$$

Note that this is the same function integrated in Figure 4.1 (but without the factor $L/2$). The function being integrated is a second-order polynomial and, therefore, can be exactly integrated using Gauss–Legendre quadrature with two integration points. Applying the Gauss–Legendre quadrature rule (Equation 4.8), this integral becomes

$$\frac{L}{2}\int_{-1}^{1} N_1(\xi)N_1(\xi)\,d\xi \simeq \frac{L}{2}\sum_{k=1}^{n} N_1(\xi_k)N_1(\xi_k)\,w_k \tag{4.19}$$

If we choose two integration points, Table 4.1 provides their respective positions ($\xi_{1,2} = \pm\sqrt{1/3}$) and weights ($w_{1,2} = 1$). Substituting $\xi_k = \pm\sqrt{1/3}$ into N_1 from 4.10 gives

$$N_1(\xi_1) = \frac{\left(1 - \left(-\sqrt{\frac{1}{3}}\right)\right)}{2} = 0.7887 \tag{4.20}$$

$$N_1(\xi_2) = \frac{\left(1 - \left(\sqrt{\frac{1}{3}}\right)\right)}{2} = 0.2113 \tag{4.21}$$

Using these results, the summation in 4.19 becomes

$$\frac{L}{2}\sum_{k=1}^{2} N_1(\xi_k)N_1(\xi_k)\; w_k = \frac{L}{2}\,[(0.7887\times 0.7887\times 1) + (0.2113\times 0.2113\times 1)]$$
$$= 0.333L \tag{4.22}$$

which is identical to the exact result (i.e., $L/3$) obtained in Chapter 2 (see Section 2.4 and note that $L = \Delta x$). As a second example, consider numerical integration of the term

$$\int_0^L \kappa \frac{\partial N_2}{\partial x}\frac{\partial N_2}{\partial x} dx \tag{4.23}$$

appearing in Equation 4.2. Transforming the integrand using Equation 4.15 and inserting $dx = (L/2)\, d\xi$ (from 4.13) leads to

$$\int_0^L \kappa \frac{\partial N_2}{\partial x}\frac{\partial N_2}{\partial x} dx = \frac{2}{L}\int_{-1}^{1} \kappa \frac{\partial N_2}{\partial \xi}\frac{\partial N_2}{\partial \xi} d\xi \tag{4.24}$$

Using the Gauss–Legendre quadrature rule (Equation 4.8), the integral can be evaluated as follows:

$$\frac{2}{L}\int_{-1}^{1} \kappa \frac{\partial N_2}{\partial \xi}\frac{\partial N_2}{\partial \xi} d\xi \simeq \frac{2}{L}\sum_{k=1}^{n} \kappa \frac{\partial N_2}{\partial \xi_k}\frac{\partial N_2}{\partial \xi_k}\; w_k \tag{4.25}$$

The assumption of two integration points, a constant κ and noting that $\partial N_2/\partial\xi = 1/2$ (from Equation 4.16), results in

$$\frac{2}{L}\sum_{k=1}^{2} \kappa \frac{\partial N_2}{\partial \xi_k}\frac{\partial N_2}{\partial \xi_k}\; w_k = \frac{2\kappa}{L}\,[(0.5\times 0.5\times 1) + (0.5\times 0.5\times 1)] = \frac{\kappa}{L} \tag{4.26}$$

which once again equals the exact result computed in Chapter 2 (see Section 2.4).

From these simple examples, we see that numerical integration gives very accurate results with minimal computation (in this case, only two integration points). This accuracy is due the strength of the Gauss–Legendre method and the fact that the functions being integrated are relatively smooth.

4.4 Variable Material Properties

It is straightforward to compute the given integrals for cases when material properties vary spatially. When this occurs, one needs to evaluate the material properties at the integration points, which can be done by interpolation using the shape functions. As an example, consider numerical calculation of an integral appearing in the element load vector (Equation 4.3)

$$\int_0^L H(x)\, N_2(x)\, dx \tag{4.27}$$

for the case when H varies spatially. In local coordinates, this integral can be rewritten as follows:

$$\frac{L}{2}\int_{-1}^{1} H(\xi)\, N_2(\xi)\, d\xi \tag{4.28}$$

Applying the Gauss–Legendre rule, the integral can be evaluated by computing

$$\frac{L}{2}\sum_{k=1}^{n} H(\xi_k)\,N_2(\xi_k)w_k \tag{4.29}$$

We will assume that H_1 and H_2 are the known values of H at nodes 1 and 2 of a single element. To find H at the integration points, one can interpolate from the nodal values using the shape functions, that is,

$$H(\xi_k) = N_1(\xi_k)\,H_1 + N_2(\xi_k)\,H_2 \tag{4.30}$$

To evaluate the summation in 4.29, we will assume two integration points (Table 4.1). Evaluating H at the integration point $\xi_1 = -\sqrt{1/3}$ gives

$$\begin{aligned} H(\xi_1) &= \frac{1-(-\sqrt{1/3})}{2}H_1 + \frac{1+(-\sqrt{1/3})}{2}H_2 \\ &= 0.7887\,H_1 + 0.2113\,H_2 \end{aligned} \tag{4.31}$$

and for the point $\xi_2 = \sqrt{1/3}$,

$$\begin{aligned} H(\xi_2) &= \frac{1-\sqrt{1/3}}{2}H_1 + \frac{1+\sqrt{1/3}}{2}H_2 \\ &= 0.2113\,H_1 + 0.7887\,H_2 \end{aligned} \tag{4.32}$$

Substituting these into 4.29 and evaluating leads to

$$\begin{aligned} \frac{L}{2}\sum_{k=1}^{n} H(\xi_k)\,N_2(\xi_k)w_k &= \frac{L}{2}[0.7887\,H_1 + 0.2113\,H_2] \times 0.2113 \times 1 \\ &+ \frac{L}{2}[0.2113\,H_1 + 0.7887\,H_2] \times 0.7887 \times 1 \end{aligned} \tag{4.33}$$

Note that if H is constant at the two-element nodes (i.e., $H_1 = H_2 = H$), this reduces to $LH/2$, which is the exact result found in Chapter 2 (Section 2.4). The same procedure is used for any other material properties that vary in space.

4.5 Programming Considerations

To illustrate how numerical integration is implemented in a finite element program written in Matlab, we resolve the heat flow problem treated in Chapter 3 but now we evaluate **MM** (4.1), **KM** (4.2), and **F** (4.3) using Gauss–Legendre quadrature. A complete listing of the newly modified program is provided in Section 4.6. In what follows, we assume one uses the Matlab program presented in Section 3.5 as a starting point. The main modifications that must be made to this program are as follows:

1) Define the number of Gauss integration points. Using Table 4.1, save the positions (ξ_k) and weights (w_k) for each point. Assuming two integration points, this is performed in Matlab with the lines

```
nip    = 2 ; % define number of integration points (ips)
points = [-sqrt(1/3) sqrt(1/3)] ; % define positions of ips
wts    = [1 1] ;                  % define weights of ips
```

2) Evaluate (and save for later use) the shape functions and their derivatives, defined in local coordinates, at the positions of the integration points

```
for k=1:nip % loop over integration points
  fun_s(k,:) = [(1-points(k))/2  (1+points(k))/2] ; % N1 and N2
  der_s(k,:) = [-1/2  1/2] ; %  shape function derivatives
end
```

where the shape functions and shape function derivatives are defined in Equations 4.10 and 4.16.

3) Within the existing element loop (within which the global matrix is assembled), delete the lines where **MM**, **KM**, and **F** were previously constructed, since now they must be computed numerically. Instead, initialize them as follows:

```
MM = zeros(nod,nod); % initialise MM
KM = zeros(nod,nod); % initialise KM
F  = zeros(nod,1);   % initialise F
```

It is important to repeat this step before each summation is performed to avoid accumulating results from the integration of other elements.

4) Within this element loop, start a new loop over all integration points and retrieve the shape functions and shape function derivatives evaluated at the current integration point (k) as follows:

```
fun = fun_s(k,:) ; % retrieve shape functions
der = der_s(k,:) ; % retrieve deriv. of shape functions
```

5) Compute the factor $L/2$ ($= dx/d\xi$, see Equation 4.13)—in the program called `detjac`—used to convert the integration limits from $[0, L]$ to $[-1, 1]$. In the program, because `dx` is used for the element length (not L) this becomes

```
detjac = dx/2 ; % factor to change integration limits
```

6) Convert derivatives of the shape functions from the local coordinates to the physical coordinates using Equation 4.15 as follows:

```
invjac = 2/dx ; % conversion factor
deriv  = der*invjac ; % convert derivatives to physical coords
```

Here, it should be recalled that `der` contains the shape function derivatives in local coordinates.

7) If material properties vary spatially (which they don't in the provided program), they should be interpolated to the integration points. As an example, if κ varies spatially and is defined on the global nodes (in the vector `kappav`), one could obtain the scalar `kappa` at the current integration point using

```
kappa = fun*kappav(num) ; % interpolate kappa to int. pt.
```

where `fun` is a row vector of length 2 and `kappav(num)` is a column vector containing the values of κ at the two nodes of the current element. Here, the shape functions, which are normally used to interpolate (approximate) the solution defined at nodes, are used to interpolate the nodal κ values.

8) Perform the quadrature summations defined, for example, by Equation 4.19 (for **MM**) and Equation 4.25 (for **KM**). Using matrix operations (see Equations 4.1, 4.2, and 4.3) these are as follows:

```
MM = MM + fun'*fun*detjac*wts(k) ;              % mass matrix
KM = KM + kappa*deriv'*deriv*detjac*wts(k) ; % stiffness matrix
F  = F  + H*fun'*detjac*wts(k) ;               % load vector
```

Note that each result is added to that computed for previous integration points (i.e., `MM = MM ... +...`). This marks the end of the loop over integration points. At this stage, the element matrices and load vector have been fully integrated. They can now be added to the global stiffness matrix, exactly as was done in the previous program.

4.6 Matlab Script

The following Matlab script computes the numerical solution to the transient heat conduction equation in 1D with the FEM using Gauss–Legendre quadrature to evaluate the element matrices. The program can be compared directly with that presented in Section 3.5 where the element matrices were evaluated analytically. The results are indeed identical.

```
%-----------------------------------------------------------
% Program diffusion1d_numint.m
% 1-D FEM solution of diffusion equation
% using Gauss-Legendre quadrature to integrate
% the element matrices.
% For the analytical solution see equation 7, page 130
% Carslaw and Jaeger (1959)
%-----------------------------------------------------------

clear % clear memory from current workspace

% physical parameters
seconds_per_yr = 60*60*24*365; % number of seconds in one year
lx     = 10000 ;      % length of spatial domain (m)
Cp     = 1e3  ;       % rock heat capacity (J/kg/K)
rho    = 2700 ;       % rock density (kg/m^3)
K      = 3.3  ;       % bulk thermal conductivity (W/m/K)
kappa  = K/(Cp*rho); % thermal diffusivity (m^2/s)
Tb     = 0 ;          % temperatures at boundaries (°C)
A      = 2.6e-6 ;     % heat production (W/m^3)
H      = A/(rho*Cp); % heat source term (°K/s)

% numerical parameters
dt       = 1000*seconds_per_yr ; % time step (s)
ntime    = 5000 ;         % number of time steps
nels     = 40 ;           % total number of elements
nod      = 2 ;            % number of nodes per element
nn       = nels+1         % total number of nodes
dx       = lx/nels ;      % element size
g_coord = [0:dx:lx] ;   % spatial domain (1-D mesh)
nip      = 2 ;            % no. Gauss int. pts per element

% integration data
points    =   [-sqrt(1/3) sqrt(1/3)] ; % positions of the Gauss points
wts       =   [1  1] ;                 % Gauss-Legendre weights

% shape functions and their derivatives (both in local coordinates)
% evaluated at the Gauss integration points
for k=1:nip
```

```
    fun_s(k,:) = [ (1-points(k))/2 (1+points(k))/2 ] ; % N1 and N2
    der_s(k,:) = [ -1/2 1/2 ] ;                        % dN1/xi  dN2/xi
end

%  define boundary conditions
 bcdof = [  1    nn ]   ; % boundary nodes
 bcval = [  Tb   Tb ]   ; % boudary values

% define connectivity and equation numbering
g_num      = zeros(nod,nels) ;
g_num(1,:) = [1:nn-1]   ;
g_num(2,:) = [2:nn]     ;

% initialise matrices and vectors
ff     = zeros(nn,1);          % system load vector
b      = zeros(nn,1);          % system rhs vector
displ  = zeros(nn,1);          % Initial temperature (°C)
lhs    = sparse(nn,nn);        % system lhs matrix
rhs    = sparse(nn,nn);        % system rhs matrix

%-------------------------------------------------------
% matrix assembly
%-------------------------------------------------------

   for iel=1:nels % loop over all elements
      num    = g_num(:,iel)  ;            % retrieve equation number
      dx     = abs(diff(g_coord(num))) ; % length of element
      MM     = zeros(nod,nod) ;
      KM     = zeros(nod,nod) ;
      F      = zeros(nod,1)    ;
      for k=1:nip % integrate element matrices
          fun = fun_s(k,:);% shape function in local coords.
          der = der_s(k,:);% deriv. of shape fun. in local coords
          detjac = dx/2   ;% factor to scale integration limits
          invjac = 2/dx   ;% factor to scale derivatives
          deriv  = der*invjac ;%shape. fun. deriv. in phys. coords.
          MM      = MM + fun'*fun*detjac*wts(k) ;        % mass matrix
          KM      = KM + deriv'*kappa*deriv*detjac*wts(k);% stiffness
          F       = F  + H*fun'*detjac*wts(k) ;          % load vector
      end % end of integration
      lhs(num,num) = lhs(num,num) + MM/dt + KM ; %  assemble lhs
      rhs(num,num) = rhs(num,num) + MM/dt      ; %  assemble rhs
      ff(num)      = ff(num)      + F          ; %  assemble load
   end    % end of element loop

%-------------------------------------------------------

% time loop
t  = 0 ; % time
k  = 1 ; % counter
ii = [100 1000 5000]; % array used for plotting
for n=1:ntime
   n
   t  = t + dt ;                       % compute time
   b = rhs*displ + ff ;                % form rhs load vector

   % impose boundary conditions
   lhs(bcdof,:) = 0 ;                  % zero the relevent equations
   tmp = spdiags(lhs,0) ;              % store diagonal
```

```
   tmp(bcdof)=1 ;                   % place 1 on stored-diagonal
   lhs=spdiags(tmp,0,lhs);          % reinsert diagonal
   b(bcdof) = bcval ;               % set rhs

   displ = lhs \ b ;                % solve system of equations

%-----------------------------------------------------------
% evaluate analytical solution
   nterms = 100 ;
   L      = lx/2 ;
   x      = linspace(-L,L,100) ;
   sumv = 0 ;
   for ni=0:nterms
      et    = exp(-kappa*(2*ni+1)^2*pi^2*t/4/L^2);
      sumv = sumv + (-1)^ni/(2*ni+1)^3*cos((2*ni+1)/2/L*pi*x)*et ;
   end
   Texact = A*L^2/(2*K)*(1-x.^2/L^2-32/pi^3.*sumv) ;
%-----------------------------------------------------------

  % plotting
    figure(1)
    plot(g_coord,displ,'o')
    hold on
    plot(x+L,Texact,'r')
    title('Comparison beteen numerical and analytical solutions')
    xlabel('Distance (m)')
    ylabel('Temperature (°C)')
    hold off
    drawnow
    pause

end % end of time loop

%-----------------------------------------------------------
```

4.7 Exercises

1) Show that

$$\int_{-1}^{1} x^3 dx = 0$$

and

$$\int_{4}^{14} x^3 dx = 9540$$

using Gauss–Legendre quadrature with two integration points.

2) Use the FEM with Gauss–Legendre quadrature to solve the boundary value problem

$$u = \frac{\partial^2 u}{\partial x^2} + x$$

with

$$0 < x < 1, \quad u(0) = u(1) = 0$$

Compare your numerical results with the exact solution

$$u(x) = x - \frac{\sinh x}{\sinh 1}$$

3) Use the FEM along with Gauss–Legendre quadrature to solve the ordinary differential equation boundary value problem

$$\frac{\partial^2 u}{\partial x^2} = u^{4x}$$

with

$$-1 < x < 1, \quad u(\pm 1) = 0$$

Check your numerical results with the exact solution

$$u(x) = \frac{e^{4x} - x\sinh(4) - \cosh(4)}{16}$$

4) Using quadratic shape functions defined as (derived as an exercise in Chapter 3)

$$N_1 = 2\frac{x^2}{L^2} - 3\frac{x}{L} + 1$$
$$N_2 = 4\frac{x}{L} - 4\frac{x^2}{L^2}$$
$$N_3 = 2\frac{x^2}{L^2} - \frac{x}{L}$$

evaluate (exactly and using Gauss–Legendre quadrature) the element matrices **MM**, **KM**, and **F** (assuming constant material properties) defined as follows:

$$\mathbf{MM} = \int_0^L \mathbf{N}^T\mathbf{N}\,dx = \int_0^L \begin{bmatrix} N_1N_1 & N_1N_2 & N_1N_3 \\ N_2N_1 & N_2N_2 & N_2N_3 \\ N_3N_1 & N_3N_2 & N_3N_3 \end{bmatrix} dx$$

$$\mathbf{KM} = \int_0^L \kappa \frac{\partial \mathbf{N}^T}{\partial x}\frac{\partial \mathbf{N}}{\partial x} dx = \int_0^L \kappa \begin{bmatrix} \frac{\partial N_1}{\partial x}\frac{\partial N_1}{\partial x} & \frac{\partial N_1}{\partial x}\frac{\partial N_2}{\partial x} & \frac{\partial N_1}{\partial x}\frac{\partial N_3}{\partial x} \\ \frac{\partial N_2}{\partial x}\frac{\partial N_1}{\partial x} & \frac{\partial N_2}{\partial x}\frac{\partial N_2}{\partial x} & \frac{\partial N_2}{\partial x}\frac{\partial N_3}{\partial x} \\ \frac{\partial N_3}{\partial x}\frac{\partial N_1}{\partial x} & \frac{\partial N_3}{\partial x}\frac{\partial N_2}{\partial x} & \frac{\partial N_3}{\partial x}\frac{\partial N_3}{\partial x} \end{bmatrix} dx$$

$$\mathbf{F} = \int_0^L H\mathbf{N}^T dx = \int_0^L H \begin{Bmatrix} N_1 \\ N_2 \\ N_3 \end{Bmatrix} dx$$

Suggested Reading

A. J. M. Ferreira, *MATLAB Codes for Finite Element Analysis: Solids and Structures (Solid Mechanics and Its Applications)*, Springer, Berlin, 2009.

Y. W. Kwong, and H. C. Bang, *The Finite Element Method using Matlab*, CRC Press, New York, 2000.

I. M. Smith, and D. V. Griffith, *Programming the Finite Element Method*, John Wiley & Sons, Ltd, Chichester, 1998.

O. C. Zienkiewicz, and R. L. Taylor, *The Finite Element Method, Volume 1, The Basis*, Butterworth-Heinemann, Oxford/Boston, MA, 2000.

5

The Finite Element Method in Two Dimensions

This chapter deals with the finite element method (FEM) in two spatial dimensions. Using the diffusion equation as an example, we consider discretization of the governing partial differential equation (PDE), numerical integration of the element matrices, assembly of the element matrices, and finally solution of the global system of equations. A program listed in Section 5.6 shows how the problem is programmed and solved using Matlab.

One of the strengths of the FEM is the relative ease with which it is possible to pass from one-dimension (1D) to two (or more) dimensions. To demonstrate the FEM in 2Ds, we once again solve the transient heat conduction equation considered in the previous chapters. Remember that the heat conduction equation is derived (see Appendix A) by combining a statement for the conservation of energy

$$\rho c \frac{\partial T}{\partial t} = -\nabla^T \cdot \mathbf{q} + A = -\begin{bmatrix} \frac{\partial}{\partial x} & \frac{\partial}{\partial y} \end{bmatrix} \cdot \begin{Bmatrix} q_x \\ q_y \end{Bmatrix} + A \tag{5.1}$$

with a constitutive relation known as Fourier's law:

$$\mathbf{q} = \begin{Bmatrix} q_x \\ q_y \end{Bmatrix} = -\begin{bmatrix} k & 0 \\ 0 & k \end{bmatrix} \begin{Bmatrix} \frac{\partial}{\partial x} \\ \frac{\partial}{\partial y} \end{Bmatrix} T = -\mathbf{k}\nabla T \tag{5.2}$$

to yield

$$\frac{\partial T}{\partial t} = \nabla^T (\mathbf{K}\nabla T) + \frac{A}{\rho c} \tag{5.3}$$

Recall that k is the thermal conductivity, ρ is the density, c is the heat capacity, and A is the rate of internal heat generation per unit volume. Note also that $\mathbf{K}$ is the thermal diffusivity tensor, with diagonal entries $k_x/\rho c$ and $k_y/\rho c$ and zeros elsewhere. If the thermal diffusivity is constant, this equation can be written as

$$\frac{\partial T}{\partial t} = \kappa \left(\frac{\partial^2 T}{\partial x^2} + \frac{\partial^2 T}{\partial y^2} \right) + H \tag{5.4}$$

where $H = A/\rho c$ and $\kappa = k/\rho c$. Nevertheless, as we will see, it is more natural and general to leave the equation in the form shown in Equation 5.3. Here, we are interested in solving Equation 5.3 with a uniform unit initial temperature distribution, that is,

$$T(x, y, t = 0) = 1 \quad \forall\, x \in [0; lx] \text{ and } y \in [0; ly] \tag{5.5}$$

and the following Dirichlet conditions on all four spatial boundaries:

$$T(x = 0, y, t) = 0,\ T(x = lx, y, t) = 0,\ T(x, y = 0, t) = 0,\ T(x, y = ly, t) = 0 \tag{5.6}$$

Practical Finite Element Modeling in Earth Science Using Matlab, First Edition. Guy Simpson.

Companion website: www.wiley.com/go/simpson

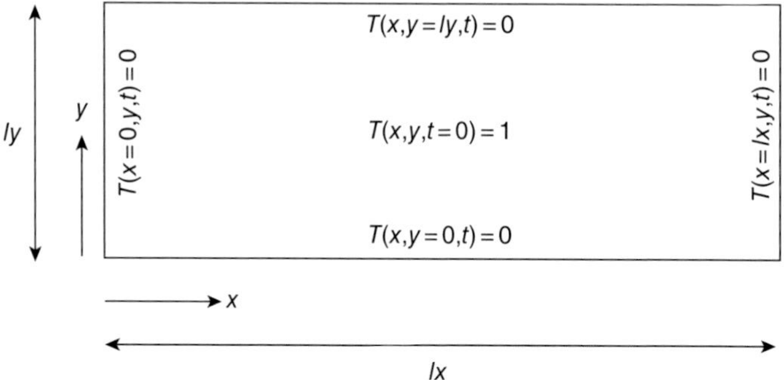

Figure 5.1 Model domain with boundary and initial conditions for problem involving heat conduction in 2Ds.

Here, the model domain extends from 0 to lx in the x-direction and from 0 to ly in the y-direction (Figure 5.1).

5.1 Discretization

As in the 1D example treated in Chapter 2, the first step of the FEM is to choose an element type where the governing equation will be discretized. An easy and common choice in 2Ds is the quadrilateral element with four nodes, one at each corner (Figure 5.2). The shape functions that go together with this type of element are linear in each direction (usually called bilinear in 2Ds). For the moment, we shall simply refer to the shape functions using the notation

$$\mathbf{N} = [N_1 \ N_2 \ N_3 \ N_4] \tag{5.7}$$

and we note that the shape functions are expressed in terms of the physical coordinate system (i.e., they are functions of x and y).

The second step of the FEM involves approximating the continuous variable T in Equation 5.3 in terms of nodal variables T_i using the shape functions just defined, that is,

$$T \simeq [N_1 \ N_2 \ N_3 \ N_4] \begin{Bmatrix} T_1 \\ T_2 \\ T_3 \\ T_4 \end{Bmatrix} = \mathbf{NT} \tag{5.8}$$

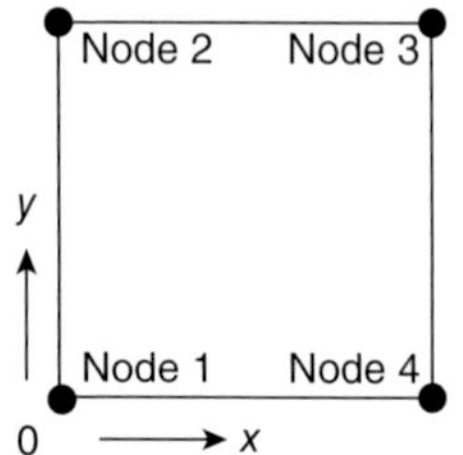

Figure 5.2 Quadrilateral element with four corner nodes.

Substituting this approximation for T into Equation 5.3 gives the residual equation

$$\frac{\partial}{\partial t}\mathbf{NT} - \nabla^T(\mathbf{K}\nabla\mathbf{NT}) - H = R \tag{5.9}$$

where R is the residual introduced by the given approximation. The term $\nabla\mathbf{N}$ stands for derivatives of the shape functions in physical coordinates, that is,

$$\nabla\mathbf{N} = \begin{Bmatrix} \frac{\partial}{\partial x} \\ \frac{\partial}{\partial y} \end{Bmatrix} [N_1\ N_2\ N_3\ N_4] = \begin{bmatrix} \frac{\partial N_1}{\partial x} & \frac{\partial N_2}{\partial x} & \frac{\partial N_3}{\partial x} & \frac{\partial N_4}{\partial x} \\ \frac{\partial N_1}{\partial y} & \frac{\partial N_2}{\partial y} & \frac{\partial N_3}{\partial y} & \frac{\partial N_4}{\partial y} \end{bmatrix} \tag{5.10}$$

The next step of the FEM requires multiplication of the residual of Equation 5.9 with weighting functions, integrating over the element and setting the result to zero. Performing these tasks and assuming that the weighting functions are identical to the shape functions (i.e., the Galerkin form of the FEM) leads to

$$\int\int \mathbf{N}^T \frac{\partial}{\partial t}\mathbf{NT}\, dxdy - \int\int \mathbf{N}^T\nabla^T(\mathbf{K}\nabla\mathbf{NT})\, dxdy = \int\int H\, \mathbf{N}^T dxdy \tag{5.11}$$

Integrating by parts where necessary (term 2, which involves double differentiation of linear shape functions) and neglecting the resulting boundary integrals (Appendix D) gives

$$\int\int \mathbf{N}^T\mathbf{N}\, dxdy \frac{\partial}{\partial t}\mathbf{T} + \int\int (\nabla\mathbf{N})^T\mathbf{K}\nabla\mathbf{N}\, dxdy\, \mathbf{T} = \int\int H\mathbf{N}^T\, dxdy \tag{5.12}$$

This equation can be simplified to

$$[\mathbf{MM}]\frac{\partial}{\partial t}\mathbf{T} + [\mathbf{KM}]\,\mathbf{T} = \mathbf{F} \tag{5.13}$$

where

$$\mathbf{MM} = \int\int \mathbf{N}^T\mathbf{N}\, dxdy \tag{5.14}$$

$$\mathbf{KM} = \int\int (\nabla\mathbf{N})^T\mathbf{K}\nabla\mathbf{N}\, dxdy \tag{5.15}$$

and

$$\mathbf{F} = \int\int H\mathbf{N}^T\, dxdy \tag{5.16}$$

Note that the element matrices and vectors have similar form to those derived in Chapter 2 in 1D. The matrix **MM** is the now familiar element mass matrix that comes from the first term in the governing PDE (5.3), **KM** is the element stiffness matrix that comes from discretizing the spatial derivatives and **F** is the element load vector related to the source term. However, the dimensions of the element matrices **MM** and **KM** are [4 x 4] for the four-node elements, whereas the element vector **F** has dimensions [4 x 1] (i.e., 4 rows and 1 column).

The final step of the discretization process involves replacing the continuous time derivative in the first term of 5.13 with a discrete approximation. Using an implicit finite difference approximation for this purpose (see Appendix E), Equation 5.13 can be written as

$$[\mathbf{MM}]\frac{\mathbf{T}^{n+1} - \mathbf{T}^{n}}{\Delta t} + [\mathbf{KM}]\ \mathbf{T}^{n+1} = \mathbf{F} \tag{5.17}$$

where Δt is the time interval between n and $n+1$, $\mathbf{T}^{n+1}$ are the unknown nodal temperatures and $\mathbf{T}^{n}$ are the known (old) nodal temperatures. Rearranging, such that the known temperatures are on the right-hand side and the unknown temperatures are on the left-hand side,

$$\left(\frac{\mathbf{MM}}{\Delta t} + \mathbf{KM}\right)\mathbf{T}^{n+1} = \frac{\mathbf{MM}}{\Delta t}\ \mathbf{T}^{n} + \mathbf{F} \tag{5.18}$$

that can be written more compactly as

$$\mathbf{L}\ \mathbf{T}^{n+1} = \mathbf{R}\ \mathbf{T}^{n} + \mathbf{F} \tag{5.19}$$

where

$$\mathbf{L} = \frac{\mathbf{MM}}{\Delta t} + \mathbf{KM} \tag{5.20}$$

and

$$\mathbf{R} = \frac{\mathbf{MM}}{\Delta t} \tag{5.21}$$

Note the close similarity between this and Equation 2.28 for the 1D version of the same problem.

5.2 Geometry and Nodal Connectivity

In Section 5.1, it was shown that discretization results in a series of element matrices (i.e., **KM**, **MM**, and **F**) that involve shape functions or their derivatives, which later must be integrated over "finite" elements. In this section, we define the geometry of these elements over which integration will be performed. This requires specification of a mesh geometry and how the various nodes and elements are interconnected. To show how this mesh geometry and nodal connectivity can be defined in 2Ds, consider a simple straight-sided mesh with nx nodes in the x-direction and ny nodes in the y-direction (Figure 5.3). Each node in the mesh has a unique x, y position and a unique global number. The nodal coordinates for this particular mesh can be generated in Matlab with the snippet

```
g_coord = zeros(ndim,nn) ; % initialisation of nodal coordinates
n = 1 ; % initialise node counter
for i=1:nx % loop over nodes in x-direction
  for j=1:ny % loop over nodes in y-direction
    g_coord(1,n) = (i-1)*dx ;
    g_coord(2,n) = (j-1)*dy ;
    n = n + 1 ; % increment nodal counter
  end
end
```

where `dx` and `dy` are the nodal spacings (element lengths) in the x- and y-directions, respectively, and `ndim` is the number of spatial dimensions (`ndim=2`). Note that because the inner loop here is in the

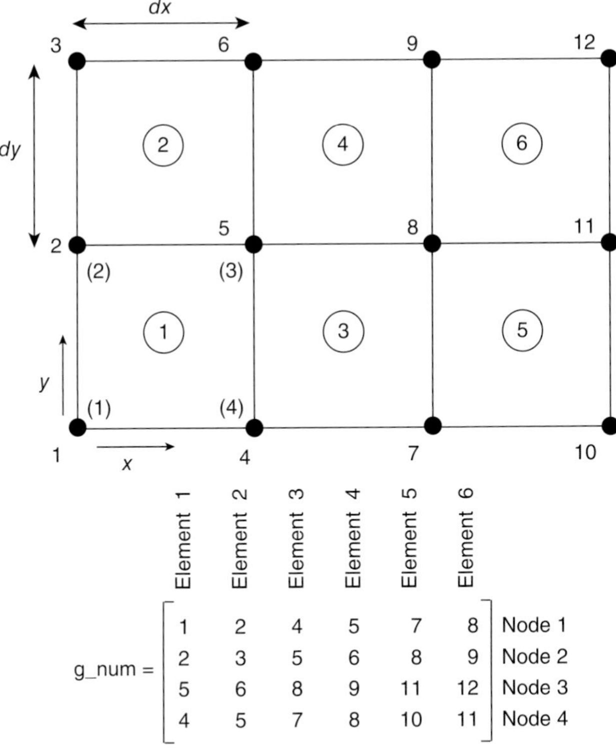

Figure 5.3 Two-dimensional mesh showing global node numbers (unbracketed), local node numbering (in parentheses), and element numbers (circled).

y-direction, nodes are numbered in this direction first. Although this choice is arbitrary, one should always choose to number first the direction that contains the least nodes, since this leads to global matrices that have a narrower bandwidth, which can be solved more efficiently.

Each element in the mesh contains four corner nodes. At an element level, these nodes are numbered using clockwise ordering, beginning from the bottom left-hand corner of each element (Figures 5.2 and 5.3). The global numbers of these same nodes are defined by the order of the node coordinates in `g_coord`. The connection between the local and global nodes for any given element (`g_num`) can be computed in Matlab with the lines

```
g_num = zeros(nod,nels); % initialisation of connectivity matrix
gnumbers = reshape(1:nn,[ny nx]) ; % grid of global node numbers
iel = 1 ;        % element counter
for i=1:nxe    % loop over elements in x-direction
  for j=1:nye % loop over elements in y-direction
    g_num(1,iel) = gnumbers(j,i) ;     % local node 1
    g_num(2,iel) = gnumbers(j+1,i) ;   % local node 2
    g_num(3,iel) = gnumbers(j+1,i+1); % local node 3
    g_num(4,iel) = gnumbers(j,i+1);   % local node 4
    iel          = iel + 1 ; % increment element counter
  end
end
```

where nxe and nye are the number of elements in each direction, nels is the total number of elements, nod is the number of nodes per element (4), and nn is the total number of nodes in the mesh. An example of g_num for a small mesh is shown in Figure 5.3. Once g_num has been created, the global node numbers (num) and global node coordinates (coord) for any given element iel can be easily obtained using the lines

```
num    = g_num(:,iel)      ;   % global node numbers for element 'iel'
coord = g_coord(:,num)' ;   % global node coords. for element 'iel'
```

5.3 Integration of Element Matrices

We now turn our attention to integrating the element matrices and load vector over the now-defined finite elements. Although this could be done analytically for simple cases, we follow the more standard approach of doing this using Gauss–Legendre quadrature (see Chapter 4), which is accurate, fast, and easy to implement. As we will see, the procedure in 2Ds is similar to that already encountered in 1D.

The formula for Gauss–Legendre quadrature in 2Ds is

$$\int_{-1}^{1}\int_{-1}^{1} f(\xi,\eta)\, d\xi d\eta \simeq \sum_{i=1}^{nipx}\sum_{j=1}^{nipy} f(\xi_i,\eta_j)\, w_i w_j = \sum_{k=1}^{nip} f(\xi_k,\eta_k)\, w_k \tag{5.22}$$

where *nipx* and *nipy* are number of integration points in each direction, ξ_i and η_j are the local spatial coordinates of the integration points in each direction, and w_i and w_j, or w_k, are the weights. The quadrature weights and locations of the integration points are once again obtained from Table 4.1. Note that the double summation in term 2 of Equation 5.22 can be evaluated as a single summation (term 3) over *nip* (= *nipx* x *nipy*) points for a rectangle. For 2D quadrature, a third-order polynomial can be integrated exactly using two integration points in each direction (i.e., $nip = 4$; see Figure 5.4).

Before evaluating the integrals in the element matrices, two transformations are necessary. First, since the shape functions (and their derivatives) in the integrals are expressed in terms of physical coordinates, it is necessary to devise some means of expressing them in terms of local coordinates (as required by Gauss–Legendre quadrature). Second, the area over which the integration has to be carried out must be expressed in terms of local coordinates, with an appropriate change in the limits of integration.

In Chapter 4, we presented the approach whereby one initially defines shape functions (and their derivatives) in terms of physical coordinates (e.g., x) and then maps them to the local coordinate system (ξ) using a linear transformation (see Section 4.2). Although this approach is easily performed

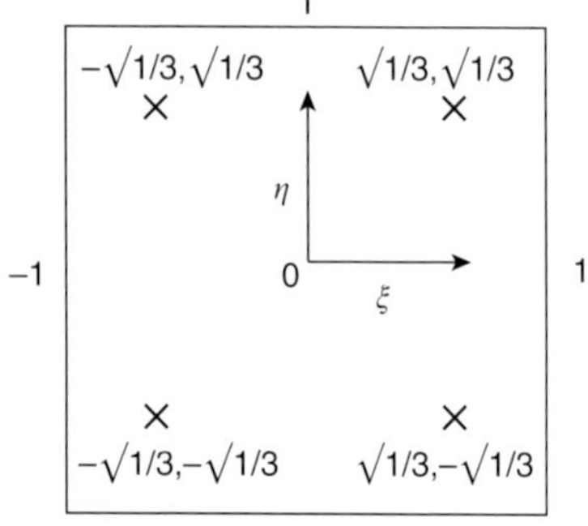

Figure 5.4 Locations of four Gauss integration points (two in each direction) in a 2D element in the local coordinate system.

in 1D, in higher dimensions and for complex elements, it becomes increasingly difficult to express the shape functions algebraically in terms of physical coordinates. A better alternative is to define the shape functions and their derivatives in local coordinates from the outset (Taig, 1961). The shape functions for four-node quadrilateral elements in local coordinates are:

$$\begin{aligned} N_1 &= \frac{1}{4}(1-\xi)(1-\eta) \\ N_2 &= \frac{1}{4}(1-\xi)(1+\eta) \\ N_3 &= \frac{1}{4}(1+\xi)(1+\eta) \\ N_4 &= \frac{1}{4}(1+\xi)(1-\eta) \end{aligned} \tag{5.23}$$

where ξ and η are local coordinates on the domain $[-1,1]\times[-1,1]$ (Figure 5.5). Because the four-node quadrilaterals are isoparametric elements, any point ξ,η in local-space can be mapped one to one to a corresponding point x, y in physical space using the following transformation:

$$\begin{aligned} x &= N_1(\xi,\eta)\,x_1 + N_2(\xi,\eta)\,x_2 + N_3(\xi,\eta)\,x_3 + N_4(\xi,\eta)\,x_4 \\ y &= N_1(\xi,\eta)\,y_1 + N_2(\xi,\eta)\,y_2 + N_3(\xi,\eta)\,y_3 + N_4(\xi,\eta)\,y_4 \end{aligned} \tag{5.24}$$

Here, $N_i(\xi,\eta)$ are given by 5.23 and x_1, y_1 are the coordinates of node 1 in physical space, and so on (Figure 5.5). This property is important because it means that the shape functions in physical coordinates appearing in the integrals (Equations 5.14 and 5.16) can simply be replaced with their normalized equivalents (5.23), as long as the change in element size is accounted for (which is dealt with in the following text). The derivatives of the given shape functions in terms of local coordinates are as follows:

$$\begin{aligned} \frac{\partial N_1}{\partial \xi} &= \frac{1}{4}(\eta-1) & \frac{\partial N_1}{\partial \eta} &= \frac{1}{4}(\xi-1) \\ \frac{\partial N_2}{\partial \xi} &= -\frac{1}{4}(\eta+1) & \frac{\partial N_2}{\partial \eta} &= \frac{1}{4}(1-\xi) \\ \frac{\partial N_3}{\partial \xi} &= \frac{1}{4}(\eta+1) & \frac{\partial N_3}{\partial \eta} &= \frac{1}{4}(\xi+1) \\ \frac{\partial N_4}{\partial \xi} &= \frac{1}{4}(1-\eta) & \frac{\partial N_4}{\partial \eta} &= -\frac{1}{4}(1+\xi) \end{aligned} \tag{5.25}$$

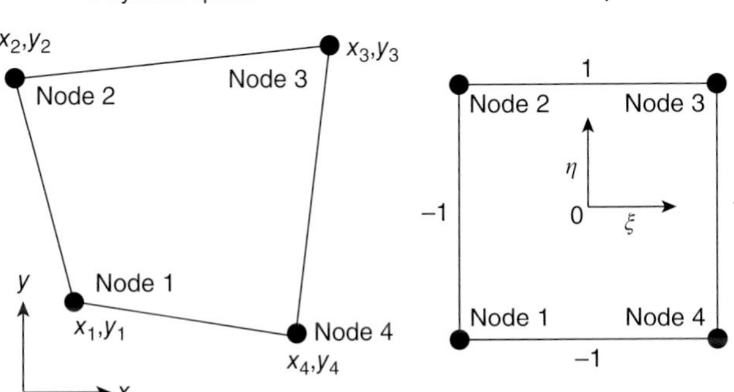

Figure 5.5 A general four-node quadrilateral element in physical space and its mapped equivalent in natural space.

Derivatives are easily converted from one coordinate system to another by means of the chain rule, expressed in matrix form as

$$\begin{pmatrix} \frac{\partial}{\partial \xi} \\ \frac{\partial}{\partial \eta} \end{pmatrix} \mathbf{N} = \begin{bmatrix} \frac{\partial x}{\partial \xi} & \frac{\partial y}{\partial \xi} \\ \frac{\partial x}{\partial \eta} & \frac{\partial y}{\partial \eta} \end{bmatrix} \begin{Bmatrix} \frac{\partial}{\partial x} \\ \frac{\partial}{\partial y} \end{Bmatrix} \mathbf{N} = \mathbf{J} \begin{Bmatrix} \frac{\partial}{\partial x} \\ \frac{\partial}{\partial y} \end{Bmatrix} \mathbf{N} \tag{5.26}$$

where $\mathbf{J}$ is the Jacobian matrix. The Jacobian matrix can be found by differentiating the global coordinates with respect to the local coordinates (i.e., multiply the coordinates of a particular element with the derivatives of the shape functions), that is,

$$\mathbf{J} = \begin{bmatrix} \frac{\partial N_1}{\partial \xi} & \frac{\partial N_2}{\partial \xi} & \frac{\partial N_3}{\partial \xi} & \frac{\partial N_4}{\partial \xi} \\ \frac{\partial N_1}{\partial \eta} & \frac{\partial N_2}{\partial \eta} & \frac{\partial N_3}{\partial \eta} & \frac{\partial N_4}{\partial \eta} \end{bmatrix} \begin{bmatrix} x_1 & y_1 \\ x_2 & y_2 \\ x_3 & y_3 \\ x_4 & y_4 \end{bmatrix} \tag{5.27}$$

where x_1 is the x-coordinate of node 1, and so on. The derivatives of the shape functions in terms of global coordinates can thus be found from

$$\begin{Bmatrix} \frac{\partial}{\partial x} \\ \frac{\partial}{\partial y} \end{Bmatrix} \mathbf{N} = \mathbf{J}^{-1} \begin{Bmatrix} \frac{\partial}{\partial \xi} \\ \frac{\partial}{\partial \eta} \end{Bmatrix} \mathbf{N} \tag{5.28}$$

where $\mathbf{J}^{-1}$ is the inverse of the Jacobian matrix. Transformation of the limits of integration is carried out using the determinant of the Jacobian (det $\mathbf{J}$) according to the following relation:

$$\int \int f(x,y)\, dxdy = \int_{-1}^{1} \int_{-1}^{1} f(\xi,\eta) \det \mathbf{J}\, d\xi d\eta \tag{5.29}$$

Combining this result with Equation 5.22 yields the final integration formula

$$\int \int f(x,y)\, dxdy \simeq \sum_{k=1}^{nip} f(\xi_k,\eta_k) \det \mathbf{J}\, w_k \tag{5.30}$$

In summary, the following steps must be carried out to integrate the element matrices:

1) Define the number of integration points (*nip*) and tabulate the integration points and weights using Table 4.1.
2) Evaluate and save the shape functions and their derivatives (defined in terms of local coordinates, see Equations 5.23 and 5.25) at the integration points.
3) Do a loop over all elements and initialize the element matrices **MM**, **KM**, and **F**. Also retrieve the global coordinates for the nodes of the current element. In Matlab, this last step can be achieved with the lines

```
num   = g_num(:,iel)      ;% nodes for current element
coord = g_coord(:,num)' ;% global nodal coords. for element
```

For the four-node quadrilateral element, `coord` has four rows (one for each node) and two columns, one for each dimension (see second term on the right-hand side of Equation 5.27).

4) Within the element loop, do a loop over each integration point k. Retrieve the shape functions and their derivatives for the current integration point

```
fun = fun_s(k,:)    ; % shape functions in local coordinates
der = der_s(:,:,k); % shape fun. derivs. in local coordinates
```

5) Calculate the Jacobian matrix by performing the operation in Equation 5.27 and invert it.

```
jac    = der*coord ;    % jacobian matrix
invjac = inv(jac)  ;    % inverse of jacobian matrix
```

6) Convert derivatives of the shape functions from local to global coordinates by performing the operation in Equation 5.28, that is,

```
deriv = invjac*der ; % transform deriv. of N to physical-space
```

7) Calculate the determinant of the Jacobian.

```
detjac = det(jac) ; % determinant of the jacobian
```

8) Perform the vector multiplications to form the function being integrated (e.g., $\mathbf{N}^T\mathbf{N}$ for the **MM** matrix, see Equation 5.14). Multiply the result by the determinant of the Jacobian and the weights (as in Equation 5.30), adding the result to that computed previously. This can be done in Matlab for the matrices **MM**, **KM**, and **F** with the lines

```
MM = MM + fun'*fun*detjac*wt(k) ;                % mass matrix
KM = KM + deriv'*kappa*deriv*detjac*wt(k) ; % stiffness matrix
F  = F  + H*fun'*detjac*wt(k) ;                  % load vector
```

Once the loop over integration points is complete, the element matrices have been completely integrated.

5.4 Multielement Assembly

In Section 5.1, we showed that discretization of the transient heat conduction equation in 2Ds led to the system of element-level equations

$$\mathbf{L}\,\mathbf{T}^{n+1} = \mathbf{R}\,\mathbf{T}^{n} + \mathbf{F} \tag{5.31}$$

where **L** and **R** are the left- and right-hand side element matrices, respectively, and **F** is the element load vector. For a single four-node quadrilateral element, these equations take the following form:

$$\begin{bmatrix} L_{11} & L_{12} & L_{13} & L_{14} \\ L_{21} & L_{22} & L_{23} & L_{24} \\ L_{31} & L_{32} & L_{33} & L_{34} \\ L_{41} & L_{42} & L_{43} & L_{44} \end{bmatrix} \begin{Bmatrix} T_1 \\ T_2 \\ T_3 \\ T_4 \end{Bmatrix}^{n+1} = \begin{bmatrix} R_{11} & R_{12} & R_{13} & R_{14} \\ R_{21} & R_{22} & R_{23} & R_{24} \\ R_{31} & R_{32} & R_{33} & R_{34} \\ R_{41} & R_{42} & R_{43} & R_{44} \end{bmatrix} \begin{Bmatrix} T_1 \\ T_2 \\ T_3 \\ T_4 \end{Bmatrix}^{n} + \begin{Bmatrix} F_1 \\ F_2 \\ F_3 \\ F_4 \end{Bmatrix} \tag{5.32}$$

We subsequently showed how to evaluate **L**, **R**, and **F** using Gauss–Legendre quadrature. Thus, at this point, the coefficients such as L_{11}, R_{21}, and F_3 are known.

Our task now is to assemble each set of element-level equations into a global multielement system. To illustrate how assembly is carried out in 2Ds, we consider the specific example of a small mesh consisting of three four-node quadrilateral elements and a total of eight global nodes (Figure 5.6).

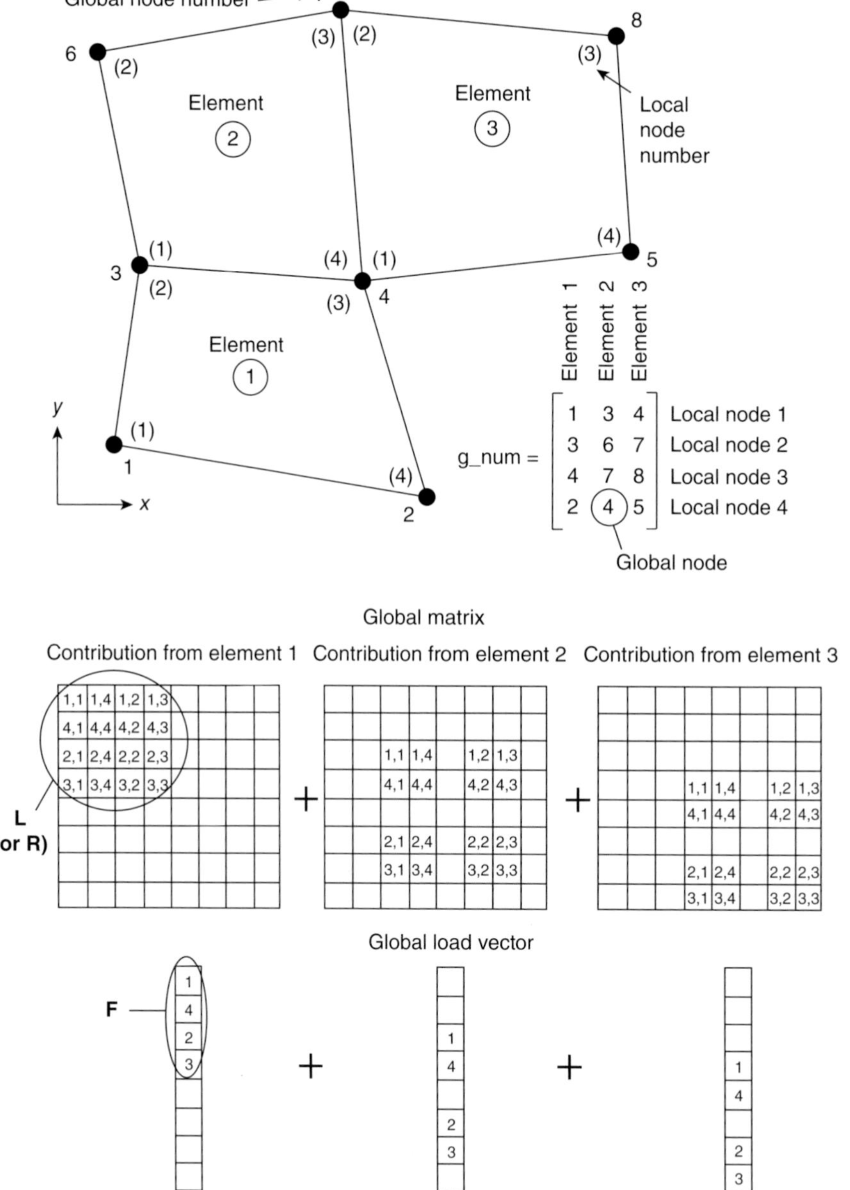

Figure 5.6 Two-dimensional finite element mesh consisting of three four-node quadrilaterals. Nodal connectivity (defined by `g_num`) determines where coefficients in the element matrices (**L** or **R**; see Equation 5.32) and element load vector **F** contribute to their global counterparts.

Because each node has only one degree of freedom (i.e., temperature), there are also a total of eight unknowns (and equations). The nodal connectivity array for this mesh is

$$\mathsf{g\_num} = \begin{bmatrix} 1 & 3 & 4 \\ 3 & 6 & 7 \\ 4 & 7 & 8 \\ 2 & 4 & 5 \end{bmatrix} \tag{5.33}$$

It is this connectivity that determines how the various element-level equations contribute to the global system of equations. To see this, consider, for example, the second element (i.e., column 2 of 5.33) that contains the global nodes 3, 6, 7, and 4, which correspond to local nodes 1, 2, 3, and 4, respectively. For this case, a coefficient in the local element matrix **L** or **R** with the subscript indices 1,2 (row 1, column 2) adds to position 3,6 in the global matrix (see Eq. 5.32), while term F_3 in the element load vector contributes to position 7 in the global load vector (Figure 5.6). Repeating this procedure for the entire mesh (Figure 5.6) leads to the following global system of equations:

$$\tilde{\mathbf{L}}\,\tilde{\mathbf{T}}^{n+1} = \tilde{\mathbf{R}}\,\tilde{\mathbf{T}}^{n} + \tilde{\mathbf{F}} = \tilde{\mathbf{b}}^{n} \tag{5.34}$$

Here,

$$\tilde{\mathbf{L}} = \begin{bmatrix} L^1_{11} & L^1_{14} & L^1_{12} & L^1_{13} & 0 & 0 & 0 & 0 \\ L^1_{41} & L^1_{44} & L^1_{42} & L^1_{43} & 0 & 0 & 0 & 0 \\ L^1_{21} & L^1_{24} & L^1_{22}+L^2_{11} & L^1_{23}+L^2_{14} & 0 & L^2_{12} & L^2_{13} & 0 \\ L^1_{31} & L^1_{34} & L^1_{32}+L^2_{41} & L^1_{33}+L^2_{44}+L^3_{11} & L^3_{14} & L^2_{42} & L^2_{43}+L^3_{12} & L^3_{13} \\ 0 & 0 & 0 & L^3_{41} & L^3_{44} & 0 & L^3_{42} & L^3_{43} \\ 0 & 0 & L^2_{21} & L^2_{24} & 0 & L^2_{22} & L^2_{23} & 0 \\ 0 & 0 & L^2_{31} & L^2_{34}+L^3_{21} & L^3_{24} & L^2_{32} & L^2_{33}+L^3_{22} & L^3_{23} \\ 0 & 0 & 0 & L^3_{31} & L^3_{34} & 0 & L^3_{32} & L^3_{33} \end{bmatrix} \tag{5.35}$$

$$\tilde{\mathbf{R}} = \begin{bmatrix} R^1_{11} & R^1_{14} & R^1_{12} & R^1_{13} & 0 & 0 & 0 & 0 \\ R^1_{41} & R^1_{44} & R^1_{42} & R^1_{43} & 0 & 0 & 0 & 0 \\ R^1_{21} & R^1_{24} & R^1_{22}+R^2_{11} & R^1_{23}+R^2_{14} & 0 & R^2_{12} & R^2_{13} & 0 \\ R^1_{31} & R^1_{34} & R^1_{32}+R^2_{41} & R^1_{33}+R^2_{44}+R^3_{11} & R^3_{14} & R^2_{42} & R^2_{43}+R^3_{12} & R^3_{13} \\ 0 & 0 & 0 & R^3_{41} & R^3_{44} & 0 & R^3_{42} & R^3_{43} \\ 0 & 0 & R^2_{21} & R^2_{24} & 0 & R^2_{22} & R^2_{23} & 0 \\ 0 & 0 & R^2_{31} & R^2_{34}+R^3_{21} & R^3_{24} & R^2_{32} & R^2_{33}+R^3_{22} & R^3_{23} \\ 0 & 0 & 0 & R^3_{31} & R^3_{34} & 0 & R^3_{32} & R^3_{33} \end{bmatrix} \tag{5.36}$$

$$\tilde{\mathbf{F}} = \begin{bmatrix} F_1^1 \\ F_4^1 \\ F_2^1 + F_1^2 \\ F_3^1 + F_4^2 + F_1^3 \\ F_4^3 \\ F_2^2 \\ F_3^2 + F_2^3 \\ F_3^3 \end{bmatrix} \quad \text{and} \quad \tilde{\mathbf{T}} = \begin{bmatrix} T_1 \\ T_2 \\ T_3 \\ T_4 \\ T_5 \\ T_6 \\ T_7 \\ T_8 \end{bmatrix} \tag{5.37}$$

Here, the superscripts indicate the element number from which a coefficient comes. This procedure of summing the element matrices and vector into the global matrix is easily achieved within a computer program. For example, once the element matrix **L** for a certain element has been integrated, it can be added to the left-hand side global matrix $\tilde{\mathbf{L}}$ (`lhs`) in Matlab using the commands

```
num = g_num(:,iel)    ; % global node numbers for current element
lhs(num,num)  = lhs(num,num) + L ; %assembly of the lhs global matrix
```

where `num` is the list of global nodes for the current element (i.e, a single column of Equation 5.33).

5.5 Boundary Conditions and Solution

Now that the global equations have been constructed, all that remains is to apply boundary conditions and solve the system of equations. Both are carried out in the same manner as for the 1D problem considered in Chapter 4. Recall from the beginning of the chapter that we are interested in applying the fixed temperature boundary condition $T = Tb$ at all four boundaries of the model domain (where Tb will be set to zero in this example; see Figure 5.1). For the straight-sided mesh illustrated in Figure 5.3, the nodes that fall on each boundary can easily be located in Matlab using the following commands:

```
eps = 0.01*min(dx,dy) ;                % small fraction of dx or dy
bx0 = find(g_coord(1,:)≤0+eps)    ; % nodes on x=0 boundary
bxn = find(g_coord(1,:)≥lx-eps)   ; % nodes on x=lx boundary
by0 = find(g_coord(2,:)≤0+eps)    ; % nodes on y=0 boundary
byn = find(g_coord(2,:)≥ly-eps)   ; % nodes on y=ly boundary
```

Note here that, in principle, one should be able to use lines such as `bx0 = find(g_coord(1,:)==0)` to find the boundary nodes. However, in some cases small rounding errors may mean that nodes do not fall exactly at the expected positions (e.g., $x = 0.0$) and, therefore, would not be located using this approach. For this reason, the given script introduces the small parameter (`eps`) to avoid this potential problem.

Following this, one can create the vectors containing the boundary node indices and their fixed temperature values as follows:

```
bcdof = [bx0 bxn by0 byn] ;             % boundary nodes to be fixed
bcval = Tb*ones(1,length(bcdof)) ; % fixed T at boundaries
```

The actual implementation of boundary conditions is carried out in an identical manner to done earlier in 1D. This must be performed within the time loop just after forming the latest right-hand side global vector $\tilde{\mathbf{b}}^n$ (see 5.34) and can be achieved using the following commands:

```
lhs(bcdof,:)    = 0 ;   % zero rows of boundary equations
tmp             = spdiags(lhs,0) ;        % store diagonal
tmp(bcdof)      = 1  ; % place 1 on entries of boundary equations
lhs             = spdiags(tmp,0,lhs)  ; % reinsert the diagonal
b(bcdof)        = bcval  ;                % modify rhs vector
```

Finally, the global system of linear equations can be solved using the Matlab backslash command, as done in Chapter 3:

```
displ = lhs \ b ; % solve global system of equations
```

where `displ` is the column vector of new temperatures at the mesh nodes.

5.6 Matlab Script

The Matlab script listed in the following text uses the FEM to solve the 2D transient heat conduction equation (Equation 5.3; see Figure 5.1). Figure 5.7 shows some graphical output from this script for the specific case $H = 0$, unit initial temperature, and zero temperature on all boundaries. The domain extends 2000 m in the x-direction (i.e., $lx = 2000$ m) and 1000 m in the y-direction. A comparison of the numerical results with the exact solution from Carslaw and Jaeger (1959, page 173) shows good agreement.

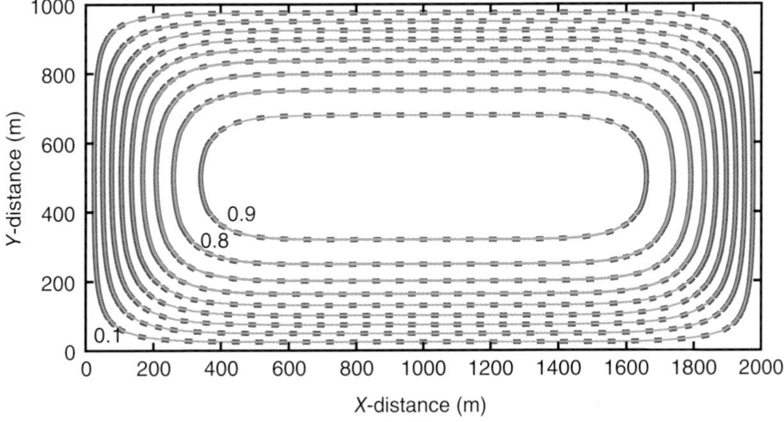

Figure 5.7 Comparison between numerical (dashed line) and exact (solid line; Carslaw and Jaeger, 1959) solutions to the 2D heat conduction equation with no internal heat source, unit initial temperature, and zero temperature on the boundaries after 600 years (with $\kappa = 10^{-6}$ m^2 s^{-1}). The numerical solution was computed with 50 bilinear elements in each direction and a time step of 10 years. Contours are the temperature isotherms 0.1, 0.2, … , 0.9°C.

```
%-----------------------------------------------
% Program: diffusion2d.m
% 2D FEM - diffusion equation
% Discretisation with 4-node quadrilaterals
%-----------------------------------------------

clear

seconds_per_yr = 60*60*24*365 ; % seconds in 1 year

% physical parameters
kappa = 1e-6   ; % thermal diffusivity, m^2/s
lx    = 1e3*2  ; % width of domain, m
ly    = 1e3    ; % depth of domain, m
H     = 0*1e-9 ; % heat source, °K/s
Tb    = 0 ;      % fixed boundary temperature
Ti    = 1 ;      % initial temperature

% numerical parameters
ndim  = 2   ;    % number of spatial dimensions
nod   = 4   ;    % number of nodes per element
nelx  = 50  ;    % number of elements in x-direction
nely  = 50   ;    % number of elements in y-direction
nels  = nelx*nely ; % total number of elements
nx    = nelx+1 ;    % number of nodes in x-direction
ny    = nely+1 ;    % number of nodes in y-direction
nn    = nx*ny ;     % total number of nodes
nip   = 4   ;       % number of integration points
dx    = lx/nelx ;   % element length in x-direction
dy    = ly/nely ;   % element length in y-direction
dt    = 10*seconds_per_yr ; % time step (s)
ntime = 60 ;        % number of time steps to perform
kay   = eye(ndim)*kappa ; % thermal diffusivity tensor

% define mesh (numbering in y direction)
g_coord = zeros(ndim,nn) ;
n = 1 ;
for i=1:nx
  for j=1:ny
    g_coord(1,n) = (i-1)*dx ;
    g_coord(2,n) = (j-1)*dy ;
    n = n + 1 ;
  end
end

% establish elem-node connectivity
gnumbers = reshape(1:nn,[ny nx]) ;
iel = 1 ;
for i=1:nelx
  for j=1:nely
    g_num(1,iel) = gnumbers(j,i) ;     % node 1
    g_num(2,iel) = gnumbers(j+1,i) ;   % node 2
    g_num(3,iel) = gnumbers(j+1,i+1);  % node 3
    g_num(4,iel) = gnumbers(j,i+1);    % node 4
    iel          = iel + 1 ;
  end
end
```

```
% find boundary nodes
eps = 0.01*min(dx,dy) ;     % small fraction of dx or dy
bx0 = find(g_coord(1,:)≤0+eps) ; % nodes on x=0 boundary
bxn = find(g_coord(1,:)≥lx-eps) ; % nodes on x=lx boundary
by0 = find(g_coord(2,:)≤0+eps) ; % nodes on y=0 boundary
byn = find(g_coord(2,:)≥ly-eps) ; % nodes on y=ly boundary

% define boundary conditions
bcdof = unique([bx0 bxn by0 byn]) ; % boundary nodes
bcval = Tb*ones(1,length(bcdof))  ; % boundary temperatures

% gauss integration data
points = zeros(nip,ndim); % location of points
root3  = 1./sqrt(3);
points(1,1)=-root3; points(1,2)= root3;
points(2,1)= root3; points(2,2)= root3;
points(3,1)=-root3; points(3,2)=-root3;
points(4,1)= root3; points(4,2)=-root3;
wts = ones(1,4) ; % weights

% save shape functions and their derivatives in local coordinates
% evaluated at integration points
for k=1:nip
  xi  = points(k,1) ;
  eta = points(k,2) ;
  etam = 0.25*(1-eta); etap = 0.25*(1+eta) ;
  xim  = 0.25*(1-xi) ; xip  = 0.25*(1+xi)  ;
  fun  = 4*[xim*etam  xim*etap  xip*etap  xip*etam ] ;
  fun_s(k,:) = fun ; % shape functions
  der(1,1)=-etam; der(1,2)=-etap; der(1,3)=etap; der(1,4)=etam ;
  der(2,1)=-xim;  der(2,2)=xim;   der(2,3)=xip;  der(2,4)=-xip ;
  der_s(:,:,k) = der ; % derivatives of shape function
end

% initialise arrays
ff    = zeros(nn,1);     % global load vector
b     = zeros(nn,1);     % global rhs vector
lhs   = sparse(nn,nn);   % global lhs matrix
rhs   = sparse(nn,nn);   % global rhs matrix

% x and y grids for plotting
xgrid = reshape(g_coord(1,:),ny,nx) ;
ygrid = reshape(g_coord(2,:),ny,nx) ;

%---------------------------------------------------
% matrix integration and assembly
%---------------------------------------------------

for iel=1:nels  % sum over elements
  num    = g_num(:,iel)      ; % element nodes
  coord  = g_coord(:,num)' ; % element coordinates
  KM     = zeros(nod,nod)   ; % initialisation
  MM     = zeros(nod,nod)   ;
  F      = zeros(nod,1)     ;
  for k = 1:nip % integration loop
    fun    = fun_s(k,:) ; % shape functions
    der    = der_s(:,:,k) ; % der. of shape functions in local coordinates
    jac    = der*coord  ;    % jacobian matrix
    detjac = det(jac) ;      % det. of jacobian
```

```
      invjac = inv(jac) ;      % inv. of jacobian
      deriv  = invjac*der ;    % der. of shape fun. in physical coords.
      KM     = KM + deriv'*kay*deriv*detjac*wts(k) ; % stiffness matrix
      MM     = MM + fun'*fun*detjac*wts(k) ;          % mass matrix
      F      = F  + fun'*H*detjac*wts(k) ;            % load vector
    end
    % assemble global matrices and vector
    lhs(num,num) = lhs(num,num) + MM/dt + KM ;
    rhs(num,num) = rhs(num,num) + MM/dt      ;
    ff(num)      = ff(num)      + F          ;
end

%---------------------------------------------------
% time loop
%---------------------------------------------------

displ = Ti*ones(nn,1) ;% initial conditions
time  = 0 ; % initialise time

for n=1:ntime

   time = time + dt ;                  % update time
   b = rhs*displ + ff  ;               % form rhs global vector

   % impose boundary conditions
   lhs(bcdof,:) = 0 ;                  % zero the boundary equations
   tmp = spdiags(lhs,0) ;              % store diagonal
   tmp(bcdof)=1 ;                      % place 1 on stored-diagonal
   lhs=spdiags(tmp,0,lhs);             % reinsert diagonal
   b(bcdof) = bcval ;                  % set boundary values

   displ = lhs \ b ;                   % solve system of equations

%---------------------------------------------------
   % exact solution from Carslaw and Jaeger (1959)
   % page 173, eqns 6 and 11
   nterms = 100 ;
   t=time; x=xgrid-lx/2; y=ygrid-ly/2 ;
   l = lx/2 ; b = ly/2;
   sumx = 0 ; sumy = 0 ;
     for m=0:nterms
       eterm    = exp(-kappa*(2*m+1)^2*pi^2*t/(4*l^2));
       costerm = cos((2*m+1)*pi*x/(2*l)) ;
       sumx     = sumx + (-1)^m/(2*m+1)*eterm*costerm ;
       eterm    = exp(-kappa*(2*m+1)^2*pi^2*t/(4*b^2));
       costerm = cos((2*m+1)*pi*y/(2*b)) ;
       sumy     = sumy + (-1)^m/(2*m+1)*eterm*costerm ;
     end
     psix  = 4/pi*sumx ;
     psiy  = 4/pi*sumy ;
     exact = psix.*psiy ;
%---------------------------------------------------

   % plot numerical and exact solutions
   figure(1)
   solution  = reshape(displ,ny,nx) ;
   isotherms = [0.1 0.2 0.3 0.4 0.5 0.6 0.7 0.8 0.9] ;
   contour(xgrid,ygrid,solution,isotherms,'b.')
   hold on
```

```
    contour(xgrid,ygrid,exact,isotherms,'r')
    xlabel('x-distance (m)')
    ylabel('y-distance (m)')
    title('Temperature isotherms')
    axis equal
    hold off
    drawnow

end

%-------------------------------------------------------------
% end of time loop
%-------------------------------------------------------------
```

5.7 Exercises

1) Using the bilinear shape functions (5.23) and their derivatives (5.25), approximate the following integrals using Gauss–Legendre quadrature with four points for an element with nodes at $x_1 = 0$, $y_1 = 0$, $x_2 = 0$, $y_2 = 2$, $x_3 = 3$, $y_3 = 2$, and $x_4 = 3$, $y_4 = 0$.

$$\int\int N_1 \, dx \, dy \quad \left(= \frac{3}{2}\right)$$

$$\int\int N_1 N_2 \, dx \, dy \quad \left(= \frac{1}{3}\right)$$

$$\int\int \frac{\partial N_1}{\partial x} \frac{\partial N_4}{\partial y} \, dx \, dy \quad \left(= \frac{1}{4}\right)$$

The exact solutions are indicated in brackets.

2) Use the bilinear shape functions to approximate the function

$$f(x,y) = \left(1 - 0.8\cos\left(\frac{\pi x}{2}\right)\right)\cos\left(\frac{\pi y}{2}\right)$$

at the point $x = 0.5$, $y = 0.5$ within an element with the four nodes positioned as follows: $x_1 = 0$, $y_1 = 0$, $x_2 = 0$, $y_2 = 1$, $x_3 = 1$, $y_3 = 1$, and $x_4 = 1$, $y_4 = 0$. Compare the interpolated solution with the exact solution (i.e., obtained by substituting $x = 0.5$, $y = 0.5$ into the given function). Hint: First, convert the coordinates $x = 0.5$ and $y = 0.5$ to local coordinates ξ and η. Second, use the shape functions evaluated at ξ and η (i.e., $N_i(\xi, \eta)$) to interpolate the values of the function defined at the nodes (f_i) using the standard approximation as follows:

$$\tilde{f} = \sum_{i=1}^{4} f_i N_i(\xi, \eta)$$

3) Consider a four-node quadrilateral with nodes located at $x_1 = 0$, $y_1 = 0$, $x_2 = -0.25$, $y_2 = 1$, $x_3 = 0.8$, $y_3 = 0.9$, and $x_4 = 1$, $y_4 = 0.25$. Produce a plot to show the element nodes and the positions of the Gauss–Legendre quadrature points, assuming four points (two in each direction). Hint: To obtain the global poisitions of the integration points, use the following isoparametric relations:

$$x = \sum_{i=1}^{4} N_i(\xi, \eta) x_i$$

$$y = \sum_{i=1}^{4} N_i(\xi, \eta) y_i$$

Here, $N_i(\xi, \eta)$ $(i = 1...4)$ are the bilinear shape functions in local coordinates and x_i and y_i are the nodal positions.

4) Modify the Matlab script listed in Section 5.6 to account for the following initial conditions:

$$T(x, y, t = 0) = 0 \quad \text{when} \quad \sqrt{(x - x_0)^2 + (y - y_0)^2} > r$$
$$\text{otherwise} \quad T(x, y, t = 0) = 1$$

Thus, $T = 1$ if the mesh nodes are within a circle centered at x_0, and y_0 with radius r, whereas $T = 0$ outside the circle. Use the following parameters: $x_0 = 1000$ m, $y_0 = 500$ m, and $r = 100$ m.

5) Modify the Matlab script listed in Section 5.6 to account for different thermal diffusivities in the x- and y-directions (10^{-6} m^2 s^{-1} and 10^{-7} m^2 s^{-1}, respectively).

6) Using the FEM with bilinear shape and weighting functions, solve Poisson's equation

$$\frac{\partial^2 u}{\partial x^2} + \frac{\partial^2 u}{\partial y^2} + 2\pi^2 \sin(\pi x)\sin(\pi y) = 0$$

with

$$0 < x < 1, \quad 0 < y < 1, \quad u(x, y) = 0 \quad \text{on all boundaries.}$$

Compare the numerical results with the exact solution

$$u(x, y) = \sin(\pi x)\sin(\pi y)$$

Suggested Reading

A. J. M. Ferreira, *MATLAB Codes for Finite Element Analysis: Solids and Structures (Solid Mechanics and Its Applications)*, Springer, Berlin, 2009.

Y. W. Kwong, and H. C. Bang, *The Finite Element Method using Matlab*, CRC Press, New York, 2000.

I. M. Smith, and D. V. Griffith, *Programming the Finite Element Method*, John Wiley & Sons, Ltd, Chichester, 1998.

O. C. Zienkiewicz, and R. L. Taylor, *The Finite Element Method, Volume 1, The Basis*, Butterworth-Heinemann, Oxford/Boston, MA, 2000.

6

The Finite Element Method in Three Dimensions

For readers proceeding chapter by chapter, it should now be no surprise that the finite element method (FEM) is easily implemented in three dimensions (3Ds). In Chapters 4 and 5, we showed that the discretization process, element integration, assembly, and solution are all very similar in 1D and 2D. In this chapter, we build on this by showing how to apply the FEM in three spatial dimensions, using once again the transient heat conduction equation as an example. A program listed in Section 6.5 shows how the problem is programmed and solved using Matlab.

The objective of this chapter is to use the FEM to solve the partial differential equation

$$\frac{\partial T}{\partial t} = \nabla^T (\mathbf{K}\nabla T) + \frac{A}{\rho c} \tag{6.1}$$

where

$$\nabla = \begin{Bmatrix} \frac{\partial}{\partial x} \\ \frac{\partial}{\partial y} \\ \frac{\partial}{\partial z} \end{Bmatrix} \tag{6.2}$$

$$\mathbf{K} = \begin{bmatrix} \kappa_x & 0 & 0 \\ 0 & \kappa_y & 0 \\ 0 & 0 & \kappa_z \end{bmatrix} \tag{6.3}$$

on the spatial domain $x = [0, lx]$, $y = [0, ly]$, $z = [0, lz]$ with fixed temperature boundary conditions and an arbitrary initial temperature distribution (Figure 6.1). In the context of heat conduction, $T(x, y, z, t)$ is the temperature, κ_x is the thermal diffusitivy in the x-direction, and so on, ρ is the density, c is the heat capacity, and A is the rate of heat generation per unit volume.

6.1 Discretization

A simple means of discretizing Equation 6.1 is with hexahedra elements containing eight corner nodes (Figure 6.2). For this element, the eight nodal temperatures (T_i) can be approximated using eight trilinear shape functions (N_i) according to

Practical Finite Element Modeling in Earth Science Using Matlab, First Edition. Guy Simpson.

Companion website: www.wiley.com/go/simpson

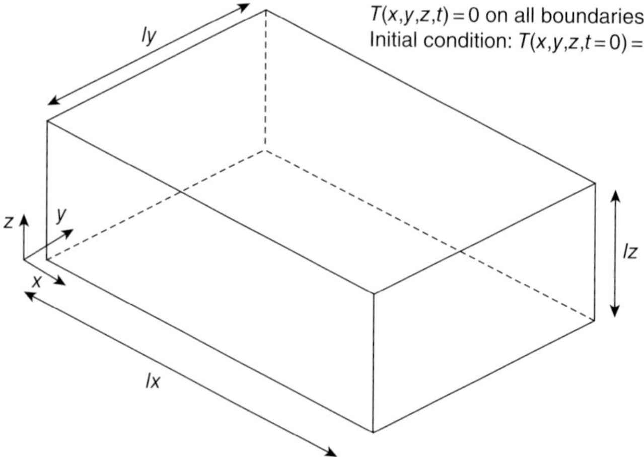

Figure 6.1 Geometry and auxiliary data for 3D heat flow problem.

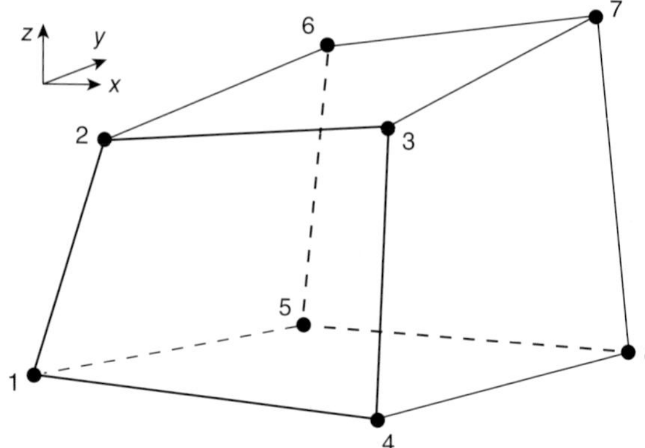

Figure 6.2 Hexahedron with eight corner nodes.

$$T \simeq [N_1\ N_2\ N_3\ N_4\ N_5\ N_6\ N_7\ N_8] \begin{Bmatrix} T_1 \\ T_2 \\ T_3 \\ T_4 \\ T_5 \\ T_6 \\ T_7 \\ T_8 \end{Bmatrix} = \mathbf{NT} \tag{6.4}$$

where the shape functions are written in terms of the physical coordinates x, y, and z. Substituting this approximation into Equation 6.1, multiplying by the weighting functions (assumed to be the same as the shape functions), integrating over the element volume and applying integration by parts to the term with second-order spatial derivatives (and ignoring the resulting boundary integrals) gives the following discrete element-level system of equations:

$$\iiint \mathbf{N}^T\mathbf{N}\, dx\, dy\, dz \frac{\partial}{\partial t}\mathbf{T} + \iiint (\nabla\mathbf{N})^T\mathbf{K}\nabla\mathbf{N}\, dx\, dy\, dz\, \mathbf{T} = \iiint H\mathbf{N}^T\, dx\, dy\, dz \tag{6.5}$$

or in compact matrix form

$$[\mathbf{MM}]\frac{\partial}{\partial t}\mathbf{T} + [\mathbf{KM}]\,\mathbf{T} = \mathbf{F} \tag{6.6}$$

where

$$\mathbf{MM} = \iiint \mathbf{N}^T\mathbf{N}\; dx\, dy\, dz \tag{6.7}$$

$$\mathbf{KM} = \iiint (\nabla\mathbf{N})^T\mathbf{K}\nabla\mathbf{N}\; dx\, dy\, dz \tag{6.8}$$

$$\mathbf{F} = \iiint H\mathbf{N}^T\; dx\, dy\, dz \tag{6.9}$$

and $H = A/\rho c$. The element mass matrix **MM** and the element stiffness matrix **KM** both have eight rows and eight columns for eight-node hexahedra, while the element load vector **F** and the vector of nodal temperatures **T** are column vectors with eight lines. Using an implicit finite difference approximation for the time derivative, Equation 6.6 becomes

$$\left(\frac{\mathbf{MM}}{\Delta t} + \mathbf{KM}\right)\mathbf{T}^{n+1} = \frac{\mathbf{MM}}{\Delta t}\,\mathbf{T}^n + \mathbf{F} \tag{6.10}$$

where Δt is the time interval between n and $n+1$. Note that the equation has exactly the same form as those derived earlier in 1D and 2D (see Equations 2.27 and 5.18).

6.2 Element Integration

The element matrices **MM**, **KM**, and the element load vector **F** contain the shape functions or their derivatives, both expressed in terms of physical coordinates, which must be integrated over the volume represented by finite hexahedra. Here, the integrals are computed numerically using Gauss–Legendre quadrature, as was also done previously in Chapters 4 and 5. The rule for Gauss–Legendre quadrature in 3Ds is

$$\int_{-1}^{1}\int_{-1}^{1}\int_{-1}^{1} f(\xi,\eta,\zeta)\; d\xi d\eta d\zeta \simeq \sum_{i=1}^{nipx}\sum_{j=1}^{nipy}\sum_{m=1}^{nipz} f(\xi_i,\eta_j,\zeta_m)\; w_i w_j w_m$$

$$= \sum_{k=1}^{nip} f(\xi_k,\eta_k,\zeta_k)\; w_k \tag{6.11}$$

where ξ, η, and ζ are local coordinates on the normalized volume $[-1,1]^3$. The positions ξ_k, η_k, ζ_k of the integration points and the weights w_k are obtained from Table 4.1. In Equation 6.11, if it is assumed that the number of integration points in each direction is the same (i.e., `nipx=nipy=nipz=n`), then one can interpret the triple summation with n points in each direction as a single summation over n^3 (`=nip`) points in 3Ds.

In order to apply the aforementioned Gauss–Legendre quadrature rule, we follow the now familiar approach by defining the shape functions and their derivatives in the local coordinate system (Figure 6.3). By applying appropriate transformations, these functions can then be rescaled to represent the original element integrals performed in physical coordinates. The trilinear shape functions for the eight-node hexahedron, written in terms of local coordinates, are

$$N_1 = \frac{1}{8}(1-\xi)(1-\eta)(1-\zeta)$$

$$N_2 = \frac{1}{8}(1-\xi)(1-\eta)(1+\zeta)$$

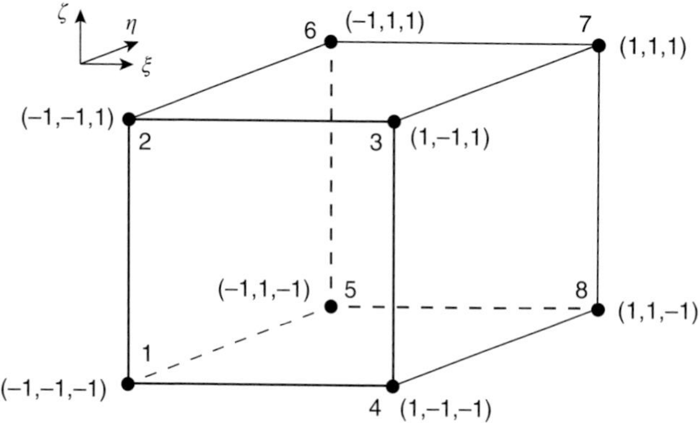

Figure 6.3 Hexahedron element with eight corner nodes in the local coordinate system (ξ, η, ζ).

$$
\begin{aligned}
N_3 &= \frac{1}{8}(1+\xi)(1-\eta)(1+\zeta)\\
N_4 &= \frac{1}{8}(1+\xi)(1-\eta)(1-\zeta)\\
N_5 &= \frac{1}{8}(1-\xi)(1+\eta)(1-\zeta)\\
N_6 &= \frac{1}{8}(1-\xi)(1+\eta)(1+\zeta)\\
N_7 &= \frac{1}{8}(1+\xi)(1+\eta)(1+\zeta)\\
N_8 &= \frac{1}{8}(1+\xi)(1+\eta)(1-\zeta)
\end{aligned}
\tag{6.12}
$$

while their first derivatives are

$$
\begin{aligned}
\frac{\partial N_1}{\partial \xi} &= -\frac{1}{8}(1-\eta)(1-\zeta) & \frac{\partial N_1}{\partial \eta} &= -\frac{1}{8}(1-\xi)(1-\zeta) & \frac{\partial N_1}{\partial \zeta} &= -\frac{1}{8}(1-\xi)(1-\eta)\\
\frac{\partial N_2}{\partial \xi} &= -\frac{1}{8}(1-\eta)(1+\zeta) & \frac{\partial N_2}{\partial \eta} &= -\frac{1}{8}(1-\xi)(1+\zeta) & \frac{\partial N_2}{\partial \zeta} &= \frac{1}{8}(1-\xi)(1-\eta)\\
\frac{\partial N_3}{\partial \xi} &= \frac{1}{8}(1-\eta)(1+\zeta) & \frac{\partial N_3}{\partial \eta} &= -\frac{1}{8}(1+\xi)(1+\zeta) & \frac{\partial N_3}{\partial \zeta} &= \frac{1}{8}(1+\xi)(1-\eta)\\
\frac{\partial N_4}{\partial \xi} &= \frac{1}{8}(1-\eta)(1-\zeta) & \frac{\partial N_4}{\partial \eta} &= -\frac{1}{8}(1+\xi)(1-\zeta) & \frac{\partial N_4}{\partial \zeta} &= -\frac{1}{8}(1+\xi)(1-\eta)\\
\frac{\partial N_5}{\partial \xi} &= -\frac{1}{8}(1+\eta)(1-\zeta) & \frac{\partial N_5}{\partial \eta} &= \frac{1}{8}(1-\xi)(1-\zeta) & \frac{\partial N_5}{\partial \zeta} &= -\frac{1}{8}(1-\xi)(1+\eta)\\
\frac{\partial N_6}{\partial \xi} &= -\frac{1}{8}(1+\eta)(1+\zeta) & \frac{\partial N_6}{\partial \eta} &= \frac{1}{8}(1-\xi)(1+\zeta) & \frac{\partial N_6}{\partial \zeta} &= \frac{1}{8}(1-\xi)(1+\eta)\\
\frac{\partial N_7}{\partial \xi} &= \frac{1}{8}(1+\eta)(1+\zeta) & \frac{\partial N_7}{\partial \eta} &= \frac{1}{8}(1+\xi)(1+\zeta) & \frac{\partial N_7}{\partial \zeta} &= \frac{1}{8}(1+\xi)(1+\eta)\\
\frac{\partial N_8}{\partial \xi} &= \frac{1}{8}(1+\eta)(1-\zeta) & \frac{\partial N_8}{\partial \eta} &= \frac{1}{8}(1+\xi)(1-\zeta) & \frac{\partial N_8}{\partial \zeta} &= -\frac{1}{8}(1+\xi)(1+\eta)
\end{aligned}
\tag{6.13}
$$

The eight-node hexahedron is an isoparametric element, meaning that the same shape functions govern both the finite element approximation (Equation 6.4) and the geometry (i.e., the shape functions can be used to map coordinates from the local space to the physical space). Thus, the shape functions (in physical coordinates) appearing in the element integrals can simply be replaced with their counterparts written in local coordinates. Derivatives of the shape functions can be converted from the local coordinate system (within which they are defined) to the physical coordinate system using the chain rule

$$\begin{Bmatrix} \frac{\partial}{\partial \xi} \\ \frac{\partial}{\partial \eta} \\ \frac{\partial}{\partial \zeta} \end{Bmatrix} \mathbf{N} = \begin{bmatrix} \frac{\partial x}{\partial \xi} & \frac{\partial y}{\partial \xi} & \frac{\partial z}{\partial \xi} \\ \frac{\partial x}{\partial \eta} & \frac{\partial y}{\partial \eta} & \frac{\partial z}{\partial \eta} \\ \frac{\partial x}{\partial \zeta} & \frac{\partial y}{\partial \zeta} & \frac{\partial z}{\partial \zeta} \end{bmatrix} \begin{Bmatrix} \frac{\partial}{\partial x} \\ \frac{\partial}{\partial y} \\ \frac{\partial}{\partial z} \end{Bmatrix} \mathbf{N} = \mathbf{J} \begin{Bmatrix} \frac{\partial}{\partial x} \\ \frac{\partial}{\partial y} \\ \frac{\partial}{\partial z} \end{Bmatrix} \mathbf{N} \tag{6.14}$$

where $\mathbf{J}$ is the Jacobian matrix. The Jacobian matrix can be found by differentiating the global coordinates with respect to the local coordinates (i.e., multiply the coordinates of a particular element with the derivatives of the shape functions), that is,

$$\mathbf{J} = \begin{bmatrix} \frac{\partial N_1}{\partial \xi} & \frac{\partial N_2}{\partial \xi} & \cdots & \frac{\partial N_8}{\partial \xi} \\ \frac{\partial N_1}{\partial \eta} & \frac{\partial N_2}{\partial \eta} & \cdots & \frac{\partial N_8}{\partial \eta} \\ \frac{\partial N_1}{\partial \zeta} & \frac{\partial N_2}{\partial \zeta} & \cdots & \frac{\partial N_8}{\partial \zeta} \end{bmatrix} \begin{bmatrix} x_1 & y_1 & z_1 \\ x_2 & y_2 & z_2 \\ \vdots & \vdots & \vdots \\ x_8 & y_8 & z_8 \end{bmatrix} \tag{6.15}$$

where x_1 is the x-coordinate of node 1, and so on. The derivatives of the shape functions in terms of physical coordinates can therefore be found from

$$\begin{Bmatrix} \frac{\partial}{\partial x} \\ \frac{\partial}{\partial y} \\ \frac{\partial}{\partial z} \end{Bmatrix} \mathbf{N} = \mathbf{J}^{-1} \begin{Bmatrix} \frac{\partial}{\partial \xi} \\ \frac{\partial}{\partial \eta} \\ \frac{\partial}{\partial \zeta} \end{Bmatrix} \mathbf{N} \tag{6.16}$$

where $\mathbf{J}^{-1}$ is the inverse of the Jacobian matrix. Transformation of the limits of integration is carried out using the determinant of the Jacobian (det $\mathbf{J}$) according to the following relation:

$$\int\int f(x,y,z)\,dx\,dy\,dz = \int_{-1}^{1}\int_{-1}^{1}\int_{-1}^{1} f(\xi,\eta,\zeta)\,\det\mathbf{J}\,d\xi\,d\eta\,d\zeta \tag{6.17}$$

Combining this result with Equation 6.11 yields the final integration formula

$$\int\int f(x,y,z)\,dx\,dy\,dz \simeq \sum_{k=1}^{nip} f(\xi_k,\eta_k,\zeta_k)\,\det\mathbf{J}\,w_k \tag{6.18}$$

Recall that $f(\xi,\eta,\zeta) = \mathbf{N}^T\mathbf{N}$ in the case of the mass matrix, $f(\xi,\eta,\zeta) = (\nabla\mathbf{N})^T\mathbf{K}\nabla\mathbf{N}$ for the stiffness matrix and $f(\xi,\eta,\zeta) = H\mathbf{N}^T$ for the element load vector. The following snippet summarizes the main steps that must be performed within a loop over all elements to integrate **KM**, **MM**, and **F** in Matlab:

```
KM      = zeros(nod,nod)    ; % initialisation
MM      = zeros(nod,nod)    ;
F       = zeros(nod,1)      ;
for k = 1:nip % integration loop
  fun     = fun_s(k,:) ; % shape functions, N
  der     = der_s(:,:,k) ; % deriv of N in local coordinates
  jac     = der*coord   ;    % jacobian matrix
  detjac = det(jac) ;        % det. of jacobian
  invjac = inv(jac) ;        % inv. of jacobian
  deriv   = invjac*der ;     % der. of shape fun. in physical coords.
  KM      = KM + deriv'*kay*deriv*detjac*wts(k) ; % stiffness matrix
```

```
    MM       = MM + fun'*fun*detjac*wts(k) ;          % mass matrix
    F        = F  + fun'*H*detjac*wts(k) ;            % load vector
end      % end of integration loop
```

One might now recognize that this code is identical to that used in 1D and 2D for the same governing equation (see scripts in Sections 4.6 and 5.6).

6.3 Assembly for Multielement Mesh

Now that the element matrices and element vectors have been integrated, they are ready to be assembled into global equivalents for a series of interconnected hexahedra. This is performed in exactly the same manner as done earlier in 1D and 2Ds. Figure 6.4 illustrates a simple 3D mesh consisting of two elements in each direction. Although this mesh has the same number of nodes in each direction, when this is not the case it's important to number in the direction with the fewest nodes first to reduce the bandwidth of the global matrix. For a simple regular mesh such as that shown in Figure 6.4, the coordinates of the global nodes can be constructed with the following Matlab commands:

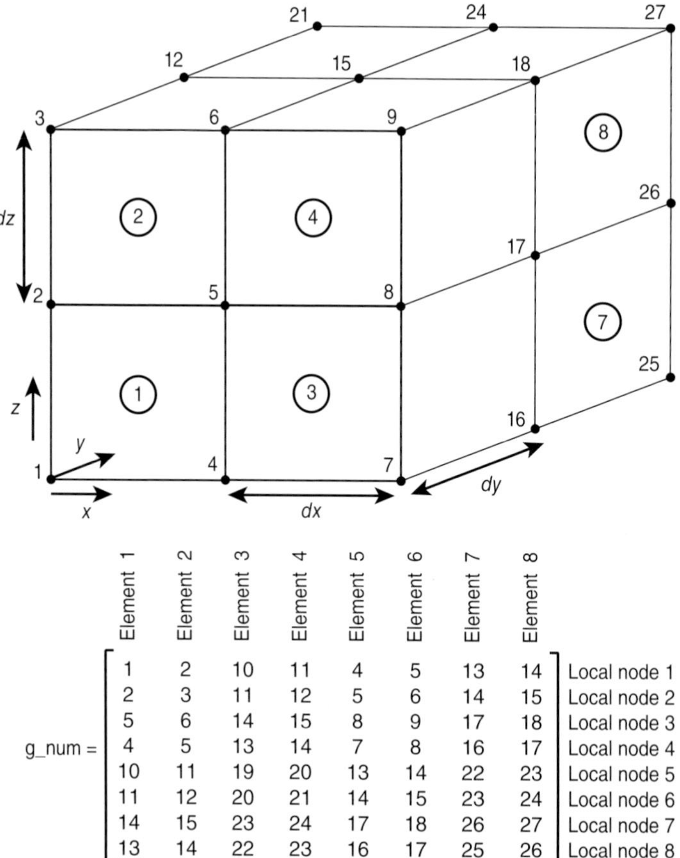

Figure 6.4 An example of a 3D mesh consisting of eight hexahedra, each with eight corner nodes. This mesh is numbered in the *z*-direction. The local node numbering within any one element is illustrated in Figure 6.2. The matrix `g_num` contains connectivity data relating the local and global nodes in each element.

```
% define mesh (numbering in z direction)
g_coord = zeros(ndim,nn) ; % global nodal coordinates
n = 1 ; % global node counter
for i=1:ny % loop over nodes in y-direction
  for j=1:nx % loop over nodes in x-direction
      for k=1:nz  % loop over nodes in z-direction
        g_coord(1,n) = (i-1)*dx ;
        g_coord(2,n) = (j-1)*dy ;
        g_coord(3,n) = (k-1)*dz ;
        n = n + 1 ; % increment node counter
    end
  end
end
```

Here, nn is the total number of nodes in the mesh (27 in Figure 6.4); ndim is the number of spatial dimensions (3); dx, dy, and dz are node spacings in the x-, y-, and z-directions, respectively; and nx is the number of nodes in the x-direction, and so on. The connectivity between local and global node numbers in each element is stored in Matlab in the matrix g_num (Figure 6.4), which can be generated with the snippet

```
% Establish local-global node connectivity for each element
gnumbers = reshape(1:nn,[nz nx ny]) ; % matrix of global node numbers
iel = 1 ; % element counter
for i=1:nelx % loop over elements in x-direction
  for j=1:nely % loop over elements in y-direction
    for k=1:nelz % loop over elements in z-direction
      g_num(1,iel) = gnumbers(k,i,j) ;       % node 1
      g_num(2,iel) = gnumbers(k+1,i,j) ;     % node 2
      g_num(3,iel) = gnumbers(k+1,i+1,j);    % node 3
      g_num(4,iel) = gnumbers(k,i+1,j);      % node 4
      g_num(5,iel) = gnumbers(k,i,j+1) ;     % node 5
      g_num(6,iel) = gnumbers(k+1,i,j+1) ;   % node 6
      g_num(7,iel) = gnumbers(k+1,i+1,j+1); % node 7
      g_num(8,iel) = gnumbers(k,i+1,j+1);    % node 8
      iel          = iel + 1 ; % increment element counter
    end
  end
end
```

Once both g_coord and g_num have been created in the preprocessing stage, and after the element matrices and load vectors have been integrated, they can be added to their global forms within an element loop using the now-familiar Matlab snippet as follows:

```
num = g_num(:,iel)   ; % global node numbers for current element
lhs(num,num) = lhs(num,num) +  MM/dt + KM ; % assemble lhs
rhs(num,num) = rhs(num,num) +  MM/dt   ; % assemble rhs
ff(num) = ff(num) +  F ;  % assemble rhs global load vector
```

6.4 Boundary Conditions and Solution

The procedure by which boundary conditions are implemented and the global system of algebraic equations is solved in 3Ds is identical to already done in 1D and 2D. As shown in Chapters 4 and 5, these can be performed within the time loop with the following Matlab commands:

```
b = rhs*displ + ff  ;                 % form rhs global vector

% impose boundary conditions
lhs(bcdof,:) = 0 ;                    % zero the boundary equations
tmp = spdiags(lhs,0) ;                % store diagonal
tmp(bcdof)=1 ;                        % place 1 on stored-diagonal
lhs=spdiags(tmp,0,lhs);               % reinsert diagonal
b(bcdof) = bcval ;                    % set fixed values in rhs vector

displ = lhs \ b ;                     % solve system of equations
```

Note however that now, due to the 3D nature of the problem, in order to obtain an accurate solution, the total number of equations to be solved can be large. For example, if one discretizes a 3D domain with 100, eight-node hexahedra in each direction, the total number of equations to solved exceeds 10^6. These systems are clearly far more challenging from a purely numerical viewpoint and may necessitate more specialized methods to make the element integration, assembly, and solution stages more efficient. There are a number of different strategies one may take to achieve this efficiency, though they are beyond the scope of the text (see Chapter 13). First, within the Matlab environment, one can use SuiteSparse (Davis, 2007), a suite of sparse matrix packages designed for the efficient solution of sparse systems of equations. Second, one can resort to iterative methods (i.e., the conjugate gradient method) to solve the global system of algebraic equations, which normally require much less memory than direct methods. Third, one can use more optimal strategies to perform element integration and assembly (e.g., see Dabrowski et al. (2008)). Fourth, one can write the finite element programs in a language that executes faster than Matlab (e.g., Fortan or C/C++) and make use of more sophisticated solver packages (e.g., PETSc).

6.5 Matlab Program

The script listed in the following text computes the numerical solution to the 3D heat conduction equation, using the approach discussed in detail before. Comparison of the results computed with the FEM with the exact solution shows excellent agreement (Figure 6.5). A more optimized version of this script is listed in Chapter 13. Here, although visualization is performed using Matlab, more advanced plotting can be performed using software such as ParaView.

```
%-----------------------------------------------------
% Program: diffusion3d.m
% FEM - Diffusion equation in 3D
% 8-node hexahedra elements
%-----------------------------------------------------

clear

seconds_per_yr = 60*60*24*365 ; % seconds in 1 year

% physical parameters
kappa = 1e-6   ; % thermal diffusivity, m^2/s
lx    = 3e3    ; % length of domain, m
ly    = 2e3    ; % width of domain, m
lz    = 1e3    ; % depth of domain, m
H     = 0*1e-9  ; % heat source, K/s
Tb    = 0      ; % fixed boundary temperature, C
```

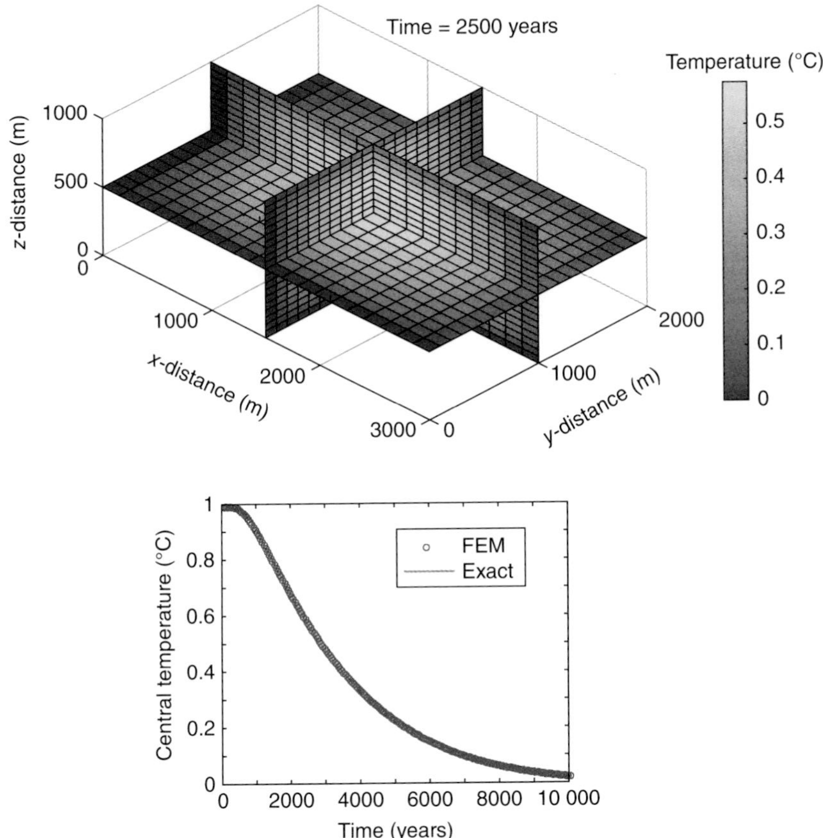

Figure 6.5 Solution of Equation 6.1 ($\kappa = 10^{-6}$ m^2 s^{-1}, $H = 0$) with unit initial temperature and zero temperatures on all boundaries for the domain $x = [0,3000$ m$]$, $y = [0,2000$ m$]$, and $z = [0,1000$ m$]$. The solution was obtained with the FEM using the program listed in Section 6.5. The lower plot shows a comparison between the FEM numerical solution (at $x = lx/2, y = ly/2, z = lz/2$) and the exact solution from Carslaw and Jaeger (1959, page 184, equations 4 and 5). The numerical solution was computed with a time step of 50 years and using 20 eight-node hexahedra in each direction.

```
Ti    = 1     ; % initial temperature, C

% numerical parameters
ntime = 200   ;    % number of time steps
ndim  = 3     ;    % number of spatial dimensions
nod   = 8     ;    % number of nodes per element
nelx  = 20    ;    % number of elements in x-direction
nely  = 20    ;    % number of elements in y-direction
nelz  = 20    ;    % number of elements in z-direction
nels  = nelx*nely*nelz ; % total number of elements
nx    = nelx+1 ;     % number of nodes in x-direction
ny    = nely+1 ;     % number of nodes in y-direction
nz    = nelz+1 ;     % number of nodes in z-direction
nn    = nx*ny*nz ;   % total number of nodes
nip   = 2^ndim   ;  % number of integration points
dx    = lx/nelx ;    % element length in x-direction
dy    = ly/nely ;    % element length in y-direction
dz    = lz/nelz ;    % element length in z-direction
```

```
dt    = 50*seconds_per_yr ; % time step (s)
kay   = eye(ndim)*kappa ;   % thermal diffusivity tensor

% define mesh (numbering in z direction)
g_coord = zeros(ndim,nn) ;
n = 1 ;
for i=1:ny
  for j=1:nx
      for k=1:nz
        g_coord(1,n) = (j-1)*dx ;
        g_coord(2,n) = (i-1)*dy ;
        g_coord(3,n) = (k-1)*dz ;
        n = n + 1 ;
    end
  end
end

% establish elem-node connectivity
gnumbers = reshape(1:nn,[nz nx ny]) ;
iel = 1 ;
for i=1:nelx
  for j=1:nely
    for k=1:nelz
      g_num(1,iel) = gnumbers(k,i,j) ;       % node 1
      g_num(2,iel) = gnumbers(k+1,i,j) ;     % node 2
      g_num(3,iel) = gnumbers(k+1,i+1,j);    % node 3
      g_num(4,iel) = gnumbers(k,i+1,j);      % node 4
      g_num(5,iel) = gnumbers(k,i,j+1) ;     % node 5
      g_num(6,iel) = gnumbers(k+1,i,j+1) ;   % node 6
      g_num(7,iel) = gnumbers(k+1,i+1,j+1);  % node 7
      g_num(8,iel) = gnumbers(k,i+1,j+1);    % node 8
      iel          = iel + 1 ;
    end
  end
end

% find boundary nodes
bx0 = find(g_coord(1,:)==0)  ;
bxn = find(g_coord(1,:)==lx) ;
by0 = find(g_coord(2,:)==0)  ;
byn = find(g_coord(2,:)==ly) ;
bz0 = find(g_coord(3,:)==0)  ;
bzn = find(g_coord(3,:)==lz) ;

% define boundary conditions
bcdof = unique([bx0 bxn by0 byn bz0 bzn]) ;
bcval = Tb*ones(1,length(bcdof))    ;

% gauss integration data
points = zeros(nip,ndim); % location of points
root3  = 1./sqrt(3);
points(1,1)= root3;points(1,2)= root3;points(1,3)= root3;
points(2,1)= root3;points(2,2)= root3;points(2,3)=-root3;
points(3,1)= root3;points(3,2)=-root3;points(3,3)= root3;
points(4,1)= root3;points(4,2)=-root3;points(4,3)=-root3;
points(5,1)=-root3;points(5,2)= root3;points(5,3)= root3;
points(6,1)=-root3;points(6,2)=-root3;points(6,3)= root3;
points(7,1)=-root3;points(7,2)= root3;points(7,3)=-root3;
points(8,1)=-root3;points(8,2)=-root3;points(8,3)=-root3;
```

```
wts = ones(1,nip) ; % weights

% save shape functions and their derivatives in local coordinates
% evaluated at integration points
for k=1:nip
  xi   = points(k,1) ;
  eta  = points(k,2) ;
  zeta = points(k,3) ;
  etam = 1-eta ;  xim=1-xi   ;  zetam=1-zeta ;
  etap = eta+1 ;  xip=xi+1   ;  zetap=zeta+1 ;
  fun=[0.125*xim*etam*zetam 0.125*xim*etam*zetap 0.125*xip*etam*zetap ...
       0.125*xip*etam*zetam 0.125*xim*etap*zetam 0.125*xim*etap*zetap ...
       0.125*xip*etap*zetap 0.125*xip*etap*zetam ] ;
  fun_s(k,:) = fun ; % shape functions
  der(1,1)=-.125*etam*zetam ; der(1,2)=-.125*etam*zetap ;
  der(1,3)=.125*etam*zetap  ; der(1,4)=.125*etam*zetam ;
  der(1,5)=-.125*etap*zetam ; der(1,6)=-.125*etap*zetap ;
  der(1,7)=.125*etap*zetap  ; der(1,8)=.125*etap*zetam ;
  der(2,1)=-.125*xim*zetam  ; der(2,2)=-.125*xim*zetap ;
  der(2,3)=-.125*xip*zetap  ; der(2,4)=-.125*xip*zetam ;
  der(2,5)=.125*xim*zetam   ; der(2,6)=.125*xim*zetap ;
  der(2,7)=.125*xip*zetap   ; der(2,8)=.125*xip*zetam ;
  der(3,1)=-.125*xim*etam   ; der(3,2)=.125*xim*etam ;
  der(3,3)=.125*xip*etam    ; der(3,4)=-.125*xip*etam ;
  der(3,5)=-.125*xim*etap   ; der(3,6)=.125*xim*etap ;
  der(3,7)=.125*xip*etap    ; der(3,8)=-.125*xip*etap  ;
  der_s(:,:,k) = der ; % derivative of shape function
end

% initialise arrays
ff    = zeros(nn,1);      % global load vector
b     = zeros(nn,1);      % global rhs vector
displ = zeros(nn,1);      % global solution vector
lhs   = sparse(nn,nn);    % global lhs matrix
rhs   = sparse(nn,nn);    % global rhs matrix

% grids for plotting
xv = linspace(0,lx,nx);
yv = linspace(0,ly,ny);
zv = linspace(0,lz,nz);
[xg,zg,yg] = meshgrid(xv,yv,zv);
%-----------------------------------------------------
% matrix integration and assembly
%-----------------------------------------------------

for iel=1:nels  % sum over elements
  num    = g_num(:,iel)     ; % element nodes
  coord  = g_coord(:,num)' ; % element coordinates
  KM     = zeros(nod,nod)   ; % initialisation
  MM     = zeros(nod,nod)   ;
  F      = zeros(nod,1)     ;
  for k = 1:nip % integration loop
    fun    = fun_s(k,:) ; % shape functions
    der    = der_s(:,:,k) ; % der. of shape functions in local coordinates
    jac    = der*coord  ;   % jacobian matrix
    detjac = det(jac) ;     % det. of jacobian
    invjac = inv(jac) ;     % inv. of jacobian
    deriv  = invjac*der ;   % der. of shape fun. in physical coords.
    KM     = KM + deriv'*kay*deriv*detjac*wts(k) ; % elem. stiffness matrix
```

```
      MM      = MM + fun'*fun*detjac*wts(k) ;        % elem. mass matrix
      F       = F  + fun'*H*detjac*wts(k) ;          % elem. load vector
    end
    % assemble global matrices and vector
    lhs(num,num) = lhs(num,num) + MM/dt + KM ;
    rhs(num,num) = rhs(num,num) + MM/dt      ;
    ff(num)      = ff(num)      + F          ;
end

%-----------------------------------------------------
% time loop
%-----------------------------------------------------

displ(1:nn) = Ti ; % initial conditions
time        = 0  ; % initial time

for n=1:ntime
   n

   time = time + dt ;                    % update time

   b = rhs*displ + ff  ;                 % form rhs global vector

   % impose boundary conditions
   lhs(bcdof,:) = 0 ;                    % zero the boundary equations
   tmp = spdiags(lhs,0) ;                % store diagonal
   tmp(bcdof)=1 ;                        % place 1 on stored-diagonal
   lhs=spdiags(tmp,0,lhs);               % reinsert diagonal
   b(bcdof) = bcval ;                    % set boundary values in rhs vector

   displ = lhs \ b ;                     % solve system of equations

%-----------------------------------------------------
   % evaluate exact solution
   % Carslaw and Jaeger (1959) eqs 4 & 5, page 184
   x = 0 ; y = 0 ; z = 0 ; % point where solution is evaluated
   nterms = 100 ;
   a = lx/2; b = ly/2; c = lz/2;
   psix = 0 ; psiy = 0 ; psiz = 0 ;
   for m=0:nterms
     nt = 2*m+1;
     t0 = (-1)^m/nt;
     cx = cos(nt*pi*x/(2*a)) ;
     ex = exp(-kappa*nt^2*pi^2*time/(4*a^2)) ;
     psix = psix + 4/pi*t0*ex*cx ;
     cy = cos(nt*pi*y/(2*b)) ;
     ey = exp(-kappa*nt^2*pi^2*time/(4*b^2)) ;
     psiy = psiy + 4/pi*t0*ey*cy ;
     cz = cos(nt*pi*z/(2*c)) ;
     ez = exp(-kappa*nt^2*pi^2*time/(4*c^2)) ;
     psiz = psiz + 4/pi*t0*ez*cz ;
   end
   exact = psix.*psiy.*psiz ;
%-----------------------------------------------------

   % plot solution
   figure(1)
   solution = reshape(displ,ny,nx,nz) ;
   xslice = lx/2 ;
```

```
    yslice = ly/2 ;
    zslice = lz/2 ;
    slice(xg,zg,yg,solution,xslice,yslice,zslice)
    xlabel('x-distance (m)')
    ylabel('y-distance (m)')
    zlabel('z-distance (m)')
    colorbar
    view([45,30])
    axis equal

    figure(2)
    ix = find(xv==x+a) ;
    iy = find(yv==y+b) ;
    iz = find(zv==z+c) ;
    numerical_s(n) = solution(ix,iz,iy) ;
    exact_s(n)     = exact ;
    time_s(n)      = time   ;
    plot(time_s/seconds_per_yr,exact_s,'r',time_s/seconds_per_yr,numerical_s,'bo')
    xlabel('Time (years)')
    ylabel('Temperature (C)')

    drawnow

end

%-----------------------------------------------------------
% end of time loop
%-----------------------------------------------------------
```

6.6 Exercises

1) Determine the derivatives of the shape functions in physical coordinates at the point with local coordinates $\xi = 1/2$, $\eta = 1/2$, $\zeta = 1/2$ for a eight-node hexahedron element with nodes at the following physical coordinates:

$$\begin{bmatrix} x_i & y_i & z_i \\ 0 & 0 & 0 \\ 0.1 & 0.1 & 1.2 \\ 1.1 & 0.05 & 1.3 \\ 1 & 0 & 0.15 \\ 0 & 1.5 & 0.1 \\ 0.1 & 1.6 & 1.2 \\ 1.8 & 1.8 & 1.8 \\ 1.7 & 1.6 & 0 \end{bmatrix}$$

Use the derivatives of the shape functions to approximate $\partial T/\partial x$, $\partial T/\partial y$, and $\partial T/\partial z$ at the point of interest assuming $T(x, y, z) = 1 + x^2/4 + y^2/5 + z^2/6$. Hint: Evaluate the trilinear shape functions at $\xi = 1/2$, $\eta = 1/2$, and $\zeta = 1/2$. Using these and the node coordinates for the element, compute the Jacobian matrix and its inverse. The derivatives of the shape functions in global coordinates are then obtained by multiplying the inverted Jacobian with the shape functions derivatives in local coordinates. Finally, obtain the partial derivatives of T by multiplying the derivatives of the shape functions in global coordinates with T_i (i.e., $T(x, y, z)$ evaluated at the eight nodes for the element).

The exact values of the derivatives at the point of interest ($x = 1.2188$, $y = 1.2914$, $z = 1.1789$) are $\partial T/\partial x = 0.6094$, $\partial T/\partial y = 0.5166$, and $\partial T/\partial z = 0.3930$, while the approximate derivatives are 0.5033, 0.4592, and 0.2954, respectively.

2) Write a Matlab script to calculate the volume of a eight-node hexahedron by performing Gauss–Legendre quadrature using eight points (two in each direction). Use the node positions given in Exercise 1. Your result should be 2.7742. Hint: The volume can be computed as $v = \int dv = \int_{-1}^{1}\int_{-1}^{1}\int_{-1}^{1} \det \mathbf{J}\, d\xi d\eta d\zeta \simeq \sum_{k=1}^{nip} \det \mathbf{J}\, W_k$. Thus, one needs to compute the Jacobian, find its determinant, and integrate using discrete rule for Gauss–Legendre quadrature.

3) Solve using the FEM Laplace's equation

$$\frac{\partial^2 \phi}{\partial x^2} + \frac{\partial^2 \phi}{\partial y^2} + \frac{\partial^2 \phi}{\partial z^2} = 0$$

in a cube of dimensions $0 \le x, y, z \le a$ with $\phi = 0$ on all boundaries except for $z = a$ where $\phi = \phi_0$. Verify your numerical results with the exact solution

$$\phi = \sum_{n,m=1}^{\infty} \frac{16\phi_0}{nm\pi^2 \sinh(\gamma a)} \sin(\alpha x)\sin(\beta y)\sinh(\gamma z)$$

where $\alpha = n\pi/a$, $\beta = m\pi/a$, and $\gamma = \sqrt{\alpha^2 + \beta^2}$.

Suggested Reading

A. J. M. Ferreira, *MATLAB Codes for Finite Element Analysis: Solids and Structures (Solid Mechanics and Its Applications)*, Springer, Berlin, 2009.

Y. W. Kwong, and H. C. Bang, *The Finite Element Method using Matlab*, CRC Press, New York, 2000.

I. M. Smith, and D. V. Griffith, *Programming the Finite Element Method*, John Wiley & Sons, Ltd, Chichester, 1998.

O. C. Zienkiewicz, and R. L. Taylor, *The Finite Element Method, Volume 1, The Basis*, Butterworth-Heinemann, Oxford/Boston, MA, 2000.

7

Generalization of Finite Element Concepts

The purpose of this chapter is to generalize some of the concepts encountered in the previous chapters before tackling various specific applications in the following chapters. We look at some of the other terms commonly encountered in partial differential equations (PDEs) before showing how the finite element method (FEM) is applied to elliptic and hyperbolic problems. We also consider how to solve systems of equations involving more than one unknown. Other features dealt with for the first time in this chapter are different element types (including triangles, three-node linear elements, and nine-node quadrilaterals, Figure 7.1), mesh generation with the Matlab pdetool GUI, and implementation of Neumann (flux) boundary conditions. Three complete Matlab scripts are listed to demonstrate how these features are incorporated in practice.

In the previous few chapters, we have shown that finite element discretization of the same PDE in one dimension (1D), two dimension (2D), and three dimension (3D) leads to very similar systems of algebraic equations for individual elements. Moreover, we saw that once the mesh and nodal connectivity are appropriately defined, the stages involving element matrix integration, assembly, implementation of boundary conditions, and solution are performed in a very similar manner in 1D, 2D, and 3D. These points make the FEM relatively easy to program and to apply in different dimensions.

Up to this point, all of our attention has been focused on different forms of the following PDE:

$$\frac{\partial T}{\partial t} = \nabla^T \kappa \nabla T + H \tag{7.1}$$

Within the context of heat transfer, T is the temperature, κ is the thermal diffusivity, and H is a source term accounting for internal heat generation. This is a parabolic (diffusion) equation with a source term. The first term is entirely responsible for transient effects, the second term accounts for dissipation (which typically smoothens the solution), while the third term alters the "mass" of the system. Diffusion-type equations such as this appear in numerous other contexts in Earth science. For example, fluid flow in the shallow subsurface is governed by diffusion (dissipation) of excess fluid pressure (i.e., the pore pressure in excess of the hydrostatic pore pressure) through porous sediments (see Chapter 10). In this case, the excess fluid pressure (P) is governed by the equation

$$\phi\beta\frac{\partial P}{\partial t} = \nabla^T \left(\frac{k}{\mu}\nabla P\right) + H \tag{7.2}$$

where ϕ is the porosity, β is the bulk compressibility (Pa^{-1}), k is the permeability (m^2), μ is the fluid viscosity (Pa s), and H accounts for any fluid pressure sources or sinks (e.g., due to dehydration reactions). The ease with which over-pressures are dissipated is controlled by the ratio of the permeability over the fluid viscosity, which plays the same role as the thermal diffusivity in heat conduction.

Practical Finite Element Modeling in Earth Science Using Matlab, First Edition. Guy Simpson.

Companion website: www.wiley.com/go/simpson

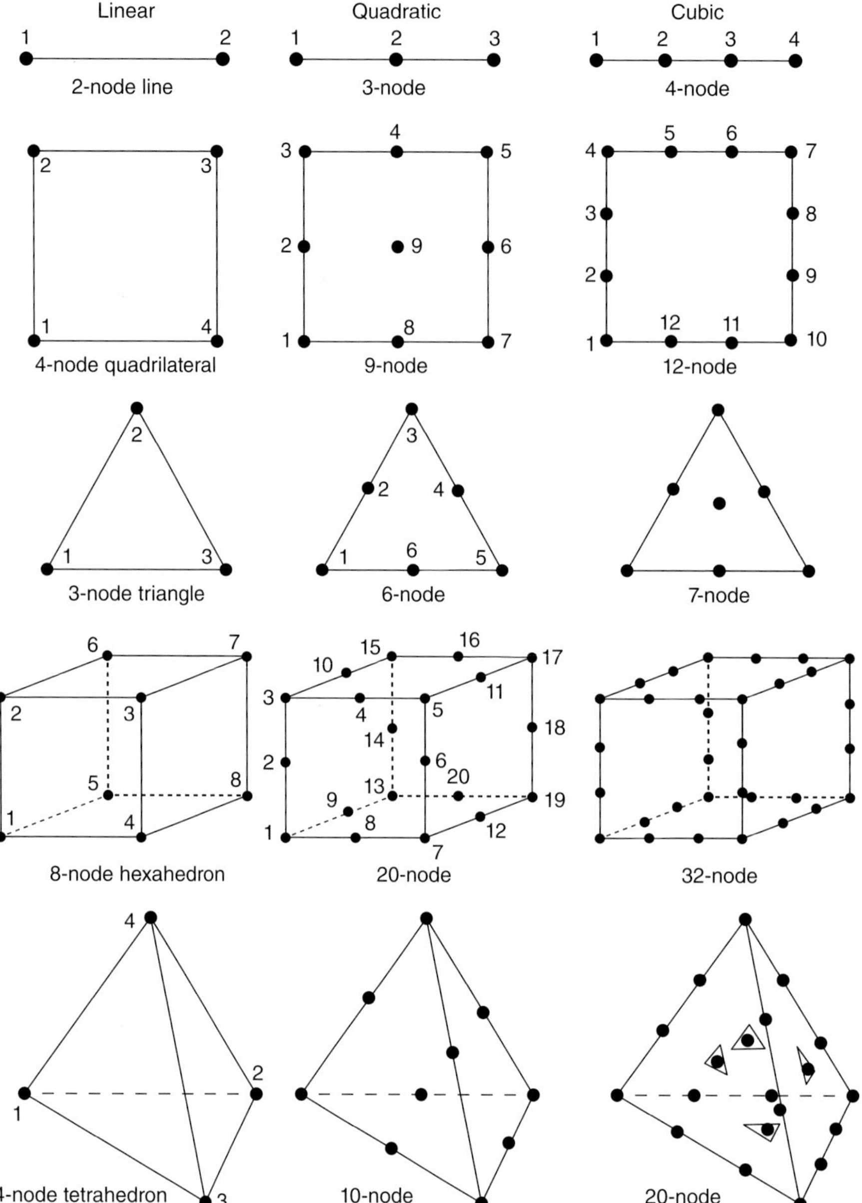

Figure 7.1 Common linear, quadratic, and cubic finite elements in 1D, 2D, and 3D. Shape functions for these and other elements can be found in numerous texts (e.g., see Hughes et al. (1989), Kwong and Bang (2000), Lewis et al. (2004), Smith and Griffiths (1998), and Zienkiewicz and Taylor (2000a)).

The transient response is controlled by compressibility of the combined fluid and rock matrix. If there was no compressibility (i.e., $\beta = 0$, the transient term on the left-hand side of 7.2 disappears and the pore pressure distribution would always be in a steady state. The important point here is that because

Equation 7.2 has exactly the same form as that governing conductive heat transfer, it can be solved by essentially the same program, once the different variables and material parameters have been correctly redefined.

In the second part of the text we will encounter other PDEs that, when discretized, will lead to element matrices that we have not yet encountered. In anticipation of this, consider the following general PDE in 1D:

$$\underbrace{A\frac{\partial u^2}{\partial t^2}}_{1} + \underbrace{B\frac{\partial u}{\partial t}}_{2} + \underbrace{Cu}_{3} + \underbrace{D\frac{\partial^4 u}{\partial x^4}}_{4} + \underbrace{E\frac{\partial^2 u}{\partial x^2}}_{5} + \underbrace{F\frac{\partial u}{\partial x}}_{6} + \underbrace{G}_{7} = 0 \tag{7.3}$$

which contains many of the commonly encountered terms of relevance in Earth science, all lumped into one "model" equation. In this equation, $u(x, t)$ is the dependent variable, while the independent variables are time (t) and space (x). The first term commonly appears in force equilibrium equations when inertia is important (i.e., Newton's law of motion) and (when combined with term 5) leads to wave-like solutions characteristic of hyperbolic equations. The second term is a classic first-order transient term. The third term occurs in problems where there is dampening (e.g., radioactive decay and isostatic restoring force in flexural problems) or amplification (e.g., population growth), depending on the sign of the coefficient C and the magnitude of the solution. The fourth term commonly arises in continuum mechanics, including elasticity, Stokes flows, and flexure. The fifth term is common in parabolic, elliptic, and hyperbolic equations and occurs in numerous different contexts where diffusion is important, but also appears in fluid- and solid mechanics problems. The sixth term is an advection term (with velocity F) and appears whenever there is flow of material relative to a fixed (Eulerian) frame of reference. The term also typically features in hyperbolic equations. Finally, the seventh term is a source or sink, which is also very commonly encountered in many different contexts (e.g., internal heat generation, gravity, and chemical reactions).

Table 7.1 shows the element matrices and vectors that result from spatial finite element discretization of each of the terms in Equation 7.3. The first three terms all lead to symmetrical mass matrices because the dependent variable in each case involves no spatial derivatives. Note that

Table 7.1 Finite element discretization of common terms appearing in PDEs (see Equation 7.3).

	Term in PDE	Term in element matrix/vector	Structure
1	$A\frac{\partial^2 u}{\partial t^2}$	$\int A\mathbf{N}^T\mathbf{N}\,dx$	Symmetrical matrix
2	$B\frac{\partial u}{\partial t}$	$\int B\mathbf{N}^T\mathbf{N}\,dx$	Symmetrical matrix
3	Cu	$\int C\mathbf{N}^T\mathbf{N}\,dx$	Symmetrical matrix
4	$D\frac{\partial^4 u}{\partial x^4}$	$\int D\frac{\partial^2\mathbf{N}^T}{\partial x^2}\frac{\partial^2\mathbf{N}}{\partial x^2}\,dx$	Symmetrical matrix
5	$E\frac{\partial^2 u}{\partial x^2}$	$-\int E\frac{\partial\mathbf{N}^T}{\partial x}\frac{\partial\mathbf{N}}{\partial x}\,dx$	Symmetrical matrix
6	$F\frac{\partial u}{\partial x}$	$\int F\mathbf{N}^T\frac{\partial\mathbf{N}}{\partial x}\,dx$	Asymmetrical matrix
7	G	$\int G\mathbf{N}^T\,dx$	Vector

The dependent variable is u, while the independent variables are time (t) and space (x). Note the negative sign appearing in front of term 5, which arises from the application of integration by parts (see Section 2.3).

the time derivatives in the first and second terms are normally discretized with finite difference approximations. These so-called mass matrices, involving products of the shape functions, have already been encountered repeatedly in the previous chapters (denoted **MM**). The fourth term leads to a symmetrical matrix involving the product of the shape functions twice differentiated. As will be seen later in the text (Chapter 11), this imposes certain limitations on the order of the shape functions that must be used for the finite element approximation, since any function with an order of less than 3 would disappear following double differentiation. The fifth term leads to a symmetrical matrix involving products of the shape functions once differentiated, after application of integration by parts (see Section 2.3). This element matrix was also encountered in the previous chapters within the context of diffusion, where it was denoted **KM**. The sixth "advection" term, involving a first-order spatial derivative, leads to an asymmetrical element matrix involving the shape function (weighting function) multiplied by its first-order derivative. The presence of this odd first-order operator makes the problem non-self-adjoint, which is not optimally solved with the standard Galerkin FEM (Zienkiewicz and Taylor, 2000c). In Chapter 10, we show how hyperbolic problems involving strong advection can be solved with a modified FEM. Finally, finite element discretization of the last term in 7.3 involving a simply a constant (i.e., a source or sink) leads to a vector involving the shape functions. This "load" vector was repeatedly encountered in the previous chapters where it was denoted **F**.

In the remainder of the chapter, we advance from parabolic problems to consider examples of the other two main classes of PDEs not yet considered: elliptic and hyperbolic equations. Other examples will be treated in the following chapters. We also consider a third (parabolic) problem, which is governed by a system of PDEs rather than a single (scalar) equation, such as considered previously. The objective is to demonstrate how the FEM is applied to solve different types of PDEs (in 2Ds), but also to consider some new aspects including incorporation of Neumann (flux) boundary conditions, different element types, and coupling.

7.1 The FEM for an Elliptic Problem

In general, the solutions to parabolic equations such as the diffusion equation

$$\frac{\partial \phi}{\partial t} = k\nabla^2\phi + S \tag{7.4}$$

change with time. However, if the boundary conditions and parameters are time-independent, we might expect the solution to eventually reach a steady-state situation when the solution remains constant. If we are not interested in the initial transient response but only in the final steady-state solution, we could actually have set the time derivative in 7.4 to 0 and solved the steady equation from the outset, that is,

$$0 = k\nabla^2\phi + S \tag{7.5}$$

This equation, which is known as Poisson's equation, is no longer parabolic but elliptic. Elliptic equations such as 7.5 are commonly encountered in near-equilibrium situations, and they arise in a number of different contexts in Earth science including steady diffusion (as just noted), flexure (see Chapter 11), and slowly flowing fluids. Unlike parabolic equations, with elliptic equations, there is no notion of initial conditions and of information propagating forward in time.

To demonstrate how the FEM is easily applied to solve elliptic equations, consider an incompressible fluid which, in 2Ds, is governed by the continuity equation

$$\frac{\partial u}{\partial x} + \frac{\partial v}{\partial y} = 0 \tag{7.6}$$

where u, v are the velocity components in the x- and y-directions, respectively. In the case that the flow is irrotational (which is reasonable when viscous effects are negligible), one can write the velocity in terms of a scalar velocity potential ϕ, defined as follows:

$$u = \frac{\partial \phi}{\partial x} \quad \text{and} \quad v = \frac{\partial \phi}{\partial y} \tag{7.7}$$

Substituting these into the continuity equation (7.6) gives

$$\frac{\partial^2 \phi}{\partial x^2} + \frac{\partial^2 \phi}{\partial y^2} = 0 \tag{7.8}$$

which is Laplace's equation. This is the potential formulation for flow in an inviscid, irrotational fluid. Note the close similarity between this and the equation governing steady-state diffusion (Equation 7.5). The potential flow model can be used to study the flow field around embedded objects in a simple manner, without solving for the individual velocity components. It is, therefore, computationally very efficient. As an example, say we want to compute the flow field around a solid circular object in 2Ds (Figure 7.2). Far upstream and downstream of the object, we assume that the flow enters and exits the domain at a constant velocity in the x-direction, denoted U, implying that

$$\frac{\partial \phi}{\partial x} = U \quad \text{at} \quad x = 0 \quad \text{and} \quad x = lx \tag{7.9}$$

Note that these are Neumann boundary conditions. No flow is assumed across the model boundaries in the y-direction (i.e., $v = 0$), implying

$$\frac{\partial \phi}{\partial y} = 0 \quad \text{at} \quad y = 0 \quad \text{and} \quad y = ly \tag{7.10}$$

Finally, we will assume no flow conditions on the boundaries of the (solid) object,

$$\mathbf{v} \cdot n = 0 \quad \rightarrow \quad \nabla \phi \cdot n = 0 \tag{7.11}$$

where $\mathbf{v}$ is the velocity vector and $\mathbf{n}$ is the vector of normals on the surface of the object. The model domain is taken to have a width and height of 1 and 0.5, respectively, while the circular object has a diameter of 0.2 (Figure 7.2).

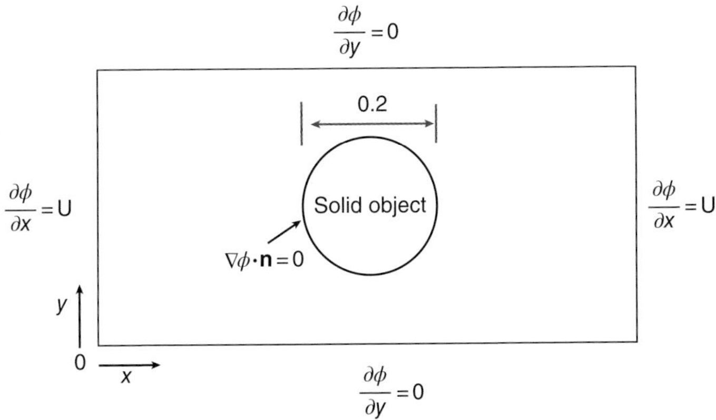

Figure 7.2 Model domain and boundary conditions for elliptic, potential flow problem. Flow enters from the left and exits from the right at a constant (imposed) velocity, U. No flow is imposed normal to the boundaries of the solid inclusion, implying $\nabla \phi \cdot \mathbf{n} = 0$ where $\mathbf{v} = \nabla \phi$.

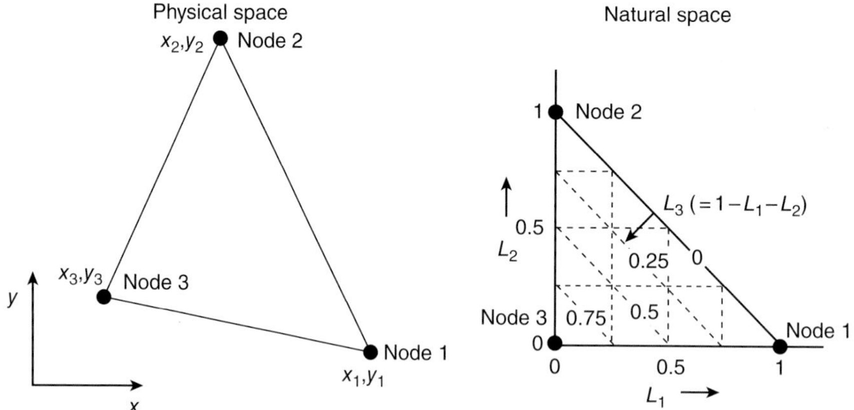

Figure 7.3 A three-node triangular element in the physical- and local (natural)-coordinate spaces.

To discretize the governing equation, we apply the Galerkin FEM, as done previously. However, because this problem involves a somewhat complicated geometry (due to the presence of the circular object), we will adopt three-node (linear) triangular elements (Figure 7.1), which offer considerable flexibility, especially for nonrectilinear geometries. We proceed by introducing the standard approximation

$$\phi = \mathbf{N}\boldsymbol{\phi} \tag{7.12}$$

where ϕ is the continuous variable in the governing PDE, $\boldsymbol{\phi}$ are the unknown velocity potentials defined at the vertices (nodes) of a triangle and the linear shape functions are defined as

$$\begin{aligned} N_1 &= \frac{x\,y_2 - x\,y_3 + x_2\,y_3 - x_2\,y - x_3\,y_2 + x_3\,y}{d} \\ N_2 &= -\frac{x\,y_1 - x\,y_3 + x_1\,y_3 - x_1\,y - x_3\,y_1 + x_3\,y}{d} \\ N_3 &= \frac{x\,y_1 - x\,y_2 + x_1\,y_2 - x_1\,y - x_2\,y_1 + x_2\,y}{d} \end{aligned} \tag{7.13}$$

where

$$d = x_1\,y_2 - x_1\,y_3 - x_2\,y_1 + x_2\,y_3 + x_3\,y_1 - x_3\,y_2$$

Here x_1, y_1 are the global coordinates of node 1, and so on (Figure 7.3). Substituting this approximation into the governing PDE (7.8) leads to the residual equation

$$\frac{\partial^2 \mathbf{N}\boldsymbol{\phi}}{\partial x^2} + \frac{\partial^2 \mathbf{N}\boldsymbol{\phi}}{\partial y^2} = R \tag{7.14}$$

Weighting the residual equation with the shape functions, integrating over an element, and setting the result to zero, we obtain the following:

$$\int\int \mathbf{N}^T \frac{\partial^2 \mathbf{N}}{\partial x^2} dxdy\ \boldsymbol{\phi} + \int\int \mathbf{N}^T \frac{\partial^2 \mathbf{N}}{\partial y^2} dxdy\ \boldsymbol{\phi} = 0 \tag{7.15}$$

Finally, applying integration by parts (see Section 2.3 and Appendix D) leads to

$$\int\int \mathbf{N}^T \frac{\partial^2 \mathbf{N}}{\partial x^2} dxdy\,\boldsymbol{\phi} + \int\int \mathbf{N}^T \frac{\partial^2 \mathbf{N}}{\partial y^2} dxdy\,\boldsymbol{\phi} = -\int\int \frac{\partial \mathbf{N}}{\partial x}^T \frac{\partial \mathbf{N}}{\partial x} dxdy\,\boldsymbol{\phi} - \int\int \frac{\partial \mathbf{N}^T}{\partial y}\frac{\partial \mathbf{N}}{\partial y} dxdy\,\boldsymbol{\phi} + \int_s \mathbf{N}^T\left(\frac{\partial \phi}{\partial x}n_x + \frac{\partial \phi}{\partial y}n_y\right) ds = 0 \tag{7.16}$$

The terms to the right of the first equal sign can be written compactly as

$$\mathbf{KM}\boldsymbol{\phi} = \mathbf{Fb} \tag{7.17}$$

where

$$\mathbf{KM} = \int\int (\nabla\mathbf{N})^T \nabla\mathbf{N}\, dxdy \tag{7.18}$$

and

$$\mathbf{Fb} = \int_s \mathbf{N}^T\left(\frac{\partial \phi}{\partial x}n_x + \frac{\partial \phi}{\partial y}n_y\right) ds \tag{7.19}$$

Here, **KM** is the familiar element stiffness matrix encountered numerous times in diffusion problems, while **Fb** is the element vector that contains a surface integral (s is the surface of an element, and n_x, n_y are direction cosines of the boundary normals). In all previous problems, we neglected the boundary term because it was overwritten with Dirichlet conditions. In the following text, rather than omitting **Fb**, we will use it to impose Neumann (flux) boundary conditions. Note that Equation 7.17 is very similar to a discretized parabolic equation, differing only in the absence of the mass matrix **MM** (see Equation 5.13).

The matrix **KM** contains first-order spatial derivatives of the shape functions, expressed in terms of physical coordinates. Although these could be obtained directly by differentiating 7.13, a more common approach is to define the shape functions and their derivatives in the natural (local) coordinate system and to map them back to the physical coordinate system. Figure 7.3 shows a general three-node triangle in physical space and its mapped equivalent in the local coordinate system (see Zienkiewicz and Taylor (2000a)). The coordinates L_1 and L_2 are assumed to be parallel to the x- and y-coordinate directions, while the third coordinate L_3 is a linear function of the other two, that is,

$$L_3 = 1 - L_1 - L_2$$

In the natural coordinate system, the shape functions listed in 7.13 can be written as

$$\begin{aligned} N_1 &= L_1 \\ N_2 &= L_2 \\ N_3 &= L_3 = 1 - L_1 - L_2 \end{aligned} \tag{7.20}$$

so their first derivatives are simply

$$\nabla\mathbf{N} = \begin{Bmatrix} \frac{\partial}{\partial L_1} \\ \frac{\partial}{\partial L_2} \end{Bmatrix} [N_1\ N_2\ N_3] = \begin{bmatrix} 1 & 0 & -1 \\ 0 & 1 & -1 \end{bmatrix} \tag{7.21}$$

As for other element types, the derivatives of the shape functions can be transformed to the physical coordinate system by performing

$$\begin{bmatrix} \frac{\partial N_1}{\partial x} & \frac{\partial N_2}{\partial x} & \frac{\partial N_3}{\partial x} \\ \frac{\partial N_1}{\partial y} & \frac{\partial N_2}{\partial y} & \frac{\partial N_3}{\partial y} \end{bmatrix} = \mathbf{J}^{-1} \begin{bmatrix} \frac{\partial N_1}{\partial L_1} & \frac{\partial N_2}{\partial L_1} & \frac{\partial N_3}{\partial L_1} \\ \frac{\partial N_1}{\partial L_2} & \frac{\partial N_2}{\partial L_2} & \frac{\partial N_3}{\partial L_2} \end{bmatrix} \tag{7.22}$$

where $\mathbf{J}^{-1}$ is the inverse of the Jacobian matrix that can be computed from

$$\mathbf{J} = \begin{bmatrix} \frac{\partial N_1}{\partial L_1} & \frac{\partial N_2}{\partial L_1} & \frac{\partial N_3}{\partial L_1} \\ \frac{\partial N_1}{\partial L_2} & \frac{\partial N_2}{\partial L_2} & \frac{\partial N_3}{\partial L_2} \end{bmatrix} \begin{bmatrix} x_1 & y_1 \\ x_2 & y_2 \\ x_3 & y_3 \end{bmatrix} \tag{7.23}$$

For triangles, an integral in physical space can be computed in normalized space according to the following relation:

$$\int\int f(x,y)\,dxdy = \int_0^1 \int_0^{1-L_1} f(L_1, L_2)\,dL_1 dL_2 \,\det\mathbf{J} \tag{7.24}$$

Here, $\det\mathbf{J}$ is the determinant of the Jacobian matrix. Gauss–Legendre quadrature approximates this last integral as

$$\int_0^1 \int_0^{1-L_1} f(L_1, L_2)\,dL_1 dL_2 \,\det\mathbf{J} \simeq \sum_{k=1}^{nip} f(L_1, L_2)_k \; w_k \,\det\mathbf{J} \tag{7.25}$$

where nip is the number of integration points in an element, L_1 and L_2 are the positions of the integration points in local coordinates, and w_k are the Gauss weights for each point. Typical values for the integration points and the weights for triangles are shown in Table 7.2.

Table 7.2 Gauss quadrature points (L_k) and weights (w_k) for triangles.

n	L_1	L_2	w_k	Degree
1	$\frac{1}{3}$	$\frac{1}{3}$	1	1
3	$\frac{1}{2}$	$\frac{1}{2}$	$\frac{1}{3}$	2
	$\frac{1}{2}$	0	$\frac{1}{3}$	
	0	$\frac{1}{2}$	$\frac{1}{3}$	
4	$\frac{1}{3}$	$\frac{1}{3}$	$-\frac{27}{48}$	3
	0.6	0.2	$\frac{25}{48}$	
	0.2	0.6	$\frac{25}{48}$	
	0.2	0.2	$\frac{25}{48}$	
7	$\frac{1}{3}$	$\frac{1}{3}$	0.225	4
	0.797426985353087	0.101286507323456	0.1259391805	
	0.101286507323456	0.797426985353087	0.1259391805	
	0.101286507323456	0.101286507323456	0.1259391805	
	0.470142064105115	0.059715871789770	0.1323941527	
	0.059715871789770	0.470142064105115	0.1323941527	
	0.470142064105115	0.470142064105115	0.1323941527	

Note that it is traditional to tabulate quadrature rules for triangles and tetrahedra such that the weights sum to 1. These weights should therefore be multiplied by $c = \frac{1}{2}$ for triangles.

Figure 7.4 Three-node triangular element on the boundary of a finite element mesh. Numbers refer to local node numbers. Also shown are unit normal vectors with direction cosines on the two sides of the element.

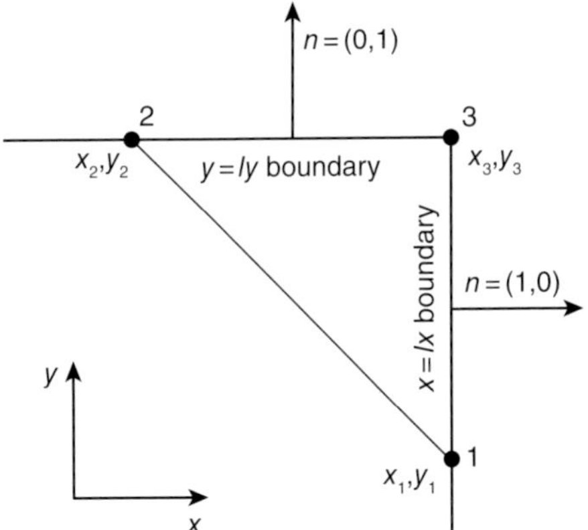

We now turn our attention to computing the boundary element vector **Fb** (7.19). Note first that the terms $\frac{\partial \phi}{\partial x} n_x$ and $\frac{\partial \phi}{\partial y} n_y$ can be identified with fluxes across the boundaries of an element. These boundary fluxes can generally be ignored for internal elements because they cancel due to mass conservation between adjacent elements. So far, we have also ignored these fluxes on boundary elements, since we have overridden them by imposing Dirichlet conditions. However, when flux boundary conditions are desired, the until-now-neglected boundary integral term can be retained.

Consider implementation of the zero flux condition (7.10) on $y = 0$ or $y = ly$ (or on the boundary of the object; see Figure 7.4). In this case, by neglecting the boundary integral in 7.16 and by not imposing any Dirichlet condition, then zero flux boundary conditions (i.e., Equation 7.10) are automatically satisfied. Such a condition is known as a natural boundary condition.

Consider now how to enforce a nonzero flux condition (e.g., 7.9) on the $x = lx$ boundary for the element shown in Figure 7.4. Inserting $\partial \phi / \partial x = U$ into 7.19, using **N** defined in 7.13, and noting that $n_x = 1$ and $n_y = 0$ on the $x = lx$ boundary, the following can be obtained:

$$\mathbf{Fb} = \int_{y1}^{y3} \mathbf{N}^T \frac{\partial \phi}{\partial x} dy = \int_{y1}^{y3} \mathbf{N}^T U \, dy = \frac{U b}{2} \begin{Bmatrix} 1 \\ 0 \\ 1 \end{Bmatrix} \tag{7.26}$$

Here, b is the distance $y_3 - y_1$ (Figure 7.4). Note that only the two nodes on the boundary contribute to the load vector (in this case, nodes 1 and 3). This vector needs to be added to the right-hand-side load vector for all elements on the $x = lx$ boundary.

To create the triangular mesh for this problem, we utilize the "pdetool" mesh-generator in Matlab. Using this, a basic mesh corresponding to the geometry illustrated in Figure 7.2 can be generated by performing the following steps:

1) With a Matlab command window, open the pdetools GUI by typing "pdetool" at the command prompt (Figure 7.5). Turn on the grid by selecting Grid from the Options menu.
2) Draw a rectangle (R1) with a width of 1.0 and a height of 0.5 (Figure 7.5). Double clicking on the rectangle opens a dialogue box, which enables you to assign its exact dimensions and position. Ensure the lower left corner of the rectangle is at the position $x = 0$, $y = 0$.

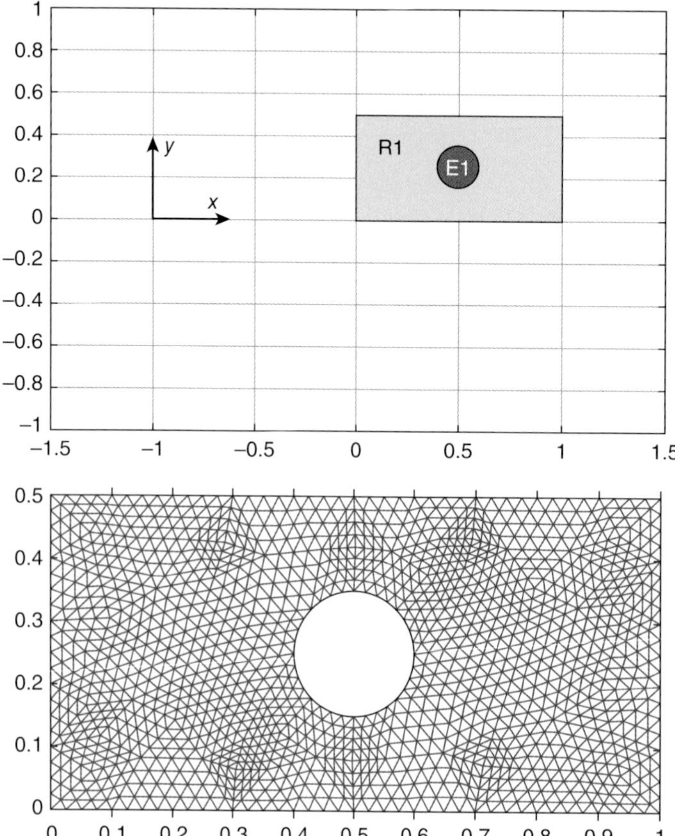

Figure 7.5 Generation of a triangular mesh around a circular object using the Matlab pdetool GUI. The top panel shows the rectangle (R1) and circle (E1) within the pdetool environment. The mesh, defined by the area R1−E1, is shown in the bottom panel of the figure.

3) Draw a circle (E1) with a diameter of 0.2. Place the circle so that its center is at the position $x = 0.5$, $y = 0.25$ (Figure 7.5). Once again, the dialogue box can be used to refine its exact dimensions and position.
4) We are now going to create a mesh defined by the formula R1−E1, that is, the area defined by the rectangle minus the area of circle (since we don't want the area within the circle in our mesh). This formula can be specified in the "Set formula" line near the top of the GUI. Once the formula is assigned, select "Initialize Mesh" from the Mesh menu. A more refined mesh can be obtained by repeating clicking "Refine Mesh" from the Mesh menu.
5) Once the desired resolution has been obtained, the mesh can be exported to the Matlab workspace by choosing Export Mesh under the Mesh menu. The three variables saved by default are `p`, `e`, and `t`. The variable `p` contains the coordinates of the nodes in the mesh (i.e., `g_coord` in this text), `e` is an array used for plotting (called `edge` in the program presented), and `t` contains element nodal connectivity in the first three lines (i.e., `g_num`), one line for each node of a particular triangle. Line 4 of `t` is a "phase" index referring to whether a particular element is within the region

R1−E1 (i.e., outside the circle, with the phase index=1) or within the region E1−R1. In our problem, all elements are, by definition, outside the circle and so all elements have a phase index of 1. Renaming of the variables can be achieved within Matlab with the following snippet:

```
g_coord = p ;          % nodal coordinates
edge    = e ;          % edge array
g_num   = t(1:3,:) ;   % element nodal numbering
phase   = t(4,:) ;     % phase indices
```

6) These variables can be saved in a .mat file using the Save command (e.g., by typing `save 'mesh_cylinder'`) and loaded into the finite element program during the preprocessor stage.

A complete Matlab script to solve the elliptic potential flow problem in 2Ds using linear triangular elements is listed in the following text. The program shares many similarities with the program to solve the diffusion equation in 2Ds, presented in Chapter 5. The following points are noteworthy to the program listed here:

1) The variables `g_coord` (nodal coordinates), `edge` (mesh edge data used only for plotting), `g_num` (node-element connectivity), and `phase` (element material indices) are read into the Matlab session from a single .mat file (mesh_cylinder[1]). This file was generated with pdetool, as described before. The file can be loaded and the mesh can be visualized with the following snippet:

```
load mesh_cylinder % load triangular mesh
pdemesh(g_coord,edge,g_num) % visualize mesh
axis equal % ensure plot has same scale in x- and y-directions
```

2) In order to perform Gauss–Legendre integration of the element matrix **KM** (defined in Equation 7.18), the positions and weights for triangles must be provided from Table 7.2. Assuming three integration points (i.e., `nip=3`), this can be done with the following lines:

```
% Gauss-Legendre integration data
points = zeros(nip,ndim); % location of points
points(1,1)=0.5 ; points(1,2)=0.5 ;
points(2,1)=0.5 ; points(2,2)=0 ;
points(3,1)=0 ; points(3,2)=0.5 ;
c = 0.5 ; % triangle factor
wts = c*1/3*ones(1,nip) ; % weights
```

Here, `ndim` is the number of spatial dimensions (here 2).

3) The values of the shape functions and their spatial derivatives for the three-node (linear) triangle, defined in terms of local triangular coordinates L_1, L_2, and L_3 (see Figure 7.3), must be evaluated at the integration points (and saved for later use). This can be achieved with the following snippet:

```
% Shape functions and their derivatives in local
% coordinates, evaluated at integration points
for k=1:nip
 L1=points(k,1); L2=points(k,2); L3=1.-L1-L2 ;
 fun = [L1 L2 L3] ;
 fun_s(k,:) = fun ; % shape functions
 der(1,1)=1 ; der(1,2)=0 ; der(1,3)=-1 ;
 der(2,1)=0 ; der(2,2)=1 ; der(2,3)=-1 ;
 der_s(:,:,k) = der ; % derivative of shape function
end
```

1 This file can be obtained by contacting the author and on the publisher website.

4) To impose nonzero ϕ-gradients on the $x = 0$ and $x = lx$ boundaries (7.9), one needs a list of elements on each boundary (defined as an element with two nodes on the boundary). The list of boundary elements (`belx0` for $x = 0$, etc.) along with the two boundary nodes for each element (`iix0` for $x = 0$) can be computed with the following snippet:

```
% data required to impose non-zero fluxes (Neumann boundary conditions)
% establish elements (with 2 nodes) on x=0 and x=lx boundaries
% save the local node indices of the 2 boundary nodes
eps = 0.001; % small number compared to element width
ii0 = 0 ; iin = 0 ;  % indices
for iel=1:nels       % loop over all elements in mesh
 num = g_num(:,iel);  % nodes of element
 x = g_coord(1,num)'; % x for nodes
 y = g_coord(2,num)'; % y for nodes
 ii = find(x≤0+eps);  % find nodes roughly on x=0 boundary
 if length(ii)==2     % 2 nodes on x=0
  ii0 = ii0 + 1 ;      % increment counter
  belx0(ii0) = iel;    % save boundary element
  iix0(:,ii0) = ii ;   % save local nodes on boundary
 end
 ii = find(x≥lx-eps); % find nodes roughly on x=lx boundary
 if length(ii)==2     % 2 nodes on x=lx
  iin = iin + 1 ;      % increment counter
  belxn(iin) = iel ;   % save boundary element
  iixn(:,iin) = ii ;   % save local nodes on boundary
 end
end
```

Here, `eps` is a small number compared to a typical element size that is used to ensure points that do not fall precisely on the boundaries are correctly located.

5) The element stiffness matrix (`KM`) is integrated and assembled into the global stiffness matrix (`lhs`) in exactly the same manner as was done in previously presented programs. Note, however, that now there is no mass matrix (`MM`) and no right-hand-side matrix (`rhs`), as in transient diffusion problems (which contain time derivatives). In the elliptic problem, there is also no time loop and so the solution is obtained in a single step.

6) The zero flux boundary conditions on the boundary of the object and on the boundaries $y = 0$ and $y = ly$ are automatically satisfied by neglecting the boundary load vector **Fb** (7.19). This vector must however be computed (see 7.26) and added to the right-hand-side vector `bv` for elements on the $x = 0$ and $x = lx$ boundaries to satisfy $\partial\phi/\partial x = U$. The snippet to implement the Neumann condition on $x = 0$ can be achieved with the following lines:

```
% compute boundary vector for Neumann condition on x=0
for i=1:length(belx0)  % loop over elements on x=0
 iel = belx0(i) ;        % extract element number
 num = g_num(:,iel);     % nodes of the element
 y = g_coord(2,num)';    % y-coordinates of the element
 ln = iix0(:,i) ;        % list of the 2 boundary nodes
 mv = zeros(nod,1);      % initialise nodal-vector
 mv(ln) = 1 ;            % place 'ones' on the bdy. nodes
 b = abs(diff(y(ln))) ; % node spacing along boundary
 Fb = U*b/2*mv ;         % element load vector
 bv(num) = bv(num) - Fb;% assemble to global load vector
end
```

Similar lines should be used to add an equivalent vector for elements on the $x = lx$ boundary.

7) Once the solution has been computed, it can be visualized using the in-built pdeplot Matlab function using a line such as the following:

```
% visualize the solution
pdeplot(g_coord,edge,g_num,'xydata',displ,'mesh','off','contour','on')
```

```
%-------------------------------------------
% Program: elliptic2d
% Potential flow in 2D
% Mesh consists of 3-node triangles
% Requires the file mesh_cylinder.mat (generated using pdetool)
%-------------------------------------------

clear

% physical parameters
U     = 1    ; % imposed boundary velocity (=phi/dx)

% numerical parameters
nod   = 3    ; % number of nodes in 1 element
nip   = 3    ; % number of Gauss integration points
ndim  = 2    ; % number of spatial dimensions in problem

% load triangular mesh (e.g., generated using pdetool)
% the following variables are loaded
% g_coord - mesh nodes
% edge     - mesh edge data
% g_num    - node-element connectivity
% phase    - index for material type
load 'mesh_cylinder'

% plot mesh
figure(1) , clf
pdemesh(g_coord,edge,g_num)
axis equal

nn     = length(g_coord) ;  % total number of nodes
nels   = length(g_num)   ;  % total number of elements
lx     = max(g_coord(1,:)); % x-dimension of domain
ly     = max(g_coord(2,:)); % y-dimension of domain

% gauss integration data
points = zeros(nip,ndim); % location of points
points(1,1)=0.5  ; points(1,2)=0.5 ;
points(2,1)=0.5  ; points(2,2)=0   ;
points(3,1)=0    ; points(3,2)=0.5 ;
c   = 0.5 ; % triangle factor
wts = c*1/3*ones(1,nip) ; % weights

% save shape functions and their derivatives in local coordinates
% evaluated at integration points
for k=1:nip
   L1=points(k,1); L2=points(k,2); L3=1.-L1-L2  ;
   fun = [L1 L2 L3] ;
   fun_s(k,:) = fun ; % shape functions
   der(1,1)=1 ; der(1,2)=0 ; der(1,3)=-1  ;
   der(2,1)=0 ; der(2,2)=1 ; der(2,3)=-1  ;
   der_s(:,:,k) = der ; % derivative of shape function
end
```

```
% data required to impose Neumann boundary conditions
% establish elements (with 2 nodes) on x=0 and x=lx boundaries
% save the local node indices of the 2 boundary nodes
eps = 0.001;
ii0 = 0 ; iin = 0 ;
for iel=1:nels
    num   = g_num(:,iel);
    x     = g_coord(1,num)';
    y     = g_coord(2,num)';
    ii    = find(x≤0+eps);
    if length(ii)==2 % 2 nodes on x=0
        ii0      = ii0 + 1 ;
        belx0(ii0)   = iel;
        iix0(:,ii0) = ii ;
    end
    ii    = find(x≥lx-eps);
    if length(ii)==2 % 2 nodes on x=lx
        iin      = iin + 1 ;
        belxn(iin) = iel ;
        iixn(:,iin) = ii ;
    end
end

% initialise global matrices and vectors
bv   = zeros(nn,1);      % global rhs vector
lhs  = sparse(nn,nn);    % global lhs matrix

%-----------------------------------------------------
% matrix integration and assembly
%-----------------------------------------------------

for iel=1:nels  % loop over elements
  num     = g_num(:,iel)      ; % element nodes
  coord   = g_coord(:,num)'   ; % element coordinates
  KM      = zeros(nod,nod)    ; % initialise element stiffness matrix
  for k = 1:nip % integration loop
    der     = der_s(:,:,k) ; % der. of shape functions in local coordinates
    jac     = der*coord  ;    % jacobian matrix
    detjac = det(jac) ;       % det. of jacobian
    invjac = inv(jac) ;       % inv. of jacobian
    deriv  = invjac*der ;     % der. of shape fun. in physical coords.
    KM     = KM + deriv'*deriv*detjac*wts(k) ; % stiffness matrix
  end
  % assemble global matrix
  lhs(num,num) = lhs(num,num) + KM ;
end
 % compute boundary vector for Neumann condition on x=0
   for i=1:length(belx0)
      iel = belx0(i) ;
      num = g_num(:,iel);
      y   = g_coord(2,num)';
      ln  = iix0(:,i) ;
      mv  = zeros(nod,1);
      mv(ln)   = 1 ;
      b        = abs(diff(y(ln))) ;
      Fb       = U*b/2*mv ;
      bv(num)  = bv(num) - Fb ;
   end
```

```
 % compute boundary vector for Neumann condition on x=lx
 for i=1:length(belxn)
    iel = belxn(i) ;
    num = g_num(:,iel);
    y   = g_coord(2,num);
    ln  = iixn(:,i) ;
    mv  = zeros(nod,1);
    mv(ln)   = 1 ;
    b        = abs(diff(y(ln))) ;
    Fb       = U*b/2*mv ;
    bv(num)  = bv(num) + Fb ;
 end

 displ = lhs \ bv ;  % solve system of equations

% postprocessing to compute velocities
for iel=1:nels  % sum over elements
num     = g_num(:,iel)      ; % element nodes
coord   = g_coord(:,num)'   ; % element coordinates
KM      = zeros(nod,nod)    ; % initialisation
F       = zeros(nod,1)      ;
for k = 1:nip % integration loop
  fun    = fun_s(k,:) ; % shape functions
  der    = der_s(:,:,k) ; % der. of shape functions in local coordinates
  jac    = der*coord  ;   % jacobian matrix
  detjac = det(jac) ;     % det. of jacobian
  invjac = inv(jac) ;     % inv. of jacobian
  deriv  = invjac*der ;   % der. of shape fun. in physical coords.
  u      = deriv(1,:)*displ(num); % x-vel at integration point
  v      = deriv(2,:)*displ(num); % y-vel at integration point
  ugp(k,iel) = u ; % save integration point velocity
  vgp(k,iel) = v ; % save integration point velocity
  xgp(k,iel) = fun*coord(:,1);% save integration points in global coords.
  ygp(k,iel) = fun*coord(:,2);% save integration points in global coords.
end

end

 % visualisation
 figure(1)
 pdeplot(g_coord,edge,g_num,'xydata',displ,'mesh','off','contour','on');
 axis equal

 xv = linspace(0,lx,50);
 yv = linspace(0,ly,50);
 [xg,yg] = meshgrid(xv,yv);
 xm = (g_coord(1,g_num(1,:))+g_coord(1,g_num(2,:))+g_coord(1,g_num(3,:)))/3;
 ym = (g_coord(2,g_num(1,:))+g_coord(2,g_num(2,:))+g_coord(2,g_num(3,:)))/3;
 hold on
 ug = griddata(xm,ym,mean(ugp),xg,yg);
 vg = griddata(xm,ym,mean(vgp),xg,yg);
 startx = 0.05*ones(1,20);
 starty = linspace(0,ly,20);
 streamline(stream2(xg,yg,ug,vg,startx,starty));
 colormap('jet')
 axis equal
%-----------------------------------------------
% end of time loop
%-----------------------------------------------
```

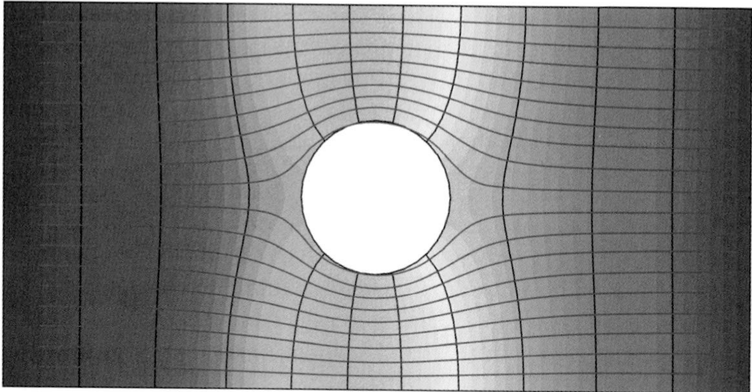

Figure 7.6 Potential flow around an circular inclusion, based on Equation 7.8 (see Figure 7.1). Results were computed with the FEM using the Matlab script listed. Flow enters from the left boundary (at a constant velocity) and exits from the right. Colors and near-vertical lines are equipotentials (constant-ϕ), whereas the nearly horizontal lines are flow directions (stream lines).

Figure 7.6 shows numerical results for the potential flow problem computed with the listed program. The results clearly show how the object strongly perturbs the flow, deflecting both the equipotentials and stream lines. Note, however, the absence of any boundary layer near the object (since flow is assumed to be inviscid) and any eddy effects (because flow is assumed to be irrotational).

7.2 The FEM for a Hyperbolic Problem

Hyperbolic PDEs describe a wide range of wave propagation and transport phenomena and appear in nearly every scientific and engineering discipline. In Earth science, hyperbolic equations appear in numerous different contexts, including porous flow, surface water flow, landsliding, faulting, and convection. Hyperbolic problems differ from both parabolic and elliptic equations in that disturbances propagate with finite time (as opposed to instantaneously) and that the solutions may propagate without decay or they may even be sharpening in time, forming what are known as shock waves. These phenomena create a number of challenges for numerical methods.

To illustrate how the FEM can be applied to solve hyperbolic problems, consider the following first-order wave (or advection) equation in 1D:

$$\frac{\partial \phi}{\partial t} + c\frac{\partial \phi}{\partial x} = 0 \tag{7.27}$$

Here, $\phi(x, t)$ is some passive scalar (e.g., the concentration of an inert chemical species) that is transported with velocity c (e.g., the mean flow velocity in a river) (Figure 7.7). This equation has the remarkable property that, once an initial condition is provided, it simply becomes propagated laterally in space at the velocity c, without any change in shape. Mathematically, this is extremely simple, though it is surprisingly challenging from a numerical point of view. On the one hand, as we will see, it is difficult to compute advection without artificially smoothing the solution. On the other hand, the fact that there is no physical diffusion in the problem is tough, since diffusion tends to dampen numerical instabilities, which would stabilize the solution. A successful numerical scheme is, therefore, one that finds the right balance between smoothing, accuracy, and stability.

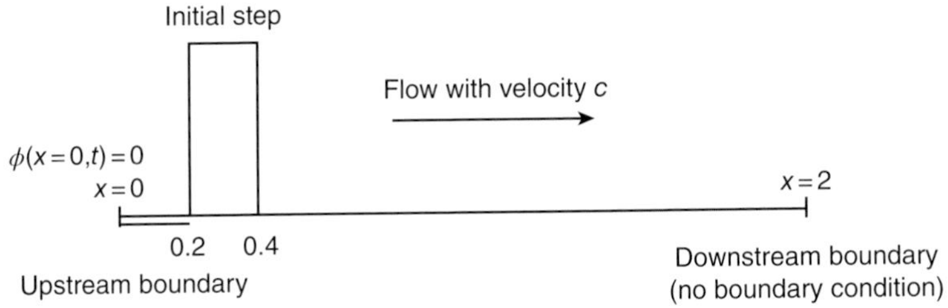

Figure 7.7 Setup for the hyperbolic advection problem.

To solve Equation 7.27, we need to specify initial conditions and boundary conditions. As an initial condition, we will consider a rather severe test case, consisting of a step function (Figure 7.7), that is,

$$\begin{aligned} &\text{If} \quad 0.2 \leq x \leq 0.4 \quad \text{then} && \phi(x, t=0) = 1 \\ &\text{otherwise} && \phi(x, t=0) = 0 \end{aligned} \tag{7.28}$$

Note that it would be far easier to compute an accurate solution with a relatively smooth initial condition such as a sine wave. Equation 7.27 contains a first-order spatial derivative indicating that only one boundary condition should be applied. This boundary condition should be located at the end from where flow is arriving (i.e., at the upstream end of the model domain). Placing the boundary condition where flow is exiting the model makes no physical sense because the exiting flow cannot influence the solution upstream. Here, because we take the flow velocity (c) to be positive, implying flow from left to right, the boundary condition should be located on the left side of the domain. At this position, we consider the following Dirichlet condition:

$$\phi(x = 0, t) = 0 \tag{7.29}$$

We now proceed with finite element discretization, introducing the standard approximation

$$\phi = \mathbf{N}\boldsymbol{\phi} \tag{7.30}$$

where $\boldsymbol{\phi}$ is the vector of nodal unknowns and $\mathbf{N}$ are a set of shape functions expressed in terms of physical coordinates (that will be defined later). For the moment, we simply remark that discretization will be performed with three-node quadratic elements of length L (Figure 7.8). Thus, $\mathbf{N}$ is a line vector with three columns, while $\boldsymbol{\phi}$ is a column vector with three rows. Substituting 7.30 into 7.27, and applying the Galerkin form of the method of weighted residuals (see Section 2.3), results in the element equations

$$\mathbf{MM}\frac{\partial}{\partial t}\boldsymbol{\phi} + \mathbf{CM}\,\boldsymbol{\phi} = 0 \tag{7.31}$$

where

$$\mathbf{MM} = \int_0^L \mathbf{N}^T\mathbf{N}\,dx = \int_0^L \begin{bmatrix} N_1N_1 & N_1N_2 & N_1N_3 \\ N_2N_1 & N_2N_2 & N_2N_3 \\ N_3N_1 & N_3N_2 & N_3N_3 \end{bmatrix} dx \tag{7.32}$$

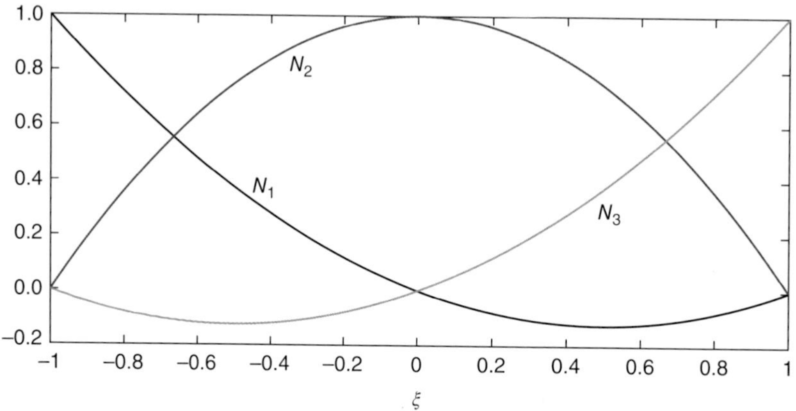

Figure 7.8 Quadratic shape functions for a 1D, three-node element with nodes at the local coordinates $\xi = -1$, $\xi = 0$, and $\xi = 1$.

and

$$\mathbf{CM} = \int_0^L c\,\mathbf{N}^T \frac{\partial \mathbf{N}}{\partial x}\,dx = \int_0^L c \begin{bmatrix} N_1 \frac{\partial N_1}{\partial x} & N_1 \frac{\partial N_2}{\partial x} & N_1 \frac{\partial N_3}{\partial x} \\ N_2 \frac{\partial N_1}{\partial x} & N_2 \frac{\partial N_2}{\partial x} & N_2 \frac{\partial N_3}{\partial x} \\ N_3 \frac{\partial N_1}{\partial x} & N_3 \frac{\partial N_2}{\partial x} & N_3 \frac{\partial N_3}{\partial x} \end{bmatrix} dx \tag{7.33}$$

Here, **MM** is the standard mass matrix encountered numerous times previously with diffusion, while **CM** is the new advection matrix. Note that **CM** is seen to be asymmetrical, which occurs whenever there are odd-ordered derivatives in the governing PDE. The presence of odd-ordered derivatives renders the equations non-self-adjoint, for which the standard Galerkin FEM is not optimal, though can be used all the same. Replacing the time derivative in 7.31 with an implicit finite difference approximation leads to the discrete element equations

$$\left(\frac{\mathbf{MM}}{\Delta t} + \mathbf{CM}\right)\boldsymbol{\phi}^{n+1} = \frac{\mathbf{MM}}{\Delta t}\boldsymbol{\phi}^{n} \tag{7.34}$$

where Δt is the time interval between n and $n+1$. We will integrate **MM** and **CM** using Gauss–Legendre quadrature, as done previously (Chapter 4). Recall that the rule for Gauss–Legendre quadrature in 1D is

$$\int_{-1}^{1} f(\xi)\,d\xi \simeq \sum_{k=1}^{n} f(\xi_k)\,w_k \tag{7.35}$$

where $f(\xi)$ is the function being integrated, n is the number of integration points, ξ_k is the spatial coordinate of the kth integration point, and w_k is the weight of the kth integration point (obtained from Table 4.1). To perform Gauss–Legendre quadrature, the function being integrated needs to be expressed in terms of local coordinates (ξ) on the domain $[-1\ 1]$. The three quadratic shape functions **N** written in the local coordinate system are

$$\begin{aligned} N_1 &= -\frac{\xi(1-\xi)}{2} \\ N_2 &= (1+\xi)(1-\xi) \\ N_3 &= \frac{\xi(1+\xi)}{2} \end{aligned} \tag{7.36}$$

from which their first derivatives are seen to be

$$\frac{\partial N_1}{\partial \xi} = -\frac{1}{2} + \xi$$
$$\frac{\partial N_2}{\partial \xi} = -2\xi \tag{7.37}$$
$$\frac{\partial N_3}{\partial \xi} = \frac{1}{2} + \xi$$

These derivatives can be mapped back to physical coordinates by multiplication with the factor $2/L$ (which is equivalent to multiplying by the inverse of the Jacobian, see Equation 7.22), that is,

$$\frac{\partial N_i}{\partial x} = \frac{\partial N_i}{\partial \xi}\frac{2}{L} \tag{7.38}$$

Finally, the integration limits can be converted from the interval $x = [0\ L]$ to $\xi = [-1\ 1]$ by multiplying by $L/2$ (which is equivalent to multiplying by the determinant of the Jacobian, see Equation 2.24). This leads to the final quadrature rule as follows:

$$\int_0^L f(x)dx = \int_{-1}^{1} f(\xi)\, d\xi \frac{L}{2} \simeq \sum_{k=1}^{n} f(\xi_k)\, w_k \frac{L}{2} \tag{7.39}$$

What follows is the complete Matlab script to solve the advection equation using this approach. The program is seen to be very similar to the script discussed earlier to solve the diffusion equation in 1D (see Section 4.6). The program given in the following text differs with respect to the following points:

1) Each element contains three nodes, one at each end of the element and one at the mid-distance. If the element spacing is assumed to be constant (denoted `dx`), the coordinates of the nodes for the entire 1D mesh can be generated with the following snippet:

```
dx      = lx/nels ;        % element size
g_coord = [0:dx/2:lx] ; % node coordinates
```

Here, `nels` is the total number of elements and `lx` is the length of the model domain.

2) The array storing node connectivity (`g_num`) for the three-node elements can be created with the following lines:

```
g_num        = zeros(nod,nels) ;
g_num(1,:)  = [1:2:nn-2] ; % node 1
g_num(2,:)  = [2:2:nn-1] ; % node 2
g_num(3,:)  = [3:2:nn] ;    % node 3
```

Here, `nn` is the total number of nodes in the mesh (`nn=2*nels+1`) and `nod` is the number of nodes per element (here 3).

3) Three integration points are used to evaluate the integrals. The weights and point positions in local coordinates (obtained from Table 4.1) are provided with the following lines:

```
% Gauss integration data
points = [-sqrt(3/5) 0 sqrt(3/5)] ; % positions of the points
wts    = [5/9 8/9 5/9] ; % weights
```

4) The quadratic shape functions (7.36) and their first derivatives (7.37) can be defined, evaluated at the integration points, and saved for later use with the following snippet:

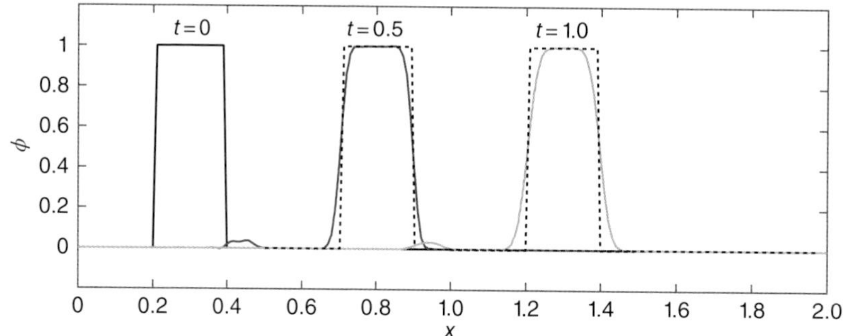

Figure 7.9 Finite element solution of the linear advection equation (7.27 using $c = 1$) with an impulse initial condition along with the exact solution (dashed line). Numerical calculation was performed with 100, three-node elements ($dx = 0.02$) and $dt = 5 \times 10^{-4}$.

```
% Define shape functions and their first derivatives
for k=1:nip % loop over integration points
 xi = points(k) ; % coordinate of point in local coordinates
 xim = 1-xi ;
 xip = 1+xi ;
 fun_s(k,:) = [ -xi*xim/2 xip*xim xi*xip/2] ; % N1, N2, N3
 der_s(k,:) = [ -1/2+xi -2*xi 1/2+xi] ; % dN1/xi dN2/xi dN3/xi
end
```

5) The element matrices **MM** (7.32) and **CM** (7.33) can be computed with the following lines:

```
MM = MM + fun'*fun*detjac*wts(k) ; % mass matrix
CM = CM + c*fun'*deriv*detjac*wts(k);% advection matrix
```

Here, `c` is the velocity, `fun` are the shape functions, `deriv` are the first derivatives of the shape functions in physical coordinates, `wts(k)` is the weight for the current integration point k, and `detjac` is $dx/2$ in 1D.

Figure 7.9 illustrates some numerical results produced with the listed program using a velocity (c) of 1, 100 elements and a time step of 5×10^{-4}. Shown for comparison is the exact solution, which is simply

$$\phi(x,t) = \phi_0(x - c\,t)$$

where ϕ_0 is the initial condition defined in 7.28. For these parameter values, one observes that the FEM advects the solution with the correct average velocity, but it introduces significant smoothing (numerical diffusion) along with producing some numerical oscillations. Calculations performed with larger time steps increase the amount of numerical diffusion which dampen the oscillations, while the converse is true if the time step is reduced. In Chapter 10, we present a modified FEM that is more successful in dealing with convection-dominated problems such as this.

```
%-----------------------------------------------------------
% Program advection1d.m
% 1-D FEM solution of linear advection equation
% using 3 node quadratic shape function
%-----------------------------------------------------------

clear % clear memory from current workspace

% physical parameters
```

```
lx     = 2 ;  % length of spatial domain
c      = 1 ;  % advection velocity

% numerical parameters
dt     = 0.001/2 ; % time step
ntime = 2000   ;    % number of time steps
nels  = 100    ;    % total number of elements
nod   = 3 ;         % number of nodes per element
nn    = 2*nels+1; % total number of nodes
dx    = lx/nels ; % element size
g_coord = [0:dx/2:lx] ; % node coordinates
nip     = 3 ;       % number of Gauss integration points

% Gauss integration data
points = [-sqrt(3/5) 0 sqrt(3/5)] ; % positions of the points
wts    = [5/9  8/9  5/9] ;           % weights

% shape functions and their derivatives (both in local coordinates)
% evaluated at the Gauss integration points
for k=1:nip
  xi  = points(k) ;
  xim = 1-xi ;
  xip = 1+xi ;
  fun_s(k,:) = [ -xi*xim/2 xip*xim xi*xip/2] ; % N1, N2,N3
  der_s(k,:) = [ -1/2+xi -2*xi 1/2+xi] ;       % dN1/xi  dN2/xi dN3/xi
end

%  Dirichlet boundary conditions
bcdof = [ 1 ] ; % boundary nodes
bcval = [ 0 ] ; % boudary values

% node connectivity
g_num      = zeros(nod,nels) ;
g_num(1,:) = [1:2:nn-2]   ;
g_num(2,:) = [2:2:nn-1]   ;
g_num(3,:) = [3:2:nn]   ;

% initialise matrices and vectors
b    = zeros(nn,1);          % system rhs vector
lhs  = sparse(nn,nn);        % system lhs matrix
rhs  = sparse(nn,nn);        % system rhs matrix

% initial conditions
displ  = zeros(nn,1);          % solution vector
displ(find(g_coord>0.2&g_coord<0.4))=1;

figure(1) , clf
plot(g_coord,displ,'-')

% time loop
t  = 0 ; % intial time
for n=1:ntime
   n
   t  = t + dt ;  % increment time

%-----------------------------------------
% Matrix integration and assembly
%-----------------------------------------
ff     = zeros(nn,1);           % system load vector
```

```
b        = zeros(nn,1);            % system rhs vector
lhs      = sparse(nn,nn);          % system lhs matrix
rhs      = sparse(nn,nn);          % system lhs matrix

   for iel=1:nels % loop over all elements
      num    = g_num(:,iel)  ;
      MM     = zeros(nod,nod) ;
      CM     = zeros(nod,nod) ;
      for k=1:nip % integrate element matrices
          fun    = fun_s(k,:); % shape fun. evaluated at an int. pt
          der    = der_s(k,:); % der. of shape fun. in local coords.
          detjac = dx/2     ;  % det. of jacobian
          invjac = 2/dx     ;  % inverse of jacobian
          deriv  = der*invjac ;% der. of shape fun. in physical coords.
          MM     = MM + fun'*fun*detjac*wts(k) ; % mass matrix
          CM     = CM + c*fun'*deriv*detjac*wts(k);% advection matrix
      end % end of integration
      lhs(num,num) = lhs(num,num) + MM/dt + CM ; % assemble lhs
      rhs(num,num) = rhs(num,num) + MM/dt ; %  assemble rhs
   end     % end of element loop

%-------------------------------------------------------------------

   b = rhs*displ   ;                  % rhs load vector

   % impose boundary conditions
   lhs(bcdof,:) = 0 ;                 % zero the relevent equations
   tmp = spdiags(lhs,0) ;             % store diagonal
   tmp(bcdof)=1 ;                     % place 1 on stored-diagonal
   lhs=spdiags(tmp,0,lhs);            % reinsert diagonal
   b(bcdof) = bcval ;                 % set rhs

   displ = lhs \ b ;                  % solve system of equations

  % plotting
  if mod(n,1000)==0
    figure(1)
    hold on
    plot(g_coord,displ,'-')
    xlabel('Distance')
    ylabel('\phi')
    hold off
    drawnow
  end

end % end of time loop

%-------------------------------------------------------------------
```

7.3 The FEM for Systems of Equations

So far, this text has been concerned with the solution of single PDEs (i.e., with only one unknown, e.g., Equations 7.1 or 7.27). However, many processes are governed by systems of PDEs containing two or more unknowns. Some examples of problems governed by systems of PDEs that are considered in greater detail later in the text are coupled heat transfer and fluid flow, deformation in two or more dimensions, and flow of viscous fluids. As an introduction to the topic of how the FEM can be performed on systems of equations, we consider the following equations

$$\frac{\partial A}{\partial t} = \nabla^2 A + \gamma\,(a - A + A^2B) \tag{7.40}$$

$$\frac{\partial B}{\partial t} = d\nabla^2 B + \gamma\,(b - A^2B) \tag{7.41}$$

with the two unknowns, A and B. This system of equations is used to model the interaction between two interacting chemicals A and B (see Maini et al. (2012)): B represents the concentration of a "substrate" (inhibitor) chemical that is consumed in a reaction by some chemical "activator," with concentration A. The reaction produces the activator, which explains the A^2B terms in both equations. Both substances are also produced at some background rate (γa for A and γb for B, respectively) and A decays with first-order kinetics at a rate γ. This "substrate depletion" reaction model is known as Schnakenberg kinetics (Gierer and Meinhard, 1972; Schnakenberg, 1979). This model is most widely applied in biology, but it might also be relevant in Earth science to explain the formation of self-organizing patterns, for example, related to mineral growth, stromatolites and corals, concretions, and so on.

The first two terms in Equations 7.40 and 7.41 are recognized to be standard diffusion equations in each substance. If only these terms were present, the equations would be completely uncoupled and they could be solved independently of each other, as done previously. However, the bracketed "reaction" term in each equation depends on both A and B. These terms couple or link the equations together, since neither can be solved for without the other. The system of equations must therefore be solved simultaneously for A and B. More generally, Equations 7.40 and 7.41 are known as reaction–diffusion equations. With this particular example, the reaction terms are nonlinear (due to the terms A^2B), which leads to solutions that may be very different from the decaying solutions characteristic of linear diffusion equations.

Here, we will solve Equations 7.40 and 7.41 in 2Ds on the domain $5 \leq x, y \leq 5$ with the following conditions:

$$A(x, y, t = 0) = a + b + r \quad \text{and} \quad B(x, y, t = 0) = \frac{b}{(a + b)^2} + r \tag{7.42}$$

which is the steady-state solution with a small random perturbation r (that has a maximum amplitude of $1/100$). No flux (zero gradient) conditions are considered across all lateral boundaries (Figure 7.10).

We begin with the standard finite element approximation by assuming that A and B can be described by simple shape functions $\mathbf{N}$, that is,

$$A = \mathbf{N}\,\mathbf{A} \tag{7.43}$$

and

$$B = \mathbf{N}\,\mathbf{B} \tag{7.44}$$

where $\mathbf{A}$ and $\mathbf{B}$ are column vectors of the unknowns at node points of a single element. Note that the same shape functions are used for A and B, though this is not essential. The shape functions will be defined fully later. For the moment, we simply note that discretization is performed using nine-node quadrilateral (Lagrangian) elements (Figures 7.1 and 7.11), for which the shape functions are quadratic polynomials. Therefore, $\mathbf{N}$ is a row vector with nine columns. Substituting these approximations into the governing PDEs (7.40 and 7.41), applying the Galerkin form of the weighted residual

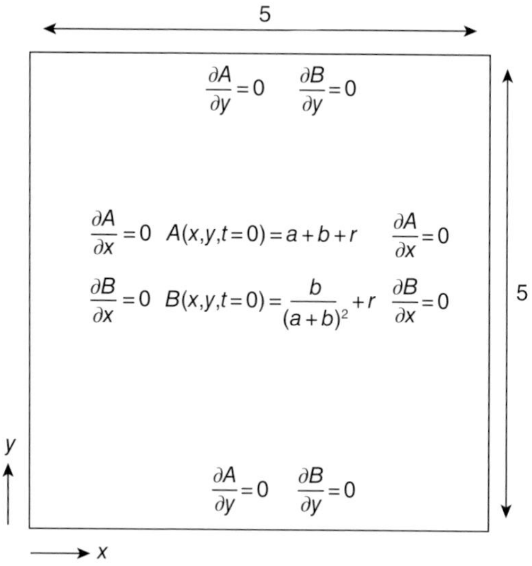

Figure 7.10 Setup for coupled reaction–diffusion problem.

method, integrating the second-order spatial derivatives by parts, and neglecting the resulting boundary integrals (consistent with the zero-flux boundary conditions), we obtain the following:

$$\begin{aligned} \mathbf{MM}\frac{\partial}{\partial t}\mathbf{A} + \gamma\mathbf{MMA} + \mathbf{KM}_A\mathbf{A} &= \mathbf{F}_A \\ \mathbf{MM}\frac{\partial}{\partial t}\mathbf{B} + \mathbf{KM}_B\mathbf{B} &= \mathbf{F}_B \end{aligned} \tag{7.45}$$

Here,

$$\mathbf{MM} = \int\int \mathbf{N}^T\mathbf{N}\, dxdy \tag{7.46}$$

$$\mathbf{KM}_A = \int\int (\nabla\mathbf{N})^T\nabla\mathbf{N}\, dxdy \tag{7.47}$$

$$\mathbf{KM}_B = \int\int d\,(\nabla\mathbf{N})^T\nabla\mathbf{N}\, dxdy \tag{7.48}$$

$$\mathbf{F}_A = \int\int \gamma\,\mathbf{N}^T\left(a + (\mathbf{NA})^2\mathbf{NB}\right)\, dxdy \tag{7.49}$$

$$\mathbf{F}_B = \int\int \gamma\,\mathbf{N}^T\left(b - (\mathbf{NA})^2\mathbf{NB}\right)\, dxdy \tag{7.50}$$

There are various ways to integrate these equations in time. Here, we will consider a fully implicit, finite difference approximation

$$\begin{aligned} \mathbf{MM}\frac{\mathbf{A}^{n+1} - \mathbf{A}^n}{\Delta t} + \gamma\mathbf{MMA}^{n+1} + \mathbf{KM}_A\mathbf{A}^{n+1} &= \mathbf{F}_A^{n+1} \\ \mathbf{MM}\frac{\mathbf{B}^{n+1} - \mathbf{B}^n}{\Delta t} + \mathbf{KM}_B\mathbf{B}^{n+1} &= \mathbf{F}_B^{n+1} \end{aligned} \tag{7.51}$$

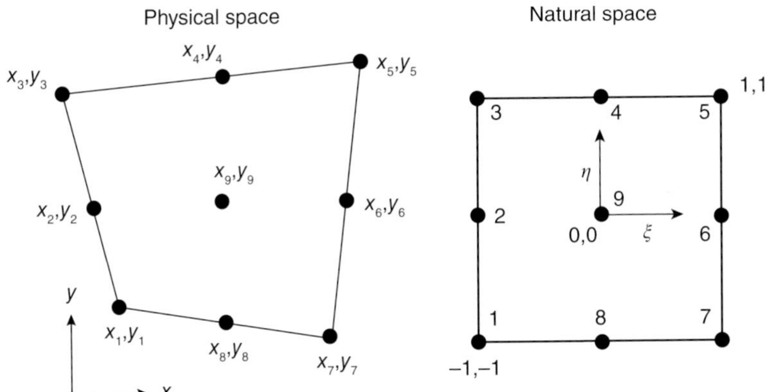

Figure 7.11 Nine-node quadrilateral element in the physical and natural coordinate systems.

that can be rearranged to give

$$\begin{aligned}\left(\frac{\mathbf{MM}}{\Delta t}+\gamma\mathbf{MM}+\mathbf{KM}_A\right)\mathbf{A}^{n+1}&=\frac{\mathbf{MM}}{\Delta t}\mathbf{A}^n+\mathbf{F}_A^{n+1}\\ \left(\frac{\mathbf{MM}}{\Delta t}+\mathbf{KM}_B\right)\mathbf{B}^{n+1}&=\frac{\mathbf{MM}}{\Delta t}\mathbf{B}^n+\mathbf{F}_B^{n+1}\end{aligned}\tag{7.52}$$

Note that the source terms $\mathbf{F}_A$ and $\mathbf{F}_B$ (which depend on A and B) are both evaluated at the $n+1$ time level, which requires an iterative solution strategy. The element matrices all have nine rows and nine columns (for the nine-node quadrilateral), while the element source term vectors are columns with nine rows. The element matrices **MM**, $\mathbf{KM}_A$, and $\mathbf{KM}_B$ are functions of the physical coordinates, while $\mathbf{F}_A$ and $\mathbf{F}_B$ are also functions of A and B. To evaluate these integrals, we follow the now-familiar approach of providing the shape functions and their derivatives in terms of local coordinates, and making the necessary transformations to perform Gauss–Legendre integration on mapped elements (Figure 7.11). The quadratic shape functions for the nine-node quadrilateral elements in local coordinates ξ and η are

$$\begin{aligned}N_1&=\frac{1}{4}\xi(\xi-1)\eta(\eta-1)\\ N_2&=-\frac{1}{2}\xi(\xi-1)(\eta+1)(\eta-1)\\ N_3&=\frac{1}{4}\xi(\xi-1)\eta(\eta+1)\\ N_4&=-\frac{1}{2}(\xi+1)(\xi-1)\eta(\eta+1)\\ N_5&=\frac{1}{4}\xi(\xi+1)\eta(\eta+1)\\ N_6&=-\frac{1}{2}\xi(\xi+1)(\eta+1)(\eta-1)\\ N_7&=\frac{1}{4}\xi(\xi+1)\eta(\eta-1)\end{aligned}\tag{7.53}$$

$$N_8 = -\frac{1}{2}(\xi + 1)(\xi - 1)\eta(\eta - 1)$$
$$N_9 = (1 + \xi)(\xi - 1)(\eta + 1)(\eta - 1)$$

while their derivatives are

$$\begin{aligned}
\frac{\partial N_1}{\partial \xi} &= \frac{1}{4}(2\xi - 1)\eta(\eta - 1) & \frac{\partial N_1}{\partial \eta} &= \frac{1}{4}\xi(\xi - 1)(2\eta - 1) \\
\frac{\partial N_2}{\partial \xi} &= -\frac{1}{2}(2\xi - 1)(\eta + 1)(\eta - 1) & \frac{\partial N_2}{\partial \eta} &= -\xi(\xi - 1)\eta \\
\frac{\partial N_3}{\partial \xi} &= \frac{1}{4}(2\xi - 1)\eta(\eta + 1) & \frac{\partial N_3}{\partial \eta} &= \frac{1}{4}\xi(\xi - 1)(2\eta + 1) \\
\frac{\partial N_4}{\partial \xi} &= -\xi\eta(\eta + 1) & \frac{\partial N_4}{\partial \eta} &= -\frac{1}{2}(\xi + 1)(\xi - 1)(2\eta + 1) \\
\frac{\partial N_5}{\partial \xi} &= \frac{1}{4}(2\xi + 1)\eta(\eta + 1) & \frac{\partial N_5}{\partial \eta} &= \frac{1}{4}\xi(\xi + 1)(2\eta + 1) \\
\frac{\partial N_6}{\partial \xi} &= -\frac{1}{2}(2\xi + 1)(\eta + 1)(\eta - 1) & \frac{\partial N_6}{\partial \eta} &= -\xi(\xi + 1)\eta \\
\frac{\partial N_7}{\partial \xi} &= \frac{1}{4}(2\xi + 1)\eta(\eta - 1) & \frac{\partial N_7}{\partial \eta} &= \frac{1}{4}\xi(\xi + 1)(2\eta - 1) \\
\frac{\partial N_8}{\partial \xi} &= -\xi\eta(\eta - 1) & \frac{\partial N_8}{\partial \eta} &= -\frac{1}{2}(\xi + 1)(\xi - 1)(2\eta - 1) \\
\frac{\partial N_9}{\partial \xi} &= 2\xi(\eta + 1)(\eta - 1) & \frac{\partial N_9}{\partial \eta} &= 2(\xi + 1)(\xi - 1)\eta
\end{aligned} \tag{7.54}$$

Using Gauss–Legendre quadrature, the integrals appearing in the various element matrices can be approximated as

$$\int\int f(x,y)\,dxdy = \int_{-1}^{1}\int_{-1}^{1} f(\xi,\eta)\,d\xi d\eta\ \det\mathbf{J} \simeq \sum_{k=1}^{n} f(\xi_k,\eta_k)\,w_k\ \det\mathbf{J} \tag{7.55}$$

where ξ_k and η_k are positions of the Gauss points, w_k are the Gauss weights (see Table 4.1), n is the number of integration points in the element, and $\det\mathbf{J}$ is the determinant of the Jacobian matrix ($\mathbf{J}$) which accounts for the change in the limits of integration. Recall from Chapter 5 that the Jacobian can be computed as

$$\mathbf{J} = \begin{bmatrix} \frac{\partial N_1}{\partial \xi} & \frac{\partial N_2}{\partial \xi} & \frac{\partial N_3}{\partial \xi} & \frac{\partial N_4}{\partial \xi} & \frac{\partial N_5}{\partial \xi} & \frac{\partial N_6}{\partial \xi} & \frac{\partial N_7}{\partial \xi} & \frac{\partial N_8}{\partial \xi} & \frac{\partial N_9}{\partial \xi} \\ \frac{\partial N_1}{\partial \eta} & \frac{\partial N_2}{\partial \eta} & \frac{\partial N_3}{\partial \eta} & \frac{\partial N_4}{\partial \eta} & \frac{\partial N_5}{\partial \eta} & \frac{\partial N_6}{\partial \eta} & \frac{\partial N_7}{\partial \eta} & \frac{\partial N_8}{\partial \eta} & \frac{\partial N_9}{\partial \eta} \end{bmatrix} \begin{bmatrix} x_1 & y_1 \\ x_2 & y_2 \\ x_3 & y_3 \\ x_4 & y_4 \\ x_5 & y_5 \\ x_6 & y_6 \\ x_7 & y_7 \\ x_8 & y_8 \\ x_9 & y_9 \end{bmatrix} \tag{7.56}$$

where x_1 is the x-coordinate of node 1, etc. (Figure 7.11). The derivatives of the shape functions can be transformed from the natural coordinate system (where they are defined) to the physical coordinate system (required in the **KM** matrices) by performing the operation

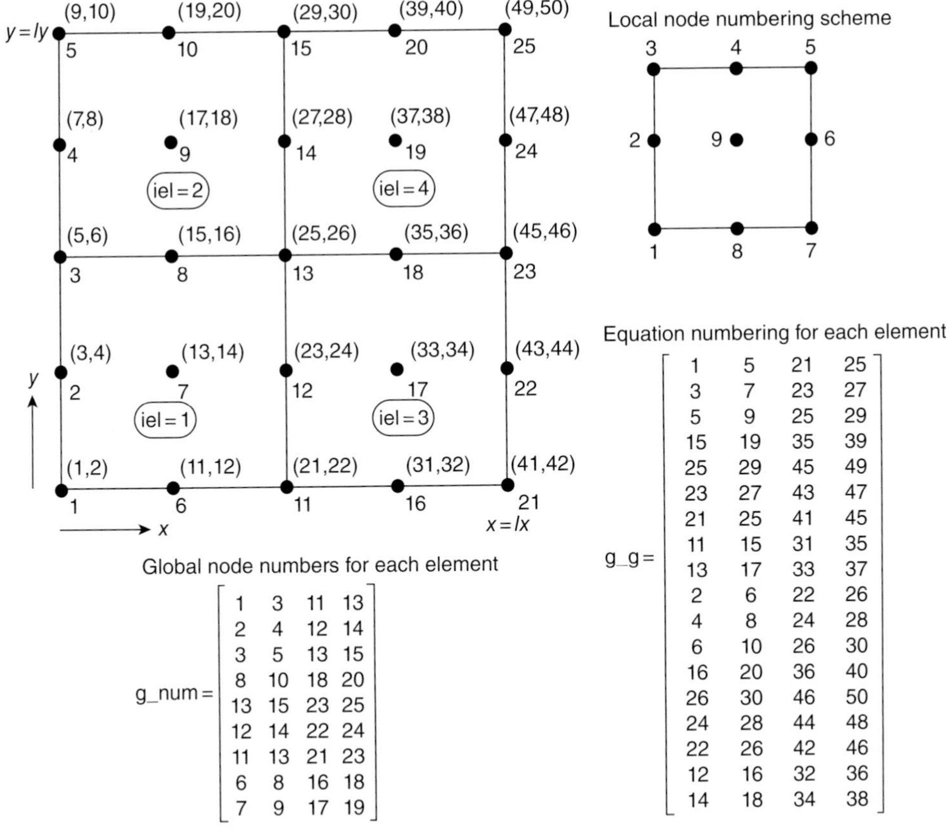

Relationship between node and equation numbers

$$\text{nf} = \begin{bmatrix} 1 & 3 & 5 & 7 & 9 & 11 & 13 & 15 & 17 & 19 & 21 & 23 & 25 & 27 & 29 & 31 & 33 & 35 & 37 & 39 & 41 & 43 & 45 & 47 & 49 \\ 2 & 4 & 6 & 8 & 10 & 12 & 14 & 16 & 18 & 20 & 22 & 24 & 26 & 28 & 30 & 32 & 34 & 36 & 38 & 40 & 42 & 44 & 46 & 48 & 50 \end{bmatrix}$$

Figure 7.12 Small 2D mesh consisting of four nine-node quadrilateral elements with two degrees of freedom on each node.

$$\left\{ \begin{matrix} \frac{\partial}{\partial x} \\ \frac{\partial}{\partial y} \end{matrix} \right\} \mathbf{N} = \mathbf{J}^{-1} \left\{ \begin{matrix} \frac{\partial}{\partial \xi} \\ \frac{\partial}{\partial \eta} \end{matrix} \right\} \mathbf{N} \tag{7.57}$$

where $\mathbf{J}^{-1}$ is the inverse of the Jacobian matrix.

Until this point, the reader might have noted that the problem has been treated in exactly the same manner as a single degree-of-freedom problem. The difference between single and multi degree-of-freedom problems really only becomes obvious once the equations are written out explicitly for one or more elements. Consider for example, a small mesh consisting of 2 nine-node quadrilateral elements in each direction (Figure 7.12). In all previous problems, because each node had only one unknown, the nodes of the mesh corresponded exactly with the equations. Conversely, now, each node has two unknowns (degrees of freedom), so it is necessary to clearly distinguish between node and equation numbers. In what follows, it is assumed that A and B are the first and second degrees of freedom, respectively. It is important to remember this order, but the choice is arbitrary. We will require the specification of three matrices, denoted `g_num`, `nf` and g_g (Figure 7.12). We have already encountered g_num; this matrix defines the relationship between elements and global node numbers. The matrix nf is used to specify the relationship between the global node numbers and the global equation numbers. Lastly, the matrix g_g defines the relationship between the elements and the global equation numbers. Examples of these matrices for a small mesh are shown in Figure 7.12. For the nine-node mesh, g_num, nf and g_g can be generated with the following Matlab snippet:

```
% establish node numbering for each element
gnumbers = reshape(1:nn,[ny nx]) ; % grid of node numbers
g_num = zeros(nod,nels);
iel = 1 ; % intialise element number
for i=1:2:nx-1 % loop over x-nodes
  for j=1:2:ny-1 % loop over y-nodes
    g_num(1,iel) = gnumbers(j,i) ;       % node 1
    g_num(2,iel) = gnumbers(j+1,i) ;     % node 2
    g_num(3,iel) = gnumbers(j+2,i) ;     % node 3
    g_num(4,iel) = gnumbers(j+2,i+1) ;   % node 4
    g_num(5,iel) = gnumbers(j+2,i+2) ;   % node 5
    g_num(6,iel) = gnumbers(j+1,i+2) ;   % node 6
    g_num(7,iel) = gnumbers(j,i+2) ;     % node 7
    g_num(8,iel) = gnumbers(j,i+1) ;     % node 8
    g_num(9,iel) = gnumbers(j+1,i+1) ;   % node 9
    iel = iel + 1 ; % increment the element number
  end
end

% establish equation number for each node
sdof = 0 ; % system degrees of freedom (dof) counter
nf = zeros(ndof,nn) ; % node degree of freedom array
for n = 1:nn            % loop over all nodes in mesh
  for i=1:ndof          % loop over each dof
    sdof = sdof + 1 ;   % increment sdof
    nf(i,n) = sdof ;    % store eqn number on each node
  end
end

% equation number for each element
g = zeros(ntot,1) ;       % equation numbers for 1 element
g_g = zeros(ntot,nels); % equation numbers for all elements
for iel=1:nels ;          % loop over elements
  num = g_num(:,iel) ;    % node numbers for this element
  inc=0 ;                 % initialise local eqn number
  % loop 2 times of nodes of an element, once for each dof
  for i=1:nod ; inc=inc+1 ; g(inc)=nf(1,num(i)) ; end
  for i=1:nod ; inc=inc+1 ; g(inc)=nf(2,num(i)) ; end
  g_g(:,iel) = g ;        % store the equation numbers
end
```

In the given script, ndof is the number of degrees of freedom in the governing PDE(s) (2); nx and ny are the number of nodes in the x- and y-directions, respectively; nn is the total number of nodes in the mesh; sdof is the total number of equations in the mesh; nxe and nye are the number of elements in the x- and y-directions, respectively; nod is the number of nodes in an element (9); and ntot is the total number of equations in a single element (18). Once these matrices have been constructed in the preprocessing stage, they can be used to specify boundary conditions, define initial conditions, and assemble the global equations. For example, the initial conditions defined in 7.42 can be assigned with the snippet

```
% initial conditions (IC)
displ          = zeros(sdof,1) ; % initialise solution vector
displ(nf(1,:)) = (a+b)*ones(nn,1) + amp*randn(nn,1); % IC for A
displ(nf(2,:)) = b/(a+b)^2*ones(nn,1) + amp*randn(nn,1); % IC for B
```

which uses `nf` to specify to which degree of freedom each initial condition is applied. The next use of the arrays `nf` and `g_g` is within the main element loop where the element matrices are integrated and assembled into their global equivalents. Here, the equation numbers for a certain element (i.e., `g`) can be obtained from `g_g` with the line

```
g = g_g(:,iel) ; % list of all equations for the current element
```

If we want only the equation numbers relating to A or B, one could use either

```
ga = g(1:nod) ;       % A-equation numbers for the current element
gb = g(nod+1:ntot) ; % B-equation numbers for the current element
```

or

```
ga = nf(1,num)' ; % A-equation numbers for the current element
gb = nf(2,num)' ; % B-equation numbers for the current element
```

where it should be remembered that the degrees of freedom for a given element are ordered as follows: $A_1, A_2, \ldots, A_9, B_1, B_2, \ldots, B_9$ (see Figure 7.12). These can be used, for example, to extract the values of A and B for the nodes of a current element. This can be achieved with the following lines:

```
An = displ(ga) ; % A-nodal values for an element
Bn = displ(gb) ; % B-nodal values for an element
```

The vectors `ga` and `gb` are also used to steer the element matrices into their correct positions in the global matrices. This, for example, can be achieved with the following lines:

```
lhs(ga,ga) = lhs(ga,ga) + MM/dt + KMa + gamma*MM ; % A contrib.
lhs(gb,gb) = lhs(gb,gb) + MM/dt + KMb ; % B contribution
bv(ga) = bv(gb) + MM/dt*displ0(ga) + FA ; % A contribution
bv(gb) = bv(gb) + MM/dt*displ0(gb) + FB ; % B contribution
```

Note here that (in comparison to some of the earlier programs) no right-hand-side matrix is formed (`rhs`). Rather, the operations `MM/dt*displ0(ga)` and `MM/dt*displ0(ga)` are performed at an element level (where `displ0` is the solution at the previous time step). Once the system of global equations is solved, the solution for the individual degrees of freedom can be extracted from the entire unknown vector (i.e., `displ`) for plotting purposes using the following lines:

```
A = displ(nf(1,:)) ; % A values for entire mesh
B = displ(nf(2,:)) ; % B values for entire mesh
```

Another noteworthy feature of the program listed in the following text is that nonlinear iterations are performed in order to evaluate the load vectors $\mathbf{F}_A$ and $\mathbf{F}_B$ at the $n+1$ time level. This requires one to loop repeatedly over the loops involving element integration and assembly to update $\mathbf{F}_A$ and $\mathbf{F}_B$ with the most recently computed values of A and B. This is performed until the solution converges—or in other words—until it stops changing within some predefined tolerance. The general overall structure of a program involving nonlinear iterations is illustrated in Figure 7.13. To evaluate the integrals in $\mathbf{F}_A$ and $\mathbf{F}_B$ during each iteration, one must interpolate A and B from nodes to the positions of the integration points. This interpolation can be performed using the shape functions, that is,

```
Ai = fun*displ(ga) ; % interp. A from nodes to an int. pt.
Bi = fun*displ(gb) ; % interp. B from nodes to an int. pt.
```

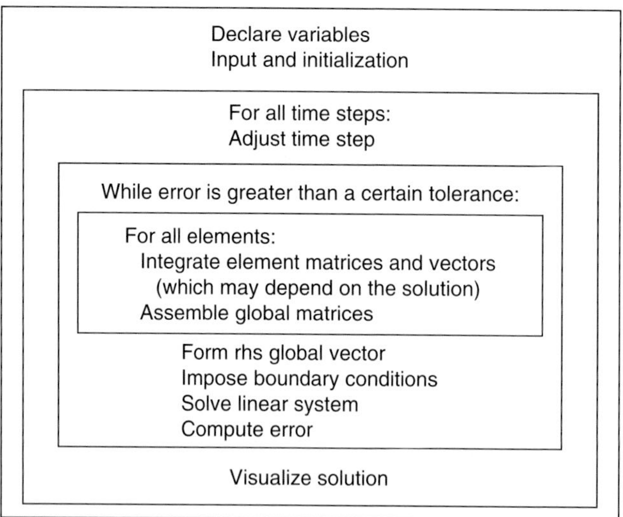

Figure 7.13 General structure of a program that includes nonlinear iterations.

Apart from these points, the program listed in the following text shares many features of the other programs presented, especially in terms of element integration, assembly, and solution.

```
%-----------------------------------------------------
% Program rxndiffn2d.m
% Solution of reaction-diffusion problem in 2D
% Unknowns are the concentrations of A (dof=1) and B (dof=2)
% FEM discretisation with 9-node quadraliterals
% Element unknowns are ordered as (A1,A2, ... A9,B1,B2,...B9)
%-----------------------------------------------------

clear

% physical parameters
lx      = 5 ;    % length of x domain
ly      = 5 ;    % length of y domain
d       = 20;    % diffusivity of species B
a       = 0.05;  % growth rate of species A
b       = 1 ;    % growth rate of species B
gamma   = 600 ;  % kinetics
amp     = 0.01 ;% max. amplitude of random noise

% numerical parameters
ntime = 200 ;       % total number of time steps to compute
nxe   = 50  ;       % number of elements in x-direction
nye   = 50  ;       % number of elements in y-direction
nels  = nxe*nye ;   % total number of elements
nx    = 2*nxe+1 ;   % number of nodes in x-direction
ny    = 2*nye+1 ;   % number of nodes in y-direction
nn    = nx*ny   ;   % total number of nodes in mesh
dx    = lx/nxe  ;   % element width in x-direction
dy    = ly/nye  ;   % element width in y-direction
```

```
dt     = 5e-4    ; % time step
nod    = 9   ;     % number of nodes in an element
ndof   = 2   ;     % number of degrees of freedom in pde
ntot   = ndof*nod ; %  total degrees of freedom in one element
nip    = 9 ;       % number of integration points in an element
ndim   = 2 ;       % number of spatial dimensions

%-----------------------------------------
% generate mesh, node and equation numbering
%-----------------------------------------

% define mesh (numbering in y direction first)
g_coord = zeros(ndim,nn) ;
n = 1 ;
for i=1:nx % loop over nodes in x-direction
    for j=1:ny % loop over nodes in y-direction
      g_coord(1,n) = (i-1)*dx/2 ;
      g_coord(2,n) = (j-1)*dy/2 ;
      n = n + 1 ;
    end
end

% reshaped mesh(for plotting only)
 xg = reshape(g_coord(1,:),ny,nx);
 yg = reshape(g_coord(2,:),ny,nx);

 % establish node numbering for each element
gnumbers = reshape(1:nn,[ny nx]) ; % grid of node numbers
g_num      = zeros(nod,nels);
iel = 1 ; % intialise element number
for i=1:2:nx-1 % loop over x-nodes
    for j=1:2:ny-1 % loop over y-nodes
      g_num(1,iel) = gnumbers(j,i)    ;     % node 1
      g_num(2,iel) = gnumbers(j+1,i)  ;     % node 2
      g_num(3,iel) = gnumbers(j+2,i)  ;     % node 3
      g_num(4,iel) = gnumbers(j+2,i+1)  ;   % node 4
      g_num(5,iel) = gnumbers(j+2,i+2)  ;   % node 5
      g_num(6,iel) = gnumbers(j+1,i+2)  ;   % node 6
      g_num(7,iel) = gnumbers(j,i+2)  ;     % node 7
      g_num(8,iel) = gnumbers(j,i+1)  ;     % node 8
      g_num(9,iel) = gnumbers(j+1,i+1)  ;   % node 9
      iel = iel + 1 ; % increment the element number
    end
end

% establish equation number for each node
sdof   = 0 ;                   % system degrees of freedom (dof) counter
nf     = zeros(ndof,nn) ;      % node degree of freedom array
for n = 1:nn                   % loop over all nodes in mesh
  for i=1:ndof                 % loop over each dof
     sdof = sdof + 1 ;         % increment sdof
     nf(i,n) =   sdof ;      % store eqn number on each node
  end
end

% equation number for each element
g     = zeros(ntot,1) ;     % equation numbers for 1 element
g_g = zeros(ntot,nels);     % equation numbers for all elements
for iel=1:nels ;            % loop over elements
```

```
    num = g_num(:,iel) ; % node numbers for this element
    inc=0 ;              % initialise local eqn number
    % loop 2 times over nodes of an element, once for each dof
    for i=1:nod ;  inc=inc+1 ; g(inc)=nf(1,num(i)) ; end
    for i=1:nod ;  inc=inc+1 ; g(inc)=nf(2,num(i)) ; end
    g_g(:,iel) = g ;     % store the equation numbers
end

%-----------------------------------------
% locate boundary nodes
%-----------------------------------------

bx0 = find(g_coord(1,:)==0)    ;
bxn = find(g_coord(1,:)==lx)   ;
by0 = find(g_coord(2,:)==0)    ;
byn = find(g_coord(2,:)==ly)   ;

%---------------------------------------------
%  integration data and shape functions
%---------------------------------------------
% local coordinates of Gauss integration points for nip=3x3
  points(1:3:7,1) = -sqrt(0.6);
  points(2:3:8,1) = 0;
  points(3:3:9,1) = sqrt(0.6);
  points(1:3,2)   = sqrt(0.6);
  points(4:6,2)   = 0 ;
  points(7:9,2)   = -sqrt(0.6);

  % Gauss weights for nip=3x3
  w   = [ 5./9. 8./9. 5./9.] ;
  v   = [ 5./9.*w ; 8./9.*w ; 5./9.*w ] ;
  wts = v(:) ;

 % evaluate shape functions and their derivatives
 % at integration points and save the results
  for k = 1:nip
     xi  = points(k,1);
     eta = points(k,2);
     etam = eta - 1; etap = eta + 1 ;
     xim  = xi - 1 ; xip  = xi + 1  ;
     x2p1 = 2*xi+1 ;   x2m1 = 2*xi-1 ;
     e2p1 = 2*eta+1 ;  e2m1 = 2*eta-1 ;
     % shape functions
     fun= [ .25*xi*xim*eta*etam -.5*xi*xim*etap*etam ...
     .25*xi*xim*eta*etap -.5*xip*xim*eta*etap ...
     .25*xi*xip*eta*etap -.5*xi*xip*etap*etam ...
     .25*xi*xip*eta*etam -.5*xip*xim*eta*etam xip*xim*etap*etam ] ;
     % derivates of shape functions
     der(1,1) = 0.25*x2m1*eta*etam  ; %dN1dxi
     der(1,2) =-0.5*x2m1*etap*etam ;  %dN2dxi, etc
     der(1,3) = 0.25*x2m1*eta*etap  ;
     der(1,4) =     -xi*eta*etap ;
     der(1,5) = 0.25*x2p1*eta*etap  ;
     der(1,6) =-0.5*x2p1*etap*etam ;
     der(1,7) = 0.25*x2p1*eta*etam  ;
     der(1,8) =     -xi*eta*etam ;
     der(1,9) = 2*xi*etap*etam    ;

     der(2,1) = 0.25*xi*xim*e2m1 ; %dN1deta
```

```
      der(2,2) =-xi*xim*eta        ;
      der(2,3) = 0.25*xi*xim*e2p1 ;
      der(2,4) =-0.5*xip*xim*e2p1    ;
      der(2,5) = 0.25*xi*xip*e2p1 ;
      der(2,6) =-xi*xip*eta        ;
      der(2,7) = 0.25*xi*xip*e2m1 ;
      der(2,8) =-0.5*xip*xim*e2m1    ;
      der(2,9) = 2*xip*xim*eta ;
      % save
      fun_s(k,:) = fun ;
      der_s(:,:,k) = der ;
   end
%-----------------------------------------
% initialisation
%-----------------------------------------

lhs   = sparse(sdof,sdof) ; % system stiffness matrix
bv    = zeros(sdof,1)     ; % system rhs vector
displ = zeros(sdof,1)     ; % solution vector

% dirichlet boundary conditions
% (zero-flux when the vectors are left empty)
bcdof = [ ] ; % fixed nodes
bcval = [ ] ; % fixed values

% initial conditions
displ            = zeros(sdof,1)  ;
displ(nf(1,:)) = (a+b)*ones(nn,1) + amp*randn(nn,1);
displ(nf(2,:)) = b/(a+b)^2*ones(nn,1) + amp*randn(nn,1);

%----------------------------------------------------
% time loop
%----------------------------------------------------

time = 0 ; % initial time
for it=1:ntime

    it
displ0 = displ ; % save old solution
time   = time + dt ; % update time

iters = 0; % initialise iteration counter
error = 1; % initialise error to arbitary large number
while error>0.001 % nonlinear iteration loop
  iters = iters + 1 % increment iteration number
  bv    = zeros(sdof,1) ; % system rhs vector
  lhs   = sparse(sdof,sdof) ; % system stiffness matrix

%-----------------------------------------------
%  element integration and assembly
%-----------------------------------------

for iel=1:nels  % sum over elements

    num       = g_num(:,iel)  ; % node numbers
    g         = g_g(:,iel)    ; % equation numbers (all)
    ga        = g(1:nod)      ; % equation numbers for A
    gb        = g(nod+1:end)  ; % equation numbers for B
    coord     = g_coord(:,num)' ; % node coordinates
```

```
    KMa      = zeros(nod,nod) ;
    KMb      = zeros(nod,nod)  ;
    MM       = zeros(nod,nod)  ;
    FA       = zeros(nod,1)    ;
    FB       = zeros(nod,1)    ;
    for k = 1:nip % loop over integration points
        fun    = fun_s(k,:) ;   % shape functions
        der    = der_s(:,:,k) ; % derivs. of N in local coords
        jac    = der*coord   ;  % Jacobian matrix
        detjac = det(jac)    ;  % determinant of Jac
        invjac = inv(jac)    ;  % inverse of Jac
        deriv  = invjac*der  ;  % derivs. of N in physical coords
        Ai     = fun*displ(ga); % interpolate A to integration pt.
        Bi     = fun*displ(gb); % interpolate B to integration pt.
        dwt    = detjac*wts(k); % multiplier
        MM     = MM  + fun'*fun*dwt; % mass matrix
        KMa    = KMa + deriv'*deriv*dwt;% A diffn matrix
        KMb    = KMb + d*deriv'*deriv*dwt; % B diffn matrix
        FA     = FA  + gamma*(a+Ai^2*Bi)*fun'*dwt ;% A load vector
        FB     = FB  + gamma*(b-Ai^2*Bi)*fun'*dwt;% B load vector
    end
      % assemble global lhs matrix and rhs vector
      lhs(ga,ga) = lhs(ga,ga) + MM/dt + KMa + gamma*MM ; % A contrib.
      lhs(gb,gb) = lhs(gb,gb) + MM/dt + KMb   ; % B contribution
      bv(ga)     = bv(ga) + MM/dt*displ0(ga) + FA  ;% A contribution
      bv(gb)     = bv(gb) + MM/dt*displ0(gb) + FB  ;% B contribution
 end

%----------------------------------------------------
 % implement boundary conditions and solve system
 %---------------------------------------------------

   % apply boundary conditions
   lhs(bcdof,:) = 0     ;
   tmp = spdiags(lhs,0) ;
   tmp(bcdof)=1         ;
   lhs=spdiags(tmp,0,lhs) ;
   bv(bcdof) = bcval ;

   displ_tmp = displ ; % save solution vector
   displ     = lhs\bv ;% solve sytem

   % check for convergence
   error = max(abs(displ-displ_tmp))/max(abs(displ))

end % end of nonlinear iteration loop

%----------------------------------------
 % visualisation
 Ag = reshape(displ(nf(1,:)),ny,nx);
 Bg = reshape(displ(nf(2,:)),ny,nx);

 figure(1)
 pcolor(xg,yg,Ag)
 shading interp
 axis equal
 title('A')
 colorbar
```

```
    figure(2)
    pcolor(xg,yg,Bg)
    shading interp
    axis equal
    title('B')
    colorbar

    drawnow

end
```

```
%-----------------------------------------------
% end of time loop
%-----------------------------------------------
```

Figure 7.14 shows an example of a numerical simulation of the coupled reaction–diffusion system. A remarkable feature of the results is that the concentrations of A and B self-organize into a regular pattern from a random initial state. This type of behavior, which is due to the nonlinear reaction kinetics between A and B, is quite different from standard linear diffusion models whose solutions progressively smoothen with increasing time.

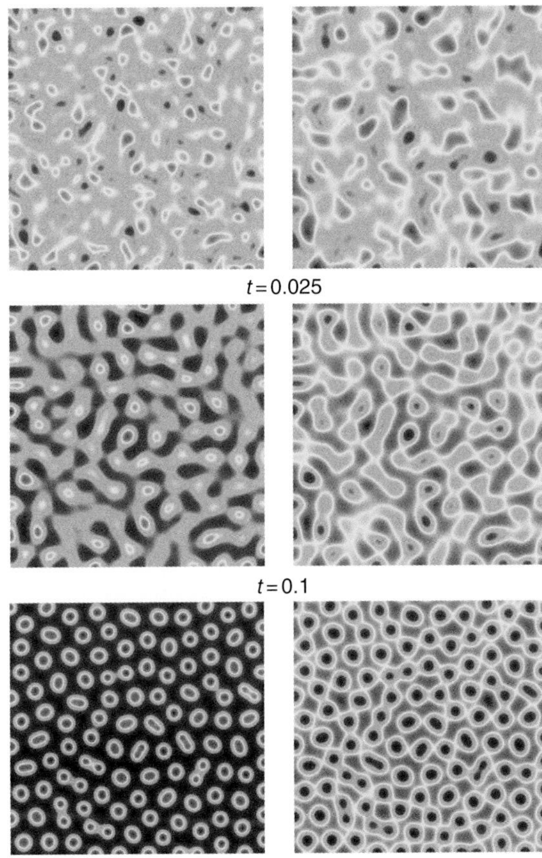

Figure 7.14 Results of a finite element simulation of a coupled reaction–diffusion problem (Equations 7.40 and 7.41). Plotted at three different times are the concentrations of two chemical species (an activator A and an inhibitor B) that diffuse and react with each other according to nonlinear kinetics. These coupled interactions lead to the formation of a self-organized pattern from random initial conditions. The simulation was performed with 50 nine-node elements in each direction ($dx = 0.1$), a domain length of 5 in each direction, a time step of $5x10^{-4}$, and the following physical parameter values: $a = 0.05$, $b = 1$, $d = 20$, and $\gamma = 600$.

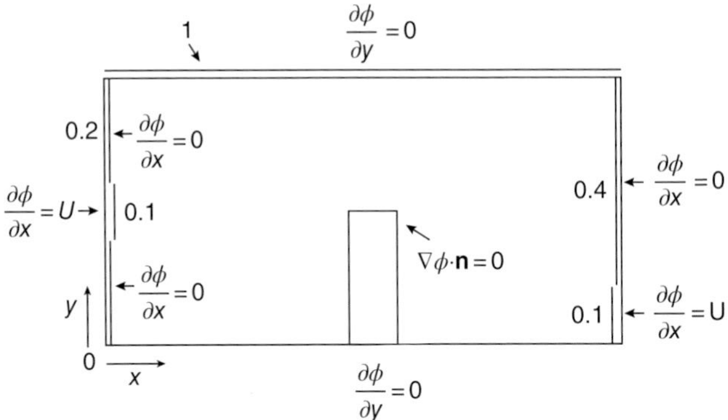

Figure 7.15 Definition of model setup for potential flow exercise.

7.4 Exercises

1) Derive the three shape functions for a linear triangle, expressed in terms of global coordinates (see 7.13). Hint: Begin by assuming that the shape functions can be expressed by an equation of the form $f(x, y) = ax + by + c$. Recall that $N_1 = 1$ at node 1, where x takes on the values $x = x_1$, $y = y_1$, and so on. Proceeding in a similar manner for the other nodes leads to three equations, which can be solved for the three unknown coefficients a, b, and c. The same approach can be used for N_2 and N_3.
2) Derive the element boundary terms (equivalent to 7.26) for the Robin boundary condition $\partial\phi/\partial\mathbf{n} = c\phi$ (where c is a constant) imposed on the $y = ly$ boundary (see Figure 7.4). You should finish up with

$$\mathbf{F}_\mathrm{b} = \frac{c\ a}{6}\begin{bmatrix} 0 & 0 & 0 \\ 0 & 2 & 1 \\ 0 & 1 & 2 \end{bmatrix}\begin{Bmatrix} \phi_1 \\ \phi_2 \\ \phi_3 \end{Bmatrix} = \mathbf{M}_\mathrm{b}\phi \tag{7.58}$$

   Note that the matrix $\mathbf{M}_\mathrm{b}$ must be added to the (left-hand side) stiffness matrix for elements falling on the boundary $y = ly$ boundary.
3) Using the Matlab pdetool, generate a triangular mesh for the setup shown in Figure 7.15. Save the mesh and use it to solve the potential flow problem (Equation 7.8).
4) Modify the Matlab script listed in Section 7.2 to solve the advection–diffusion equation

$$\frac{\partial\phi}{\partial t} + c\frac{\partial\phi}{\partial x} = k\frac{\partial^2\phi}{\partial x^2}$$

   using $c = 1$, $k = 0.01$, the initial conditions defined in 7.28, and $\phi = 0$ on the $x = 0$ and $x = 2$ boundaries.

Suggested Reading

S.-W. Cheng, T. K. Dey, and J. Shewchuk, *Delaunay Mesh Generation*, Chapman and Hall/CRC Press, New York, 2012.

C. A. J. Fletcher, *Computational Techniques for Fluid Dynamics,* Springer, Berlin, 2000.

Y. W. Kwong, and H. C. Bang, *The Finite Element Method Using Matlab,* CRC Press, New York, 2000.

R. W. Lewis, P. Nithiarasu, and K. N. Seetharamu, *Fundamentals of the Finite Element Method for Heat and Fluid Flow,* John Wiley & Sons, Ltd, Chichester, 2004.

I. M. Smith, and D. V. Griffith, *Programming the Finite Element Method,* John Wiley & Sons, Ltd, Chichester, 1998.

P. Šolín, K. Segeth, and I. Doležel, *Higher-Order Finite Element Methods,* Chapman and Hall/CRC Press, New York, 2004.

O. C. Zienkiewicz, and R. L. Taylor, *The Finite Element Method, Volume 1, The Basis,* Butterworth-Heinemann, Oxford/Boston, MA, 2000.

Part II

Applications of the Finite Element Method in Earth Science

In Part II, we will illustrate how the finite element method (FEM) introduced in Part I can be applied to solve different model equations of relevance in Earth science. In doing so, we revisit various topics treated for the first time in Chapter 7 including implementation of different boundary conditions, discretization with triangles, higher-order elements, mesh generation using the Matlab pdetool GUI, specialized methods for advection-dominated problems, coupled problems involving more than one unknown, and treatment of nonlinearities. The main disciplines considered are heat flow, landscape evolution, fluid flow in porous media, lithospheric flexure, and deformation.

Practical Finite Element Modeling in Earth Science Using Matlab, First Edition. Guy Simpson.

Companion website: www.wiley.com/go/simpson

8

Heat Transfer

This chapter deals with application of the finite element method (FEM) to modeling heat transfer in the crust. Although numerous aspects of this problem have already been treated in Chapters 1–6, here we consider several new factors that were only treated briefly for the first time in Chapter 7, including material advection, discretization with triangular finite elements, mesh generation with the Matlab pdetool GUI, flux boundary conditions, and treatment of nonlinearities. Two independent Matlab scripts are presented to show how these various aspects are implemented in practice.

The temperature within the Earth is controlled by competition between processes that generate or consume heat and processes that transfer heat (mainly conduction and advection/convection). For example, heat may be generated by radioactive decay, motion on frictional faults, and latent heating related to crystallization, while it may be consumed by endothermic metamorphic reactions and melting. Conduction is heat transfer transmitted by atom vibrations and occurs whenever there is a spatial variation in temperature. Advective or convective heat transfer is that which is caused either by motion of the rocks themselves or by relative motion between a porous fluid phase and the rock matrix. The interaction between these processes can be expressed mathematically as follows (see Carslaw and Jaeger (1959) and Turcotte and Schubert (1982)):

$$\underbrace{\frac{\partial T}{\partial t}}_{\text{Transient term}} + \underbrace{\mathbf{r}^T \nabla T}_{\text{Advection/Convection}} = \underbrace{\nabla^T \mathbf{K} \nabla T}_{\text{Conduction}} + \underbrace{\frac{A}{\rho c}}_{\text{Heat source/sink}} \tag{8.1}$$

Here, T is the temperature (K), ρ is the rock density (kg m^{-3}), c is the specific heat capacity (J kg^{-1} K^{-1}), $\mathbf{K}$ is the thermal diffusivity tensor (m^2 s^{-1}), A is the rate of internal heat production per unit volume (J s^{-1} m^{-3}), and $\mathbf{r}$ is the advection/convection velocity vector (m s^{-1}). Equation 8.1 applies to one dimension (1D), two dimension (2D), and three dimension (3D). For example, in 2Ds

$$\nabla = \begin{Bmatrix} \frac{\partial}{\partial x} \\ \frac{\partial}{\partial z} \end{Bmatrix} \tag{8.2}$$

$$\mathbf{r} = \begin{Bmatrix} u \\ v \end{Bmatrix} \tag{8.3}$$

$$\mathbf{K} = \begin{bmatrix} \kappa_x & 0 \\ 0 & \kappa_z \end{bmatrix} \tag{8.4}$$

and Equation 8.1 can be written as follows:

$$\frac{\partial T}{\partial t} + u\frac{\partial T}{\partial x} + v\frac{\partial T}{\partial z} = \frac{\partial}{\partial x}\left(\kappa_x \frac{\partial T}{\partial x}\right) + \frac{\partial}{\partial z}\left(\kappa_z \frac{\partial T}{\partial z}\right) + \frac{A}{\rho c} \tag{8.5}$$

Practical Finite Element Modeling in Earth Science Using Matlab, First Edition. Guy Simpson.

Companion website: www.wiley.com/go/simpson

Equation 8.1 is closely related to the diffusion equation studied in the preceding chapters but, due to the advective term, is typically called an advection–diffusion equation with a source/sink term. In this chapter, we numerically solve the equation applied to two different contexts.

8.1 Conductive Cooling in an Eroding Crust

Consider a crust that experiences conductive heat transfer, spatially uniform surface erosion that is constant in time and no internal heat generation or consumption. Erosion advects rocks at depth toward the surface with a constant vertical velocity that is equal to the erosion rate $\dot{e}$. The equation governing this problem is obtained from 8.5 by setting the vertical velocity to $\dot{e}$, omitting the source term, and dropping spatial derivatives in the horizontal (x) direction as follows:

$$\frac{\partial T}{\partial t} + \dot{e}\frac{\partial T}{\partial z} = \frac{\partial}{\partial z}\left(\kappa \frac{\partial T}{\partial z}\right) \tag{8.6}$$

Equation 8.6 will be solved with a constant initial temperature gradient of $-30°\mathrm{C\,km^{-1}}$, a fixed temperature of $0°\mathrm{C}$ at the surface and a constant heat flux q at the base of the crust (initially assumed to be at -30 km depth; see Figure 8.1). The flux is chosen so that it perfectly balances the initial geothermal gradient (i.e., $q = -\kappa\, \partial T/\partial z = 10^{-6} \times 0.03 = 3 \times 10^{-8}$). An important nondimensional parameter controlling the behavior of this problem is the thermal Peclet number, defined as $Pe = |\dot{e}|L/\kappa$ where L is the characteristic length scale (here 30 km). When $Pe \gg 1$ advective heat transfer is dominant over conductive heat transfer and the equation behaves mainly like a hyperbolic equation. Conversely, when $Pe \ll 1$, conductive heat transfer dominates and the equation is dominantly parabolic.

This problem includes two relatively new features. First, a Neumann (constant flux) boundary condition must be included at $z = -30$ km (see Section 7.1). Second, Equation 8.6 contains an advection term (i.e., a first-order spatial derivative) that renders the problem non-self-adjoint (see Zienkiewicz and Taylor (2000) and Section 7.2). Neither of these factors present any difficulty and the program listed in the following closely resembles those presented earlier (especially the program listed in Section 4.6).

Applying Galerkin finite element discretization to Equation 8.6 and using integration by parts on the term with a second-order spatial derivative result in the following discrete system for a single element:

$$[\mathbf{MM}]\frac{\partial \mathbf{T}}{\partial t} + [\mathbf{CM}]\,\mathbf{T} + [\mathbf{KM}]\,\mathbf{T} = \mathbf{Fb} \tag{8.7}$$

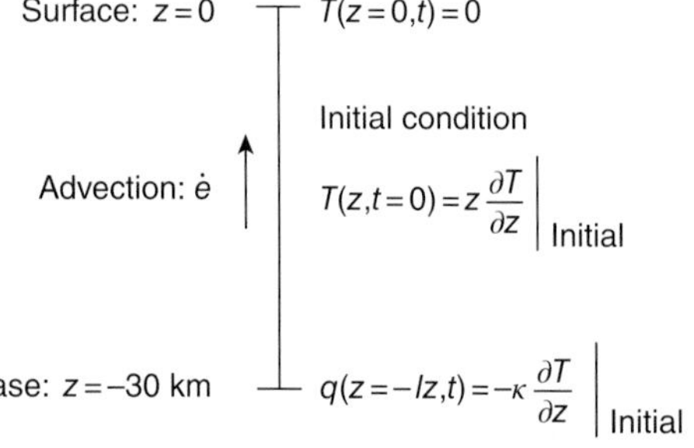

Figure 8.1 Initial and boundary conditions for 1D problem involving conductive cooling of an eroding crust.

Here, **T** is the vector of nodal temperatures for an element,

$$\mathbf{MM} = \int \mathbf{N}^T \mathbf{N}\, dz \tag{8.8}$$

$$\mathbf{CM} = \int \dot{e}\, \mathbf{N}^T \frac{\partial \mathbf{N}}{\partial z}\, dz \tag{8.9}$$

$$\mathbf{KM} = \int \kappa \frac{\partial \mathbf{N}}{\partial z}^T \frac{\partial \mathbf{N}}{\partial z}\, dz \tag{8.10}$$

$$\mathbf{Fb} = -\kappa \frac{\partial T}{\partial z}\bigg|_b \tag{8.11}$$

and **N** are the shape functions. Here, **MM** and **KM** are the familiar element mass and stiffness matrices, respectively, while **CM** is an asymmetrical element matrix resulting from discretization of the advection term in Equation 8.6 (see Section 7.2). The term **Fb** is the boundary flux at $z = -lz$, which results from integration by parts. This term is normally neglected when Dirichlet boundary conditions are applied, but here it is retained to account for nonzero flux (Neumann) conditions.

Discretizing the time derivative in 8.7 using an implicit finite difference approximation and rearranging results in

$$\left(\frac{\mathbf{MM}}{\Delta t} + \mathbf{CM} + \mathbf{KM}\right) \mathbf{T}^{n+1} = \frac{\mathbf{MM}}{\Delta t} \mathbf{T}^n + \mathbf{Fb} \tag{8.12}$$

where Δt is the time interval between the n and $n + 1$ levels. In this form, the bracketed term on the left-hand side is seen to be the total element stiffness matrix, while the entire right-hand side combines to form the element load vector, which changes through time because it depends on the old temperature $\mathbf{T}^n$. The element matrices **MM**, **KM**, and **CM** can be evaluated exactly or computed using Gauss–Legendre quadrature and assembled into the global left- and right-hand matrices, as done in the preceding chapters.

What follows is a Matlab script that computes the numerical solution to 8.12. Most aspects of the program are identical to the script discussed and listed in Chapter 4. The script presented below differs in the following respects:

1) No load vector appears in the element loop (i.e., there is no `F`).
2) The element advection matrix `CM` (see Equation 8.9) must be integrated, which is done with the snippet

```
CM   = CM + edot*fun'*deriv*detjac*wts(k) ;   % advection matrix
```

   This line appears within the loop over integration points and must be performed after `CM` is initialized for each element.
3) After the integration loop, `CM` and the other element matrices can be assembled into the left-hand side global matrix (see the left-hand side of Equation 8.12) as follows:

```
lhs(num,num) = lhs(num,num) + MM/dt + KM + CM ; % assemble lhs
```

4) After implementation of the Dirichlet boundary condition for the surface node, the basal heat flux is added to the right-hand side global vector as follows:

```
b(nbcdof) = b(nbcdof) + nbcval ;  % add heat flux on base
```

   Here, `nbcdof` are nodes where Neumann boundary conditions are applied (in this case, `nbcdof=[ nn ]`) and `nbcval` are the flux values (in this case, `nbcval=[ -kappa*dTdz]`) where nn is the last node in the finite element mesh, kappa is the thermal diffusivity, and dTdz is the initial geothermal gradient.

```
%-----------------------------------------------------------
% Program advdiffn1d.m
% 1-D FEM solution of advection-diffusion equation
%-----------------------------------------------------------

clear     % clear memory from current workspace
close all % close all figures

% physical parameters
seconds_per_yr = 60*60*24*365; % number of seconds in one year
lz     = 30000 ;          % length of spatial domain (m)
kappa = 1e-6;             % thermal diffusivity (m^2/s)
Tb     = 0 ;              % temperatures at surface (°C)
dTdz   = -0.03 ;          % initial geothermal gradient (°C/m)
q      = -kappa*dTdz ; % basal heat flux
edot   = -1e-3/seconds_per_yr ; % erosion rate (m/s)

% numerical parameters
dt     = 10000*seconds_per_yr ; % time step (s)
ntime = 2000 ;        % number of time steps
nels   = 100 ;        % total number of elements
nod    = 2 ;          % number of nodes per element
nn     = nels+1       % total number of nodes
dz     = lz/nels ; % element size
nip    = 2 ;          % number of Gauss integration points per element

g_coord = -[0:dz:lz] ; % spatial domain (1-D mesh)

% integration data
points    =   [-sqrt(1/3) sqrt(1/3)] ; % positions of the Gauss points
wts       =   [1  1] ;                 % Gauss-Legendre weights

% shape functions and their derivatives (both in local coordinates)
% evaluated at the Gauss integration points
for k=1:nip
   fun_s(k,:) = [ (1-points(k))/2 (1+points(k))/2 ] ; % N1 and N2
   der_s(k,:) = [ -1/2 1/2 ] ;                        % dN1/xi, dN2/xi
end

% define Dirichlet boundary conditions
bcdof = [  1    ]   ; % boundary nodes
bcval = [  Tb   ]   ; % boundary values

% define Neumann (flux) boundary conditions
nbcdof = [ nn ] ;% boundary nodes
nbcval = [ q ]  ;% boundary flux

% define connectivity and equation numbering
g_num        = zeros(nod,nels) ;
g_num(1,:) = [1:nn-1]   ;
g_num(2,:) = [2:nn]     ;

% initialise matrices and vectors
b        = zeros(nn,1);          % system rhs vector
lhs      = sparse(nn,nn);        % system lhs matrix
rhs      = sparse(nn,nn);        % system rhs matrix
```

```
%-------------------------------------------
% matrix assembly
%-------------------------------------------
 for iel=1:nels % loop over all elements
   num    = g_num(:,iel)   ;            % retrieve equation number
   dz     = abs(diff(g_coord(num))) ; % length of element
   MM     = zeros(nod,nod) ;
   KM     = zeros(nod,nod) ;
   CM     = zeros(nod,nod) ;
   for k=1:nip % integrate element matrices
     fun = fun_s(k,:)   ;
     der = der_s(k,:)   ;
     detjac = dz/2      ;
     invjac = 2/dz      ;
     deriv  = der*invjac ;
     MM   = MM + fun'*fun*detjac*wts(k) ;         % mass matrix
     KM   = KM + deriv'*kappa*deriv*detjac*wts(k);% diffusion matrix
     CM   = CM + edot*fun'*deriv*detjac*wts(k) ; % advection matrix
   end % end of integration
   lhs(num,num) = lhs(num,num) + MM/dt + KM + CM ; % assemble lhs
   rhs(num,num) = rhs(num,num) + MM/dt           ; % assemble rhs
 end    % end of element loop
%-------------------------------------------------------------------

% time loop
displ = dTdz*g_coord' ; % initial temperature (°C)
t  = 0 ;                % initial time

for n=1:ntime
   t  = t + dt ;                          % update time
   b  = rhs*displ  ;                      % form rhs load vector

   % impose Dirichlet boundary conditions
   lhs(bcdof,:) = 0 ;                   % zero the relevent equations
   tmp = spdiags(lhs,0) ;               % store diagonal
   tmp(bcdof)=1 ;                       % place 1 on stored-diagonal
   lhs=spdiags(tmp,0,lhs);              % reinsert diagonal
   b(bcdof) = bcval ;                   % set rhs

   b(nbcdof) = b(nbcdof) + nbcval ;   % add heat flux on base

   displ = lhs \ b ;         % solve global system of equations
   %-----------------------------------------------
   % plotting
  if mod(n,200)==0
   figure(1) ; plot(displ,g_coord/1e3,g_coord*dTdz,g_coord/1e3,'r')
   xlabel('Temperature (C)')
   ylabel('Depth (km)')
   title(['Time = ',num2str(t/seconds_per_yr/1e6), 'My'])
   drawnow
   hold on
  end
end % end of time loop

%-------------------------------------------------------------------
```

Some numerical results computed with the given script are illustrated in Figure 8.2. For the parameters values used, the Peclet number is close to unity, indicating that both advective and conductive

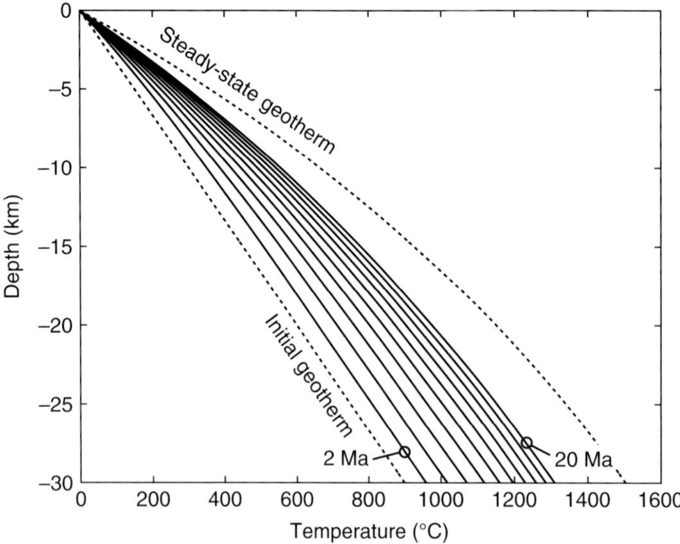

Figure 8.2 Numerical solution of Equation 8.6 with a fixed surface temperature and a constant heat flux at the base (Figure 8.1). The calculation was performed with an erosion rate ($\dot{e}$) of 1 mm year^{-1} and a thermal diffusivity (κ) of 10^{-6} m^2 s^{-1}. For these conditions, erosion efficiently advects heat toward the surface. Solid lines are plotted every 2 Ma. Also shown are the initial geothermal gradient (30°C km^{-1}) and the exact steady-state geotherm, which have not yet been attained after 20 Ma.

heat transfer are of similar importance. For this case, erosion is quite effective in advecting heat toward the surface, which increases the geothermal gradient throughout the crust.

8.2 Conductive Cooling of an Intrusion

Our task here is to compute the temperature within, and adjacent to, an instantaneously intruded spherical magma chamber that cools conductively and releases latent heat as it crystallizes. Although this problem can be solved in 3Ds using the approach presented in Chapter 6, the problem is effectively 2D due to rotational symmetry about a vertical axis passing through the center of the sphere (Figure 8.3). The 2D axisymmetric formulation presented is computationally efficient and requires only minimal modification of a standard 2D finite element program. The equation to be solved is

$$\rho c \frac{\partial T}{\partial t} = \frac{1}{r}\frac{\partial}{\partial r}\left(rk\frac{\partial T}{\partial r}\right) + \frac{\partial}{\partial z}\left(k\frac{\partial T}{\partial z}\right) + \rho L \frac{\partial \psi_s}{\partial t} \tag{8.13}$$

where r is the radial coordinate, z is the axial coordinate (Figure 8.3), k is the thermal conductivity (W m^{-1}), L is the latent heat of fusion (J kg^{-1}), c is the specific heat capacity (J kg^{-1} K^{-1}), ρ is the density (kg m^{-3}), and ψ_s is the solid (crystal) fraction in the molten rock, which varies as a function of temperature. Note that advective heat transfer (appearing in 8.1) has been neglected. We assume fixed temperatures on the upper and lower boundaries ($T = 0$ and $T = Tb = 300$°C, respectively), while zero horizontal heat flux (i.e., zero horizontal temperature gradient) is imposed on the two lateral boundaries (Figure 8.3). Recall that this latter condition is the natural boundary condition

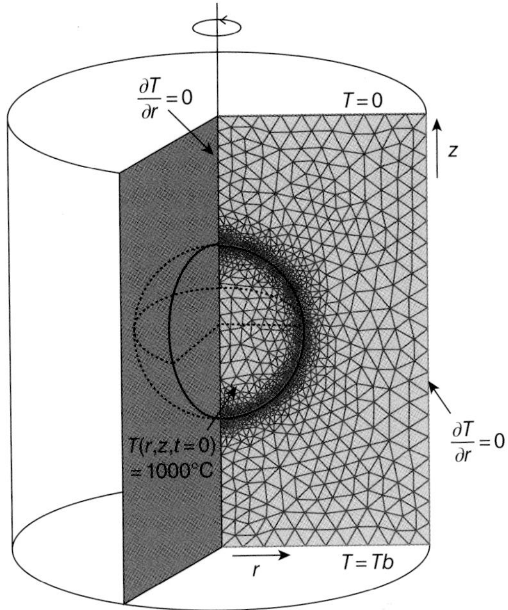

Figure 8.3 Axisymmetrical representation of a spherical intrusion. The triangular mesh for this problem was generated using the Matlab pdetool.

and requires no special treatment to be automatically satisfied. Recall also that the vertical left-hand boundary is a rotational symmetry axis (Figure 8.3). The initial temperature is assumed to increase linearly with depth from zero at the surface to $Tb = 300°\mathrm{C}$ at the base, with a gradient of $-30°\mathrm{C\,km}^{-1}$, except within the intrusion where the initial temperature is taken to be 1000°C.

The source term in 8.13 depends indirectly on the temperature in a nonlinear way and necessitates additional development. Using the chain rule, one can rewrite the source term as follows:

$$\rho L \frac{\partial \psi_s}{\partial t} = \rho L \frac{\partial \psi_s}{\partial T} \frac{\partial T}{\partial t} \tag{8.14}$$

We consider the following simple parameterization for the relationship between crystal fraction and temperature:

$$\psi_s = 1 - \frac{1}{1 + \exp \frac{c_1 - T}{c_2}} \tag{8.15}$$

Here, c_1 and c_2 are empirical fitting parameters (with dimensions of °C). A reasonable fit to experimental data of granodioritic composition (Piwinskii and Wyllie, 1968) is obtained with $c_1 = 800$ and $c_2 = 23$. Taking the derivative of ψ_s with respect to T gives

$$\frac{\partial \psi_s}{\partial T} = -\frac{\exp \frac{c_1 - T}{c_2}}{c_2 \left(1 + \exp \frac{c_1 - T}{c_2}\right)^2} = S(T) \tag{8.16}$$

Substituting 8.14 and 8.16 into 8.13 leads to

$$\rho c \frac{\partial T}{\partial t} = \frac{1}{r} \frac{\partial}{\partial r} \left(rk \frac{\partial T}{\partial r} \right) + \frac{\partial}{\partial z} \left(k \frac{\partial T}{\partial z} \right) - \rho L S(T) \frac{\partial T}{\partial t} \tag{8.17}$$

which upon rearranging becomes

$$\rho\left(LS(T)+c\right)\frac{\partial T}{\partial t}=\frac{1}{r}\frac{\partial}{\partial r}\left(rk\frac{\partial T}{\partial r}\right)+\frac{\partial}{\partial z}\left(k\frac{\partial T}{\partial z}\right) \tag{8.18}$$

Note that the source term related to latent heating has been transformed into a variable coefficient time derivative. Dependency of the parameter S on T (see Equation 8.16) means that the governing equation is no longer linear, as were most of the equations considered up to now. Nonlinear equations such as this can be solved using an iterative strategy (see Section 7.3), where the coefficient $S(T)$ is continuously updated at any given time step until the solution converges.

Applying Galerkin finite element discretization to 8.18, integration by parts to the second-order spatial derivatives (and neglecting the resulting boundary terms), and using an implicit finite difference approximation for the time derivative lead to the following system of discrete element equations:

$$\left(\frac{\mathbf{MM}^{n+1}}{\Delta t}+\mathbf{KM}\right)\mathbf{T}^{n+1}=\frac{\mathbf{MM}^{n}}{\Delta t}\mathbf{T}^{n} \tag{8.19}$$

Here, **T** is the vector of nodal temperatures and

$$\mathbf{MM}^{n+1}=\int\int r\rho\left(LS^{n+1}+c\right)\mathbf{N}^T\mathbf{N}\,drdz \tag{8.20}$$

$$\mathbf{MM}^{n}=\int\int r\rho\left(LS^{n}+c\right)\mathbf{N}^T\mathbf{N}\,drdz \tag{8.21}$$

$$\mathbf{KM}=\int\int r(\nabla\mathbf{N})^T k\nabla\mathbf{N}\,drdz \tag{8.22}$$

To discretize this problem, we use linear triangular elements (see Section 7.1). Triangles are especially good for this problem because they can accurately describe the circular intrusion margin (Figure 8.3). The mesh used for the calculation presented was generated using the "pdetool" tool box in Matlab. A basic mesh can be created by performing the following steps:

1) Open the pdetools GUI by typing "pdetool" at the command prompt (Figure 8.4). Turn on the grid by selecting Grid from the Options menu.
2) Draw a rectangle (R1) with the desired aspect ratio (Figure 8.4). Double clicking on the rectangle opens a dialogue box, which enables you to assign its exact dimensions and position.
3) Draw a circle (E1) with the desired radius. Place the circle so that it is x-centered on the left-hand boundary of the preexisting rectangle and z-centered midway between the top and bottom of rectangle R1 (Figure 8.4). Once again, the dialogue box can be used to refine its exact dimensions and position.
4) Draw a second rectangle (R2), with the same dimensions as the first and place it to the left of the first rectangle, over half the circle (Figure 8.4).
5) We are now going to create a mesh defined by the formula R1 + E1 − R2, that is, the intersection of the areas defined by the objects R1 and E1 minus the area of R2. This formula can be specified in the "Set formula" line near the top of the GUI. Once the formula is assigned, select "Initialize Mesh" from the Mesh menu. A more refined mesh can be obtained by repeating clicking "Refine Mesh" from the Mesh menu.
6) Once the desired resolution has been obtained, the mesh can be exported to the Matlab workspace by choosing Export Mesh under the Mesh menu. The three variables saved by default are `p`, `e`, and `t`. The variable `p` contains the coordinates of the nodes in the mesh (i.e., `g_coord` in this text), `e` is an array used for plotting (called `edge` in the program presented), and `t` contains element nodal

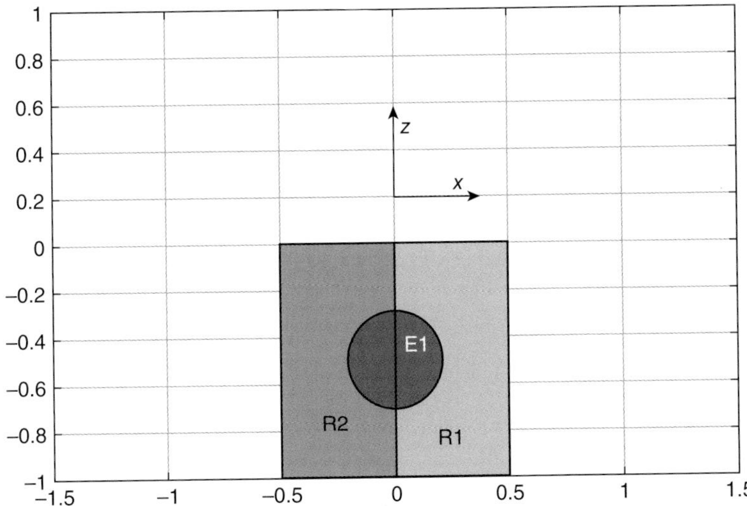

Figure 8.4 Generating a basic 2D triangular mesh within and around a circular "intrusion" using the Matlab pdetool GUI. The final mesh (shown in Figure 8.3) consists of the union between rectangle R1 and the circle E1 minus the rectangle R2.

connectivity in the first three lines (i.e., g_num): one line for each node of a particular triangle. Line 4 of t is a "phase" index referring to whether a particular element is within the region R1 − E1 (i.e., outside the intrusion, with the phase index = 1) or within the region E1 − R1 (i.e., within the intrusion, with phase index = 2). Renaming of the variables can be achieved within Matlab with the following snippet:

```
g_coord   = p ;           % nodal coordinates
edge      = e ;           % edge array
g_num     = t(1:3,:) ;    % element nodal numbering
phase     = t(4,:) ;      % phase indices
```

7) These variables can be saved in a .mat file using the Save command and loaded into the finite element program during the preprocessor stage. At this stage, the mesh can also be repositioned and rescaled to match the desired geometry and dimension of the problem. Although the positioning of the mesh is arbitrary, in the program listed in the following text, we will assume that the left boundary is located at $x = 0$, the right boundary is at $x = lx$, the upper boundary is at $z = 0$, and the lower boundary is at $z = -lz$. If this convention is not followed, the boundary conditions may either be incorrectly implemented, or they may not be applied at all.

Despite the axisymmetrical geometry, the nonlinear governing equation and use of a triangular mesh, the finite element Matlab program listed in the following text bears a strong resemblance to those presented in the preceding chapters (especially the 2D script listed in Section 5.6). The new script differs in the following respects:

1) The variables g_coord (nodal coordinates), edge (mesh edge data used only for plotting), g_num (node-element connectivity), and phase (indices, where 1 indicates the element is outside the intrusion, whereas 2 indicates the element is within the intrusion) are read into the Matlab session

from a single .mat file (mesh_axisym[1]). This file was generated with pdetool, as described earlier. The file can be loaded and the mesh can be visualized with the following snippet:

```
load mesh_axisym % load triangular mesh
pdemesh(g_coord,edge,g_num) % visualise mesh
axis equal % ensure plot has same scale in x and z directions
```

2) In order to perform Gauss–Legendre integration of the element matrices **MM**, and **KM** (defined in Equations 8.20, 8.21, and 8.22), the position and weights for triangles must be provided from Table 7.2. Assuming three integration points (i.e., `nip=3`), this can be done with the following lines:

```
% gauss integration data
points = zeros(nip,ndim); % location of points
points(1,1)=0.5   ; points(1,2)=0.5 ;
points(2,1)=0.5   ; points(2,2)=0   ;
points(3,1)=0     ; points(3,2)=0.5 ;
c   = 0.5 ; % triangle factor
wts = c*1/3*ones(1,nip) ; % weights
```

Here, `ndim` is the number of spatial dimensions (here 2).

3) The values of the shape functions and their spatial derivatives for the three-node (linear) triangle, defined in terms of local triangular coordinates L_1, L_2, and L_3 (see Figure 7.3), must be evaluated at the integration points (and saved for later use). This can be achieved with the following snippet:

```
% save shape functions and their derivatives in local coordinates
% evaluated at integration points
for k=1:nip
   L1=points(k,1); L2=points(k,2); L3=1.-L1-L2   ;
   fun = [L1 L2 L3] ;
   fun_s(k,:) = fun ; % shape functions
   der(1,1)=1 ; der(1,2)=0 ; der(1,3)=-1   ;
   der(2,1)=0 ; der(2,2)=1 ; der(2,3)=-1   ;
   der_s(:,:,k) = der ; % derivative of shape function
end
```

4) Integration of the element matrices (Equations 8.20, 8.21, and 8.22) requires the temperatures at the n and $n+1$ time levels ($T0$ and T) along with the radial coordinate r, all of which must be evaluated at the positions of the integration points. These can be computed using the shape functions (`fun`) as interpolants, that is,

```
T    = fun*displ(num) ; % T interpolated to current int. point
T0   = fun*displ0(num) ;% T0 interpolated to current int. point
r    = fun*coord(:,1) ; % distance from symmetry axis
```

where `displ(num)` and `coord(:,1)` are the nodal solution vector and nodal r-coordinates, respectively, for a given element.

5) The program uses nonlinear iterations, repeatedly solving the linear system at each time step, while the $\mathbf{MM}^{n+1}$ element matrix is updated with the newly computed temperature. This is done until the solution converges, or stops changing within some predefined limit. The basic algorithm is illustrated in Figure 7.13. The program also uses a variable time step that increases linearly with time up to a maximum of 100 years, where it remains constant thereafter. Using much smaller time steps at the beginning of the calculation, when the rate of temperature change is relatively fast, leads to faster convergence.

---

1 This file can be obtained by contacting the author.

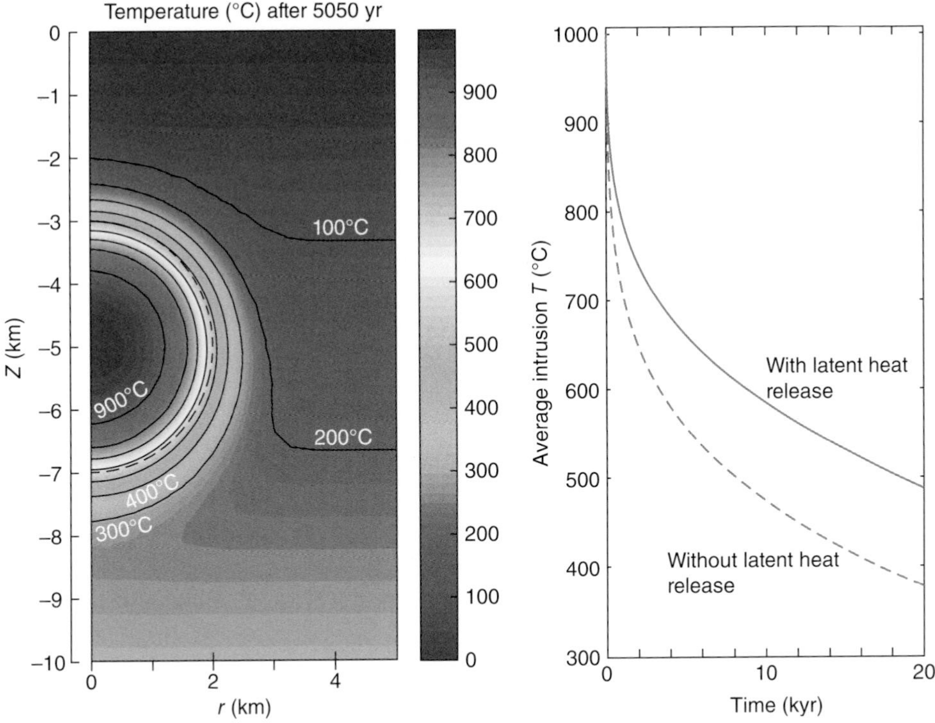

Figure 8.5 Results from axisymmetrical intrusion thermal model. Left panel: Temperature 5050 years after intrusion at 5 km depth of an axisymmetrical spherical intrusion with a radius of 2 km undergoing conductive cooling and releasing latent heat due to crystallization (computed from Equation 8.13; see Figure 8.3). Intrusion margin is shown as a dashed line. Isotemperatures are plotted for 100, 200, …, 900°C. Right panel: Mean temperature of elements within the intrusion as a function of time for simulations with (solid line) and without (dashed line) latent heating.

6) Visualization of the solution on the triangular mesh can be performed using the following Matlab pdeplot command:

```
% plot solution defined on a triangular mesh
pdeplot(g_coord,edge,g_num,'xydata',displ,'mesh','off','contour','on');
```

Various other plot options can be found in the Help menu for the command pdeplot.

An example of output from the program `diffnaxisym` is illustrated in Figure 8.5. After 5050 years, the temperature of the intrusion (initially intruded at 1000°C at a depth of 5 km) has decayed due to conductive heat transfer outward into the surrounding, relatively cool country rocks that undergo heating. As expected, latent heating due to crystallization delays cooling of the intrusion. This is most easily demonstrated by directly comparing the results of two different simulations: one where latent heat is included and a second where it is omitted (i.e., by setting $L = 0$, Figure 8.5).

```
%-----------------------------------------
% Program: diffnaxisym
% 2D FEM - axisymmetric diffusion equation
% with Latent heat release
% 3-node triangles - Mesh generated externally with pdetool
%-----------------------------------------
```

```
clear

seconds_per_yr = 60*60*24*365 ; % seconds in 1 year

% physical parameters
lx     = 5e3    ;    % width of domain (m)
lz     = 10e3   ;    % depth of domain (m)
radius = 2e3    ;    % intrusion radius (m)
px     = 0      ;    % x coord. of intrusion centroid (m)
pz     = -lz/2  ;    % z coord. of intrusion centroid (m)
dTdz   = -0.03  ;    % initial geothermal gradient (C/m)
Tz0    = 0 ;         % fixed surface temperature (C)
Tzn    = -dTdz*lz ;  % fixed basal temperature (C)
Tintrusion = 1000 ;  % initial intrusion temperature (C)
K          = 3.3   ; % thermal conductivity (W/m/K)
latent     = 350e3 ; % latent heat of fusion (J/kg)
rhov       = [2700 2300] ; % densities of rock and magma (kg/m^3)
Cp         = 1000; % specific heat capacity (J/kg/K)
c1         = 800 ; % parameter for crystallinity function (C)
c2         = 23  ; % parameter for crystallinity function (C)

% numerical parameters
ndim  = 2    ;    % number of spatial dimensions
nod   = 3    ;    % number of nodes in 1 element
nip   = 3   ;     % number of Gauss integration points
dt    = 1*seconds_per_yr ; % initial time step (s)
tolerance = 0.001 ; % error tolerance

% load triangular mesh (e.g., generated using pdetool)
% the following variables are loaded
% g_coord - mesh nodes
% edge    - mesh edge data
% g_num   - node-element connectivity
% phase   - index (1 or 2)
load mesh_axisym

nn    = length(g_coord) ;  % total number of nodes
nels  = length(g_num)   ;  % total number of elements

figure(10) , clf % visualise the mesh
pdemesh(g_coord,edge,g_num) ; axis equal

% Find boundary nodes
bx0 = find(g_coord(1,:)==0) ;   % nodes on x=0 boundary
bxn = find(g_coord(1,:)==lx) ; % nodes on x=lx boundary
bz0 = find(g_coord(2,:)==0) ;   % nodes on z=0 boundary
bzn = find(g_coord(2,:)==-lz) ;% nodes on z=-lz boundary

% define fixed boundary conditions
bcdof = [bz0 bzn ] ;  % boundary nodes
% boundary temperatures
bcval = [Tz0*ones(1,length(bz0)) Tzn*ones(1,length(bzn))]   ;

% gauss integration data
points = zeros(nip,ndim); % location of points
points(1,1)=0.5   ; points(1,2)=0.5 ;
points(2,1)=0.5   ; points(2,2)=0   ;
points(3,1)=0     ; points(3,2)=0.5 ;
```

```
c   = 0.5 ; % triangle factor
wts = c*1/3*ones(1,nip) ; % weights

% save shape functions and their derivatives in local coordinates
% evaluated at integration points
for k=1:nip
   L1=points(k,1); L2=points(k,2); L3=1.-L1-L2   ;
   fun = [L1 L2 L3] ;
   fun_s(k,:) = fun ; % shape functions
   der(1,1)=1 ; der(1,2)=0 ; der(1,3)=-1   ;
   der(2,1)=0 ; der(2,2)=1 ; der(2,3)=-1   ;
   der_s(:,:,k) = der ; % derivative of shape function
end

% initialise arrays
ff     = zeros(nn,1);      % global load vector
b      = zeros(nn,1);      % global rhs vector
lhs    = sparse(nn,nn);    % global lhs matrix
rhs    = sparse(nn,nn);    % global rhs matrix

% initial conditions
displ = dTdz*g_coord(2,:)' ; % initial const. gradient
% d = distance between all points in mesh and the intrusion center
d      = sqrt((g_coord(1,:)-px).^2+(g_coord(2,:)-pz).^2) ;
ii     = find(d≤radius) ; % indices of all nodes within the intrusion
% set nodes within the intrusion to the intrusion temperature
displ(ii) = Tintrusion ;
displ0    = displ ; % save old temperature

%-----------------------------------------------------
% time loop
%-----------------------------------------------------
time = 0 ; % initialise time
for n=1:1000

  % variable time step
  if n<100
    dt = n*seconds_per_yr    ;
  else
    dt = 100*seconds_per_yr   ;
  end

  time = time + dt ;  % update time

  % nonlinear iteration loop -------------------------
  error = 1 ;
  iters = 0 ;
  while error>tolerance
       iters = iters + 1
%-----------------------------------------------------
% matrix integration and assembly
%-----------------------------------------------------
lhs    = sparse(nn,nn);    % reinitialise global lhs matrix
rhs    = sparse(nn,nn);    % reinitialise global rhs matrix
for iel=1:nels  % sum over elements
  num      = g_num(:,iel)      ; % element nodes
  coord    = g_coord(:,num)' ; % element coordinates
  KM       = zeros(nod,nod)    ; % initialisation
  MM       = zeros(nod,nod)    ;
```

```
    MM0     = zeros(nod,nod)   ;
    F       = zeros(nod,1)      ;
    rho     = rhov(phase(iel))   ;

    for k = 1:nip                     % integration loop
      fun    = fun_s(k,:) ;           % shape functions
      der    = der_s(:,:,k) ;         % der. of shape functions
      jac    = der*coord  ;           % jacobian matrix
      ri     = fun*coord(:,1);        % distance from symmetry axis
      detjac = det(jac) ;             % det. of jacobian
      invjac = inv(jac) ;             % inv. of jacobian
      deriv  = invjac*der ;           % der. of shape fun. in physical coords.
      T      = fun*displ(num) ; % T interpolated to int. pt.
      eterm  = exp((c1-T)/c2) ;
      S      = eterm/(c2*(1+eterm)^2); % deriv of xtal fract. wtr T
      T0     = fun*displ0(num) ;       % old T interpolated to int. pt.
      eterm0 = exp((c1-T)/c2) ;
      S0     = eterm/(c2*(1+eterm)^2);% deriv of xtal. fract. wtr T0
      dwt    = detjac*wts(k) ;
      KM     = KM + ri*deriv'*K*deriv*dwt ;       % stiffness matrix
      MM     = MM + ri*rho*(latent*S+Cp)*fun'*fun*dwt ; % mass matrix
      MM0    = MM0 + ri*rho*(latent*S0+Cp)*fun'*fun*dwt; % mass matrix0
    end
    % assemble global matrices
    lhs(num,num) = lhs(num,num) + MM/dt + KM  ;
    rhs(num,num) = rhs(num,num) + MM0/dt      ;
  end
  %-------------------------------------------------------------------

    b = rhs*displ0 ;                        % form rhs global vector

    % impose boundary conditions
    lhs(bcdof,:) = 0 ;                      % zero the boundary equations
    tmp = spdiags(lhs,0) ;                  % store diagonal
    tmp(bcdof)=1 ;                          % place 1 on stored-diagonal
    lhs=spdiags(tmp,0,lhs);                 % reinsert diagonal
    b(bcdof) = bcval ;                      % set boundary values

    displ_tmp = displ ;                     % same temporary solution
    displ = lhs \ b ;                       % solve system of equations
    error = max(abs(displ-displ_tmp))/max(abs(displ)) % evaluate error

  end
  % end of nonlinear iterations %--------------------------------------

  displ0 = displ ; % update old solution

    %-------------------------------------------------------------
    % visualisation
    figure(1)
    clevels = [0.1 0.2 0.3 0.4 0.5 0.6 0.7 0.8 0.9]*Tintrusion ;
    pdeplot(g_coord/1e3,edge,g_num,'xydata',displ,'mesh','off','contour',
      'on','levels',clevels);
    title([num2str(time/seconds_per_yr),' years after intrusion'])
    xlabel('Distance (km)' )
    ylabel('Depth(km)' )
    colormap('jet')
    axis equal
    drawnow
```

```
    %-------------------------------------------------------------

end

%---------------------------------------------------
% end of time loop
%---------------------------------------------------
```

Suggested Reading

C. M. R. Fowler, *The Solid Earth: An Introduction to Global Geophysics*, Cambridge University Press, Cambridge, 2006.

M. Kaviany, *Essentials of Heat Transfer*, Cambridge University Press, Cambridge, 2011.

R. W. Lewis, P. Nithiarasu, and K. N. Seetharamu, *Fundamentals of the Finite Element Method for Heat and Fluid Flow*, John Wiley & Sons, Ltd, Chichester, 2004.

K. Stüwer, *Geodynamics of the Lithosphere*, Springer, Berlin, 2007.

D. L. Turcotte, and G. Schubert, *Geodynamics*, Cambridge University Press, Cambridge, 2002.

O. C. Zienkiewicz, and R. L. Taylor, *The Finite Element Method, Volume 1, The Basis*, Butterworth-Heinemann, Oxford/Boston, MA 2000.

9

Landscape Evolution

This chapter shows how the finite element method (FEM) can be applied to study topographic evolution in response to the transport of sediments on Earth's surface. Two scripts are listed and discussed in detail. The first investigates evolution of a one-dimensional (1D) river profile cutting into an uplifting block of crust, while the second studies coupling between surface water run-off and topographic incision in 2Ds (plan view). We show that the solutions to these problems are quite different to the "diffuse" heat conduction-type solutions, due to strong spatial variability of the effective diffusion coefficient in the case of topography.

Topography on Earth is controlled by the interaction between climate-mediated surface processes that redistribute mass (e.g., due to the action of rivers and glaciers), and internal processes (i.e., folding, faulting, plutonism, and mantle flow) that deform the surface both vertically and horizontally. These interactions can be treated in a manner analogous to those governing the evolution of temperature considered in Chapter 8, that is, by combining a statement for mass conservation with an empirical function for the mass flux (equivalent to Fourier's law). Mathematically, such a model can be expressed as (e.g., see Smith and Bretherton (1972) and Simpson and Schlunegger (2003))

$$\frac{\partial h}{\partial t} + u\frac{\partial h}{\partial x} + v\frac{\partial h}{\partial y} = -\nabla^T.(\mathbf{n}\,q_s) + w \tag{9.1}$$

where h is the elevation of the topography, u and v are the rates of horizontal tectonic motion in the x- and y-directions, respectively, w is the vertical rate of uplift or subsidence, q_s is the volumetric sediment flux per unit width (with units m^2 s^{-1}), and $\mathbf{n}$ is a unit vector directed down the slope of the surface (i.e., $\mathbf{n} = -\nabla h/S$ where S is the local surface slope defined as $|\nabla h|$). The sediment flux depends on numerous factors, most notably the local slope and the discharge of water flowing over the surface. In situations where sediment transport is controlled by the supply of material rather than detachment processes (e.g., as would be the case in an alluvial river), one can write the sediment flux as

$$q_s = c_0 S + b(q - q_c)^n S^m \tag{9.2}$$

where c_0, b, n, and m are positive constants, S is the local slope, q is the local water discharge, and q_c is a critical discharge below which there is no sediment transport. This relation includes contributions from two terms, a "hillslope" flux that depends linearly on the local slope (term 1) and a "fluvial"

Practical Finite Element Modeling in Earth Science Using Matlab, First Edition. Guy Simpson.

Companion website: www.wiley.com/go/simpson

flux that depends nonlinearly on both the water discharge and the local slope (term 2). The discharge of water flowing over the surface accumulates, for example, due to run-off from rainfall. Here, rather than computing the discharge from the equations governing the conservation of mass and momentum of the surface water (e.g., see Simpson and Castelltort (2006)), we use a simple empirically based power law relationship between the water discharge and the upstream drainage area A, that is,

$$q = kA^p \tag{9.3}$$

where k and p are positive constants. A similar relation can be introduced for the critical discharge

$$q_c = kA_c^p \tag{9.4}$$

where A_c is the critical drainage area below which there is no fluvial sediment transport. Substituting equations 9.2, 9.3, and 9.4 into 9.1 leads to

$$\frac{\partial h}{\partial t} + u\frac{\partial h}{\partial x} + v\frac{\partial h}{\partial y} = \nabla^T . \left[(c_0 + cA_e^r S^{m-1})\nabla h\right] + w \tag{9.5}$$

where $c = bk^n$, $r = pn$, and A_e is the effective drainage area defined as $A_e = A - A_c$ for $A > A_c$ and $A_e = 0$ for $A \leq A_c$. This equation bears a strong resemblance to that governing the evolution of temperature in 2Ds, considered in Chapter 8 (see Equation 8.1). However, in this case, rather than the diffusion coefficient being a constant, it is the function $\kappa = c_0 + cA_e^r S^{m-1}$ that depends nonlinearly on the upstream drainage area, which itself varies strongly in time and space as the drainage network develops. As we will see, this property leads to very different solutions that are strongly localized in space compared to the "diffuse" solutions characteristic of heat conduction problems. In what follows, we present two finite element Matlab programs to investigate the evolution of topography. The first example focuses on the evolution of a 1D uplifting river profile, while the second considers the dissection of an uplifting transport-limited landscape in 2Ds.

9.1 Evolution of a 1D River Profile

Consider a linear river system extending away from a drainage divide located at $x = 0$ (Figure 9.1). We assume that the drainage area increases downstream (e.g., due to the confluence of side rivers) and can be described by a smooth function of the downstream distance from the drainage divide, x. We will also assume that the substratum is not experiencing any horizontal motion, while it is uplifting at a constant rate relative to a base level at $x = lx = 10$ km where the elevation is maintained at zero.

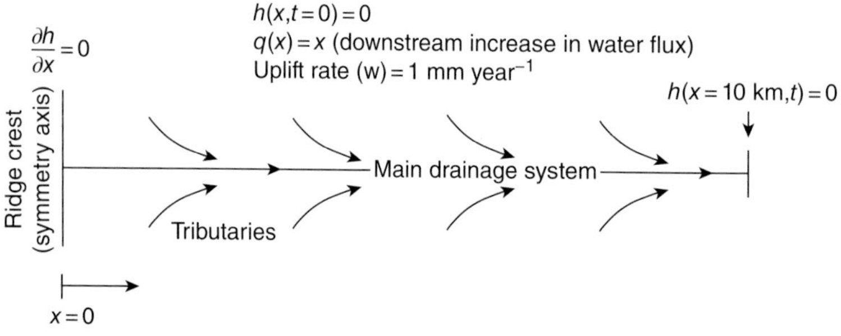

Figure 9.1 Schematic representation (in plan view) of model for topographic evolution of a 1D river profile.

For this setup, the equation governing the evolution of the river profile with time is (obtained from Equation 9.5)

$$\frac{\partial h}{\partial t} = \frac{\partial}{\partial x}\left[(c_0 + cx^r S^{m-1})\frac{\partial h}{\partial x}\right] + w \tag{9.6}$$

This parabolic equation will be solved with an initially flat surface at zero elevation. The boundary conditions considered are a zero flux (zero topographic gradient) at $x = 0$ (i.e., at the drainage divide) and an elevation fixed to zero at $x = lx$. We use the following parameter values: $w = 1$ mm year^{-1}, $c_0 = 10^{-8}$ m^2 s^{-1}, $r = 2$, $m = 1$, and $c = 10^{-13}$ s^{-1}. For these values, the governing equation has a spatially variable diffusion coefficient ($\kappa = c_0 + cx^2$) but is still linear (because it doesn't depend on h). The exact steady-state solution for this case is

$$h(x) = \frac{1}{2}\frac{w\ln(c\,l_x^2 + c_0)}{c} - \frac{1}{2}\frac{w\ln(c\,x^2 + c_0)}{c} \tag{9.7}$$

Applying Galerkin finite element discretization to Equation 9.6, using integration by parts on the second-order spatial derivatives, and neglecting the resulting boundary integral result in the following discrete system for a single element:

$$\mathbf{MM}\,\frac{\partial \mathbf{h}}{\partial t} + [\mathbf{KM}]\,\mathbf{h} = \mathbf{F} \tag{9.8}$$

Here, **h** is the vector of nodal elevations,

$$\mathbf{MM} = \int \mathbf{N}^T\mathbf{N}\,dx \tag{9.9}$$

$$\mathbf{KM} = \int \left(c_0 + c\,x^2\right)\frac{\partial \mathbf{N}}{\partial x}^T\frac{\partial \mathbf{N}}{\partial x}\,dx \tag{9.10}$$

$$\mathbf{F} = \int w\,\mathbf{N}^T\,dx \tag{9.11}$$

and **N** are the shape functions. Discretizing the time derivative using an implicit finite difference approximation and rearranging results in

$$\left(\frac{\mathbf{MM}}{\Delta t} + \mathbf{KM}\right)\mathbf{h}^{n+1} = \frac{\mathbf{MM}}{\Delta t}\,\mathbf{h}^n + \mathbf{F} \tag{9.12}$$

where Δt is the time interval between the n and $n+1$ time levels. The Matlab script used to compute the numerical solution to Equation 9.12 (that is listed in the following text) closely resembles several other 1D heat conduction programs presented earlier (e.g., see Sections 4.6 and 8.1). As the other mentioned programs, the script listed here uses linear two-node elements, a regularly spaced mesh and Gauss–Legendre quadrature with two integration points per element (see Table 4.1). The following points are worthy of special mention:

1) Although the governing equation is linear for the chosen parameter values (notably for $m=1$), nonlinear iterations are nevertheless included to be able to treat nonlinear cases. The general structure of a code with nonlinear iterations is illustrated in Figure 7.13.
2) The erosive "diffusivity" (i.e., $c_0 + cx^r S^{m-1}$) potentially depends nonlinearly on both the downstream distance from the drainage divide x and the local slope S ($= |\partial h/\partial x|$ in 1D). These variables can be computed within the loop over integration points with the following snippet:

```
xk      = fun*g_coord(num)';             % x coord. at int. pt. k
S       = abs(deriv*displ(num));         % topographic slope
kappa   = c0 + c*xk^rexp*S^(mexp-1) ;   % effective diffusivity
```

Here, `fun` are the shape functions for the current integration point (that are used as interpolants), `deriv` are the derivatives of the shape functions expressed in terms of global coordinates, `displ(num)` is the column vector containing values of the topography at the two nodes of the current element, and `g_coord(num)'` are the nodal coordinates of the current element.

3) The zero flux boundary condition at $x = 0$ (i.e., at global node number 1) is automatically satisfied by making no modification to either the global matrix or the global right-hand-side vector. The Dirichlet boundary condition at $x = lx$ (i.e., the last global node (nn) in the mesh) is treated using the approach presented in Section 2.7.

```
%--------------------------------------------------------------
% Program erosion1d.m
% FEM solution of 1D erosion problem
%--------------------------------------------------------------

clear % clear memory from current workspace
seconds_per_yr = 60*60*24*365; % number of seconds in one year

% physical parameters
lx     = 10000 ;          % length of spatial domain (m)
w      = 1e-3/seconds_per_yr; % uplift source term (m/s)
c0     = 1e-8   ;  % linear 'hillslope' erosion coefficient (m^2/s)
rexp   = 2   ;     % fluvial erosion exponent (discharge term)
mexp   = 1   ;     % fluvial erosion exponent (slope term)
c      = 1e-13 ;   % fluvial erosion coefficient(m^(2-rexp)/s)

% numerical parameters
dt      = 1000*seconds_per_yr ; % time step (s)
tolerance = 1e-3 ; % error tolerance
ntime  = 5000 ;     % number of time steps
nels   = 50 ;       % total number of elements
nod    = 2 ;        % number of nodes per element
nn     = nels+1 ;   % total number of nodes
dx     = lx/nels ;  % element size
nip    = 2 ;        % number of integration points

g_coord = [0:dx:lx] ; % spatial domain (1-D mesh)

% integration data
points    =   [-sqrt(1/3) sqrt(1/3)] ; % positions of the Gauss points
wts       =   [1  1] ;                 % Gauss-Legendre weights

% shape functions and their derivatives (both in local coordinates)
% evaluated at the Gauss integration points
for k=1:nip
   fun_s(k,:) = [ (1-points(k))/2 (1+points(k))/2 ] ; % N1 and N2
   der_s(k,:) = [ -1/2 1/2 ] ;                        % dN1/xi  dN2/xi
end

% Dirichlet boundary conditions
 bcdof = [ nn ]    ; % boundary nodes
 bcval = [ 0  ]    ; % boundary values

% define connectivity and equation numbering
g_num       = zeros(nod,nels) ;
g_num(1,:)  = [1:nn-1]   ;
g_num(2,:)  = [2:nn]     ;
```

```
% initialise matrices and vectors
ff      = zeros(nn,1);        % system load vector
b       = zeros(nn,1);        % system rhs vector
lhs     = sparse(nn,nn);      % system lhs matrix
rhs     = sparse(nn,nn);      % system rhs matrix
displ   = zeros(nn,1);        % initial solution
displ0 = displ ;              % solution at n time level

%-------------------------------------------------
% time loop
%-------------------------------------------------
time  = 0 ; % initial time

for n=1:ntime
  time  = time + dt ; % update time
  iters = 0 ;    % initialise iteration counter
  error = 1 ;    % set initial error to an arbitrary high value
  while error>tolerance % start of nonlinear iterations
  iters = iters + 1 ; % iteration counter

%-------------------------------------------------
% Matrix assembly
ff      = zeros(nn,1);          % system load vector
lhs     = sparse(nn,nn);        % system lhs matrix
rhs     = sparse(nn,nn);        % system rhs matrix
   for iel=1:nels % loop over all elements
      num    = g_num(:,iel) ;              % retrieve equation number
      dx     = abs(diff(g_coord(num))) ; % length of element
      MM     = zeros(nod,nod) ;  % reinitialise MM
      KM     = zeros(nod,nod) ;  % reinitialise KM
      F      = zeros(nod,1)    ;  % reinitialise F
      for k=1:nip % integrate element matrices
        fun = fun_s(k,:)   ; % retrieve shape functions (sf)
        der = der_s(k,:)   ; % retrieve shape fun. der.
        detjac = dx/2      ; % det. of the Jacobian
        invjac = 2/dx      ; % inv of the Jacobian
        deriv  = der*invjac ; % sf der. in global coordinates
        xk     = fun*g_coord(num)'; % x coord. at int. pt. k
        S      = abs(deriv*displ(num)); % topographic slope
        kappa  = c0 + c*xk^rexp*S^(mexp-1) ; % diffusivity
        dwt    = detjac*wts(k) ;
        MM     = MM + fun'*fun*dwt ; % mass matrix
        KM     = KM + deriv'*kappa*deriv*dwt ; % stiffness matrix
        F      = F  + w*fun'*dwt ; % load vector
      end % end of integration
      lhs(num,num) = lhs(num,num) + MM/dt + KM ; % assemble lhs
      rhs(num,num) = rhs(num,num) + MM/dt      ; % assemble rhs
      ff(num)      = ff(num)      + F  ; % assemble load vector
   end  % end of element loop

%---------------------------------------------------------------------

   b  = rhs*displ0 + ff ;            % form rhs vector

   % impose boundary conditions
   lhs(bcdof,:) = 0 ;                % zero the relevent equations
   tmp = spdiags(lhs,0) ;            % store diagonal
   tmp(bcdof)=1 ;                    % place 1 on stored-diagonal
```

```
      lhs=spdiags(tmp,0,lhs);              % reinsert diagonal
      b(bcdof) = bcval ;                   % set rhs vector

      displ_tmp = displ;% save a copy of solution from prev. iteration
      displ = lhs \ b ;                  % solve system of equations
      error = max(abs(displ-displ_tmp))/max(abs(displ)) % check error

   end % end of nonlinear iterations

      displ0 = displ ; % save solution from the last time step

     % visualisation
     if mod(n,100)==0
       figure(1)
       plot(g_coord,displ,'o-')
       title(['Time = ',num2str(time/seconds_per_yr),' years'])
       xlabel('Distance away from drainage divide (m)')
       ylabel('Surface height (m)')
       hold off
       drawnow
     end

   end % end of time loop

   %-----------------------------------------------------------------------
```

Figure 9.2 shows an example of numerical results computed with the given script using 50 finite elements and a time step of 1000 years. With increasing time, the "river profile" is seen to develop a characteristic concave upward form due to the combination of tectonic uplift and the nonlinear

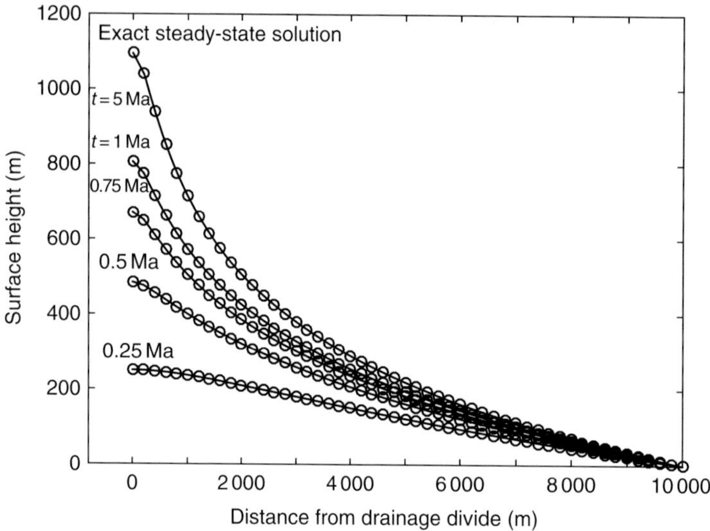

Figure 9.2 Numerical solution of Equation 9.6 with constant rock uplift rate, zero sediment flux at the left-hand boundary and right-hand boundary with elevation fixed to zero. The upper curve shows the steady-state solution, computed from Equation 9.7 superimposed on the numerical solution after 5 Ma (circles). Numerical results were computed with the following parameter values: $w = 1$ mm year^{-1}, $c_0 = 10^{-8}$ m^2 s^{-1}, $r = 2$, $m = 1$, $c = 10^{-13}$ s^{-1}, $dx = 200$ m, and $\Delta t = 1000$ years.

increase in diffusivity in the downstream direction. The numerical solution after 5 Ma is seen to be in good agreement with the exact steady-state solution computed using Equation 9.7.

9.2 Evolution of a Fluvially Dissected Landscape

Although 1D models such as that just presented are useful to study topographic evolution in certain simplified situations, erosion is a process that localizes strongly in space, leading to complex incised patterns that can only be treated with 2D models. In this section, we solve the full 2D surface evolution model (i.e., Equation 9.5) without lateral tectonic advection, that is,

$$\frac{\partial h}{\partial t} = \nabla^T . \left[(c_0 + cA_e^r S^{m-1}) \nabla h \right] + w \tag{9.13}$$

where h is the surface elevation, c_0 is the "hillslope" diffusivity, c is a fluvial transport coefficient, A_e is the effective upstream drainage area (defined as $A_e = A - A_c$ where A is the upstream drainage area and A_c is a critical drainage area below which there is no fluvial sediment transport), r and m are positive exponents, and w is the rate of rock uplift (assumed to be uniform and constant). This equation will be solved on a square 50×50 km domain that has an initial random elevation of between 0 and 10 m, while the elevation of all four boundaries is fixed to zero. Here, we use linear three-node triangular elements and generate the mesh using the Matlab pdetool GUI (as described in Sections 7.1 and 8.2).

The discretized form of Equation 9.13 has exactly the same form as for the 1D problem,

$$\left(\frac{\mathbf{MM}}{\Delta t} + \mathbf{KM} \right) \mathbf{h}^{n+1} = \frac{\mathbf{MM}}{\Delta t} \mathbf{h}^n + \mathbf{F} \tag{9.14}$$

whereas the element matrices are now defined as

$$\mathbf{MM} = \int\int \mathbf{N}^T \mathbf{N}\, dx\, dy \tag{9.15}$$

$$\mathbf{KM} = \int\int \left(c_0 + cA_e^r S^{m-1} \right) (\nabla \mathbf{N})^T \nabla \mathbf{N}\, dx\, dy \tag{9.16}$$

$$\mathbf{F} = \int\int w \mathbf{N}^T\, dx\, dy \tag{9.17}$$

where $\mathbf{h}$ is the vector of nodal elevations for an element, $\mathbf{N}$ are the shape functions, $\nabla\mathbf{N}$ are the shape function derivatives expressed in terms of physical coordinates, and Δt in the time increment. These element matrices are evaluated using Gauss–Legendre quadrature on triangular elements, as described in Section 7.1 (see also Section 8.2).

The upstream drainage area A appearing in the element matrix $\mathbf{KM}$ (recall that $A_e = A - A_c$) is computed according to the following scheme (Figure 9.3):

1) Calculate the surface area of each finite element. In the program listed in the following text, the result is saved in the array `area`.
2) Find and save the three elements adjacent to each element, saved in the array `gnei`. Elements on the boundary that have only two neighbors are assigned their own element index to the missing third neighbor. If the mesh doesn't change through time, this step and the previous one need to be performed only once before the time loop.
3) At each step in time, calculate the average elevation of each element, based on the three corner nodes. The result is saved in the array `zc`.

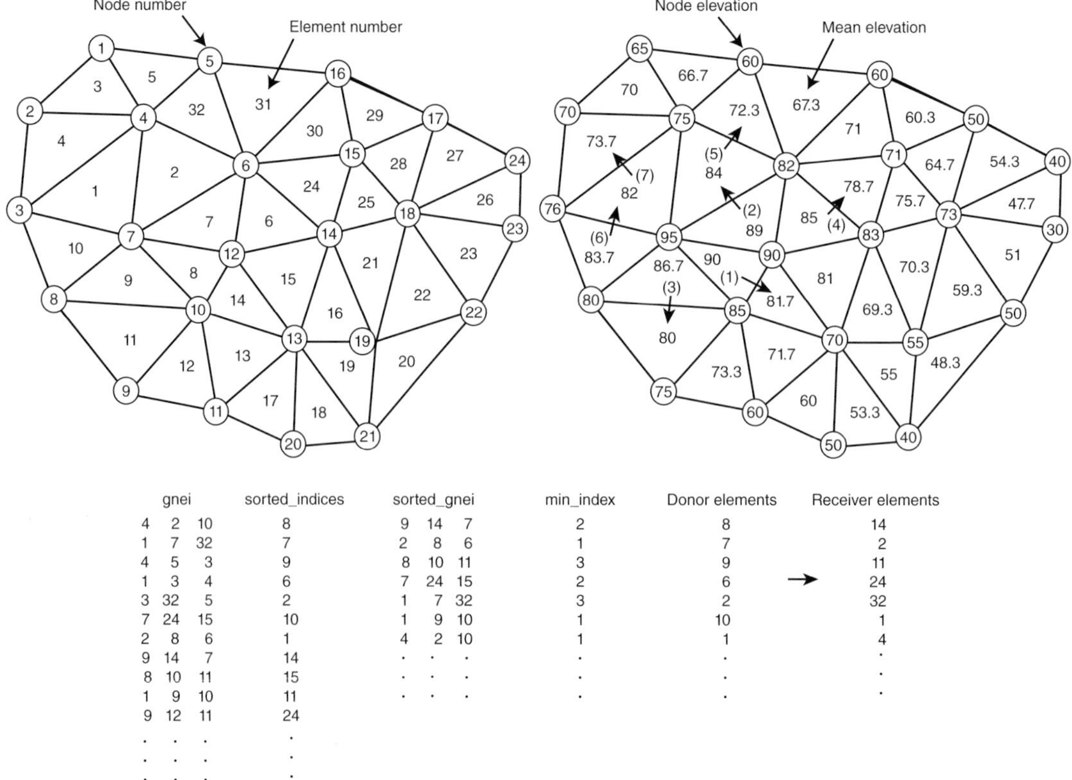

Figure 9.3 Illustration of drainage area calculation algorithm for a small triangular mesh. The mesh on the left shows global node (encircled) and element numbers. The mesh on the right shows node elevations (encircled) and element elevations computed as the average of the three triangle vertices. The arrows show the order (indicated with a bracketed number) and direction in which the cumulative drainage area is "passed" down the network (from donor to receiver) for part of the mesh. The Matlab variable `gnei` stores the three element neighbors for each element. The vector `sorted_indices` contains a list of elements, sorted on the basis of elevation, from highest to lowest. The array `sorted_gnei` contains the element neighbors, sorted according to the element list in `sorted_indices`. The vector `min_index` stores the local index (1, 2, or 3) of the lowest of the three neighboring elements for each element. The cumulative drainage area is redistributed progressively down the network from each (donor) element to its lowest neighbor (receiver), from the highest to the lowest element of the mesh.

4) Sort the average element elevation for the entire mesh from highest to lowest. The indices of the sorted elevations are saved in the array `sorted_indices`.
5) Using the sorted indices, sort the neighboring elements, the result of which is saved in `sorted_gnei`. At this stage, all elements in the mesh have been ordered, along with their three adjacent neighbors, from highest to lowest.
6) For each element, find and save the local index of the lowest of the three adjacent elements. The result (i.e., 1, 2, or 3) is saved in the array `min_index`.
7) Set the upstream drainage area (`A`) initially to the element area.
8) In a loop over all ordered elements from highest to lowest, "pass" the accumulated drainage area from each (donor) element to its lowest adjacent neighbor (receiver). The result after the loop has

been completed is the accumulated surface area that "drains" to each element in the landscape, A. Note that this procedure implicitly assumes a spatially uniform rainfall, which can easily be accounted for if desired.

Apart from this drainage area calculation, the program listed in the following text to compute the numerical solution to 9.13 bears a strong resemblance to other scripts previously presented (e.g., see Sections 5.6, 7.1, and 8.2). One noteworthy point is that the mesh used by the listed program needs to be provided from a .mat file.[1] As already mentioned, this mesh can easily be generated and saved for later use using the Matlab pdetool GUI (see Sections 7.1 and 8.2). A second remark is that in the listed program, mass lumping (or diagonalization) has been applied to the element mass matrix **MM**, which is performed to suppress numerical oscillations that can arise if the diffusivity in **KM** becomes too small. While there are no strict rules on how to perform mass lumping, here it is done by summing each column of **MM** separately, putting the result on the diagonal, and zeroing out all off-diagonal terms as follows:

```
MM = diag(sum(MM)); % lump the mass matrix
```

The reader is referred to Zienkiewicz and Taylor (2000) for more information regarding this practice.

Figure 9.4 show an example of some numerical results calculated with the program. The results show the development of a branching drainage network that progressively invades the uplifting topography from the lateral boundaries. Note the difference between this solution and that typical of 2D heat conduction investigated in Section 8.2, even though both are essentially governed by diffusion equations. This difference here is mainly due to the coupling between the diffusion coefficient and the upstream drainage area, which varies strongly in space and time.

```
%-----------------------------------------
% Program: erosion2d.m
% FEM solution of 2D erosion with diffusivity dependent
% on upstream drainage area and local slope
% 3-node triangles - mesh generated externally with pdetool
%-----------------------------------------

clear

seconds_per_yr = 60*60*24*365 ; % seconds in 1 year

% physical parameters
lx     = 50e3    ;  % x extent of domain (m)
ly     = 50e3    ;  % y extent of domain (m)
rexp   = 2 ;        % drainage area exponent
mexp   = 1 ;        % slope exponent
w      = 1e-3/seconds_per_yr ;  % rock uplift rate (m/s)
vx     = 0 ;                    % horizontal x velocity (m/s)
vy     = 0 ;                    % horizontal y velocity (m/s)
c      = 1e-11/seconds_per_yr ; % fluvial erosion coeff. (m^(2-2rexp)/s)
c0     = 0.1/seconds_per_yr ;   % linear diffusion coeff.(m^2/s)
Ac     = 0  ;  % critical drainage area (m^2)
scale = 10  ;  % max. amplitude random initial topography (m)

% numerical parameters
ntime = 3000 ;    % number of time steps
ndim  = 2    ;    % number of spatial dimensions
nod   = 3    ;    % number of nodes in 1 element
```

1 This file can be obtained from the author on request.

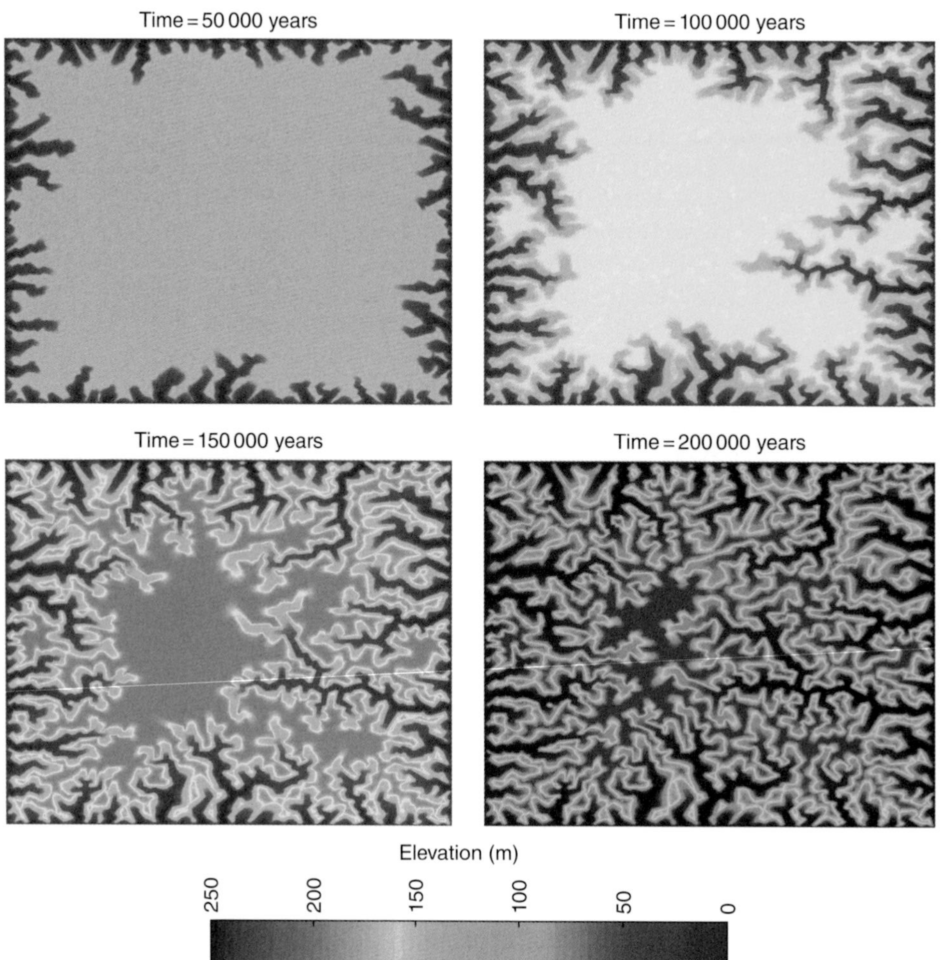

Figure 9.4 Topography after four different times based on the numerical solution of Equation 9.13 with a constant rock uplift rate and zero elevation prescribed on all boundaries. Results were computed with the script listed in Section 9.2 using a total of 19 968 linear triangular finite elements. Parameter values used were $w = 1$ mm year^{-1}, $c_0 = 3.171 \times 10^{-9}$ m^2 s^{-1}, $r = 2$, $m = 1$, $c = 3.171 \times 10^{-19}$ s^{-1}, and $\Delta t = 100$ years.

```
nip    = 3   ;      % number of Gauss integration points
dt     = 100*seconds_per_yr ; % time step (s)
tolerance = 1e-3 ; % iteration tolerance

% load triangular mesh (e.g., generated using pdetool)
% the following variables are loaded
% g_coord - mesh nodes
% edge    - mesh edge data
% g_num   - node-element connectivity
load('mesh_erosion2d')

nn   = length(g_coord)    ;  % total number of nodes
nels = length(g_num)      ;  % total number of elements
```

```
figure(10), clf  % visualise mesh
pdemesh(g_coord,edge,g_num) ; axis equal

% find boundary nodes
bx0 = find(g_coord(1,:)==0) ;  % nodes on x=0 boundary
bxn = find(g_coord(1,:)==lx) ; % nodes on x=lx boundary
by0 = find(g_coord(2,:)==0) ;  % nodes on y=0 boundary
byn = find(g_coord(2,:)==ly) ; % nodes on y=ly boundary

% specify Dirichlet boundary conditions
bcdof = [ by0 byn bx0 bxn ]       ; % nodes to be fixed
bcval = [ zeros(1,length(bcdof)) ]; % fixed values

% gauss integration data
points = zeros(nip,ndim); % location of points
points(1,1)=0.5   ; points(1,2)=0.5 ;
points(2,1)=0.5   ; points(2,2)=0   ;
points(3,1)=0     ; points(3,2)=0.5 ;
c   = 0.5 ; % triangle factor
wts = c*1/3*ones(1,nip) ; % weights

% save shape functions and their derivatives in local coordinates
% evaluated at integration points
for k=1:nip
   L1=points(k,1); L2=points(k,2); L3=1.-L1-L2   ;
   fun = [L1 L2 L3] ;
   fun_s(k,:) = fun ; % shape functions
   der(1,1)=1 ; der(1,2)=0 ; der(1,3)=-1   ;
   der(2,1)=0 ; der(2,2)=1 ; der(2,3)=-1   ;
   der_s(:,:,k) = der ; % derivative of shape function
end

% establish 3 adjacent neighbours for each element
% and compute element area
gnei = zeros(3,nels);
area = zeros(1,nels);
for iel=1:nels
  num  =  g_num(:,iel);
  numc = [num ;  num(1)] ;
  coord = g_coord(:,num)';
  area(iel) = polyarea(coord(:,1),coord(:,2));
  for i=1:3 % loop over the three edges of an element
     p1 = find(g_num(1,:)==numc(i)|g_num(2,:)==numc(i)| ...
               g_num(3,:)==numc(i)) ;
     p2 = find(g_num(1,:)==numc(i+1)|g_num(2,:)==numc(i+1)| ...
              g_num(3,:)==numc(i+1)) ;
     nei = intersect(p1,p2);
     n   = find(nei≠iel)   ;
     if n>0
         gnei(i,iel)=nei(n);
     else
         gnei(i,iel)=iel;
     end
  end
end

% initialise arrays
ff     = zeros(nn,1);      % global load vector
```

```
b      = zeros(nn,1);      % global rhs vector
lhs    = sparse(nn,nn);    % global lhs matrix
rhs    = sparse(nn,nn);    % global rhs matrix

% initial conditions
displ  = rand(nn,1)*scale ;
displ0 = displ ;
%-------------------------------------------------------
% time loop
%------------------------------------------------------
time = 0 ; % initialise time

for n=1:ntime
  n
  time = time + dt ;  % update time

%-----------------------------------------
% Compute upstream drainage area (A)
%----------------------------------------

% elevation of triangle centroids
zc = 1/3*(displ(g_num(1,:))+displ(g_num(2,:))+displ(g_num(3,:)))' ;

% sort elements from highest to lowest
% save the indices required to achieve the sort
[szc,sort_indices] = sort(-zc) ;

% sort the 3 neighbours
sorted_gnei(:,:) = gnei(:,sort_indices) ;

% elevations of the sorted neighbours
szcmap(1,:) = zc(sorted_gnei(1,:)) ;
szcmap(2,:) = zc(sorted_gnei(2,:)) ;
szcmap(3,:) = zc(sorted_gnei(3,:)) ;

% compute minima of the 3 neighbouring elements
% save the index (1, 2 or 3) of each minima
[mv,min_index] = min(szcmap) ;

% pass drainage area down network
A = area ;
for iel=1:nels
   donner       = sort_indices(iel) ;
   receiver     = sorted_gnei(min_index(iel),iel) ;
   A(receiver) = A(receiver) + A(donner) ;
end

%------------------------------------------------------
% nonlinear iterations
error = 1 ; % initial error, set to arbitrary large value
iters = 0 ; % initialise iteration counter
while error > tolerance
  iters = iters + 1  % update iteration counter
%------------------------------------------------------
% matrix integration and assembly
%------------------------------------------------------
  lhs    = sparse(nn,nn);    % reinitialise global lhs matrix
  rhs    = sparse(nn,nn);    % reinitialise global rhs matrix
  ff     = zeros(nn,1);      % reinitialise global load vector
```

```
  b      = zeros(nn,1);      % reinitialise global rhs vector
for iel=1:nels  % sum over elements
  num     = g_num(:,iel)      ; % element nodes
  coord   = g_coord(:,num)' ; % element coordinates
  KM      = zeros(nod,nod)    ; % initialisation
  MM      = zeros(nod,nod)    ;
  CM      = zeros(nod,nod)    ;
  F       = zeros(nod,1)      ;
  for k = 1:nip % integration loop
    fun     = fun_s(k,:) ;    % shape functions
    der     = der_s(:,:,k) ; % der. of shape functions
    jac     = der*coord  ;    % jacobian matrix
    detjac = det(jac) ;       % det. of jacobian
    invjac = inv(jac) ;       % inv. of jacobian
    deriv  = invjac*der ;     % der. of shape fun. in physical coords.
    dwt    = detjac*wts(k) ;
    grads  = deriv*displ(num) ; % topo. gradient
    S      = sqrt(grads(1)^2+grads(2)^2); % slope
    if A(iel)>Ac % diffusivity
    kappa  = c0 + c*(A(iel)-Ac)^rexp*S^(mexp-1);
    else
    kappa  = c0 ;
    end
    KM  = KM + deriv'*kappa*deriv*dwt ;       % stiffness matrix
    MM  = MM + fun'*fun*dwt ;                 % mass matrix
    CM  = CM + fun'*(vx*deriv(1,:)+vy*deriv(2,:))*dwt; % adv matrix
    F   = F  + w*fun'*dwt ;                   % load vector
  end
    MM = diag(sum(MM)); % lump the mass matrix
  % assemble global matrices
  lhs(num,num)  = lhs(num,num)  + MM/dt  + KM   + CM ;
  rhs(num,num)  = rhs(num,num)  + MM/dt        ;
  ff(num)       = ff(num)  + F;
end
%-----------------------------------------------

   b = rhs*displ0 + ff ;               % form rhs global vector

   % impose boundary conditions
   lhs(bcdof,:) = 0 ;                  % zero the boundary equations
   tmp = spdiags(lhs,0) ;              % store diagonal
   tmp(bcdof)=1 ;                      % place 1 on stored-diagonal
   lhs=spdiags(tmp,0,lhs);             % reinsert diagonal
   b(bcdof) = bcval ;                  % set boundary values

   displ_tmp = displ ;            % save solution from last iteration
   displ = lhs \ b ;              % solve system of equations

   error = max(abs(displ-displ_tmp))/max(abs(displ))

end % end of nonlinear iterations
%-------------------------------------------

   displ0 = displ; % save solution from this time step

   %-----------------------------------------
   % visualisation
    figure(1) , clf % Topography
    pdeplot(g_coord,edge,g_num,'xydata',displ,'mesh','off','contour','off');
```

```
    title(['Topography in meters (after ',num2str(time/seconds_per_yr),' years)'])
    colormap('jet')
    colorbar

    figure(2) % Drainage area
    pdeplot(g_coord,edge,g_num,'xydata',log10(A),'contour','off') ;
    colormap('jet')
    title('Log10 of Drainage area (m^2)')
    drawnow
   %-------------------------------------

end

%-----------------------------------------------
% end of time loop
%-----------------------------------------------
```

Suggested Reading

P. A. Allen, *Earth Surface Processes*, John Wiley & Sons, Inc., Hoboken, NJ, 2009.

P. A. Allen, and J. R. Allen, *Basin Analysis*, Wiley Blackwell, Oxford, 2013.

P. R. Bierman, and D. R. Montgomery, *Key Concepts in Geomorphology*, Freeman, New York, 2014.

D. W. Burbank, and R. S. Anderson, *Tectonic Geomorphology*, Wiley-Blackwell, Oxford, 2011.

R. J. Huggett, *Fundamentals of Geomorphology*, Routledge, London, 2007.

J. D. Pelletier, *Quantitative Modeling of Earth Surface Processes*, Cambridge University Press, Cambridge, 2008.

10

Fluid Flow in Porous Media

This chapter treats the important topic of modeling fluid flow in porous media using the finite element method (FEM). Two applications are investigated. The first concerns fluid flow around a fault, which is a problem very similar to the previously presented parabolic and/or elliptic problems. The second application involves interaction between two moving miscible fluids, which is a coupled problem that we solve with a variant of the Galerkin FEM (known as the Petrov–Galerkin FEM) that is effective in solving advection-dominated hyperbolic problems.

Most rocks in the upper few kilometers of the crust contain pore space that is filled with fluid (e.g., water, gas, and oil). These fluids exert a pressure on the enclosing solid skeleton that modulates the stress state and, therefore, influences the ability to deform the rocks. In the presence of nonequilibrium pore pressure gradients (i.e., when the gradient differs from the hydrostatic gradient), the fluids flow through the porous matrix, potentially transporting mass and heat and altering the properties of the solid skeleton. These phenomena are important in numerous different contexts, including the formation of mineral deposits, petroleum migration and extraction, groundwater exploitation, and geothermal energy (e.g., see Phillips (1991) and Ingebritsen et al. (2006)).

The simplest mathematical model for flow in porous media considers that the fluid has no influence on the solid skeleton as it flows passively through the available pore space. In this case, the equations governing the evolution of the fluid pressure and fluid flow velocities are (e.g., see Walder and Nur (1984))

$$\phi\beta\frac{\partial p}{\partial t} = -\nabla.\mathbf{u} + H \tag{10.1}$$

$$\mathbf{u} = -\frac{k}{\mu}\left(\nabla p - \rho_f g\mathbf{e}\right) \tag{10.2}$$

where p is the fluid pressure, $\mathbf{u}$ is the fluid velocity vector, ϕ is the porosity, β is the bulk compressibility (Pa^{-1}), k is the permeability (m^2), μ is the fluid viscosity (Pa s), ρ_f is the water density, g is acceleration due to gravity, $\mathbf{e}$ is a unit vector oriented in the vertical direction, and H accounts for any fluid pressure sources or sinks (e.g., due to devolatilization reactions). Equation 10.1 is a statement for the conservation of mass for the pore fluid, while Equation 10.2 is known as Darcy's law. Introducing the excess pore fluid pressure defined as the pressure in excess of the hydrostatic pressure (i.e., $p_e = p - p_h$ where $p_h = \rho_f gz$) and substituting 10.2 into 10.1 lead to a single parabolic equation for the excess fluid pressure as follows:

$$\phi\beta\frac{\partial p_e}{\partial t} = \nabla.\left(\frac{k}{\mu}\nabla p_e\right) + H \tag{10.3}$$

Practical Finite Element Modeling in Earth Science Using Matlab, First Edition. Guy Simpson.

Companion website: www.wiley.com/go/simpson

Note the close similarity between this diffusion equation and those considered in Chapters 8 and 9 in the context of surface evolution (Chapter 9) and heat transfer (Chapter 8). Equation 10.3 shows that fluid flow is driven by gradients in excess pressure, while it is resisted by k/μ. In the following text, we obtain a numerical solution to this equation in two spatial dimensions using the FEM.

The equation governing the fluid pressure evolution becomes more complicated in situations where the pore fluid exchanges heat, momentum, and/or mass either with the solid skeleton or with another fluid phase. When this occurs, the pore pressure evolution becomes coupled to the evolution of other variable(s) (e.g., temperature and concentration of a dissolved species). For such problems, one has to solve simultaneously for more than one unknown. As an example, consider coupling between a flowing fluid and the temperature, which is described mathematically by the following equations (e.g., see Turcotte and Schubert (1982)):

$$\nabla \cdot \mathbf{u} = 0 \tag{10.4}$$

$$\mathbf{u} = -\frac{k}{\mu}\left(\nabla p - \rho_f g \mathbf{e}\right) \tag{10.5}$$

$$\rho_b c_b \frac{\partial T}{\partial t} + \rho_f c_f \nabla.(\mathbf{u}\ T) = \nabla.(k_b \nabla T) \tag{10.6}$$

$$\rho_f = \rho_o(1 - \alpha\ T) \tag{10.7}$$

Here, T is the temperature (assumed to be the same in the fluid and solid matrix), p is the fluid pressure, $\mathbf{u}$ is the vector of Darcy velocities, ρ_b is the bulk density (i.e., an average of the rock and fluid in the pore space), ρ_f is density of the pore fluid, ρ_o is density of the pore fluid at a reference temperature (0°C), c_b is the bulk heat capacity, c_f is the heat capacity of the fluid, α is the thermal expansion coefficient of the fluid, k_b is the bulk heat conductivity, k is the permeability, μ is the fluid viscosity, and g is the acceleration due to gravity. Equations 10.4 and 10.5 are similar to the equations for mass conversation and Darcy's law introduced earlier except now the flow is assumed incompressible so the time derivative is absent. Equation 10.6 is a statement of energy balance for the porous media, while Equation 10.7 is an equation of state for the fluid density. This model is coupled in two respects. First, the flowing fluid that is driven by gradients in p (via Equation 10.5) advects heat, as seen by the advection term (term 2) in 10.6. Second, the fluid density, which is one of the factors driving flow (see Equation 10.5), depends on the temperature (via Equation 10.7). Under certain circumstances, this coupling can lead to convection (e.g., see Turcotte and Schubert (1982)). In the following text, we show how a coupled problem similar to this one is solved within the framework of the FEM using the approach outlined in Section 7.3. More details on how the FEM can be applied to problems involving coupling between fluid flow and deformation can be found in Lewis and Schrefler (1998) and Lewis et al. (2004).

10.1 Fluid Flow Around a Fault

To illustrate how to solve a basic uncoupled porous flow problem using the FEM, we consider the following simple scenario. A porous crust with an initially uniform and isotropic permeability has an overpressure that increases linearly with depth. At time = 0 a dipping, high permeability fault zone is suddenly introduced into the crust (Figure 10.1). Our task is to compute how the fault zone influences fluid flow in the surrounding crust in a vertical two-dimensional (2D) section. The transient evolution of the excess pore pressure is governed by

$$\phi\beta\frac{\partial p_e}{\partial t} = \nabla \cdot \left(\frac{k(x,z)}{\mu}\nabla p_e\right) \tag{10.8}$$

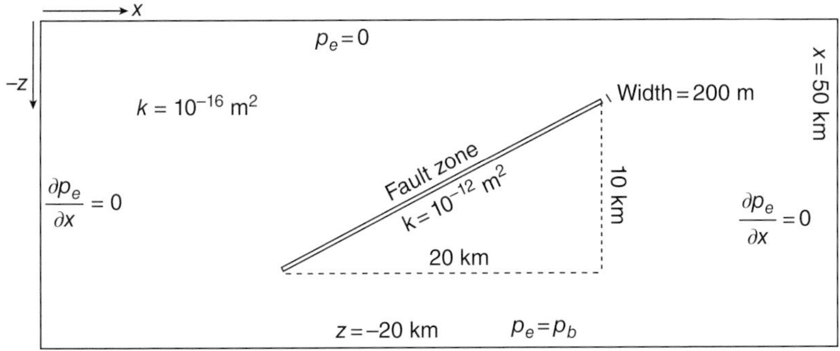

Figure 10.1 Setup for model of fluid flow around a high permeability fault zone.

subject to the boundary conditions

$$\begin{aligned} &p_e = 0 \quad \text{at} \quad z = 0 \\ &p_e = p_b \quad \text{at} \quad z = -20 \text{ km} \\ &\frac{\partial p_e}{\partial x} = 0 \quad \text{at} \quad x = 0 \quad \text{and} \quad x = 50 \text{ km} \end{aligned} \tag{10.9}$$

and the initial condition

$$p_e = -\frac{z p_b}{lz} \tag{10.10}$$

where lz is the vertical extent of the model domain (=20 km), z is the vertical coordinate (negative downward), and p_b is the positive fluid overpressure at the base of the model, which is taken to be equivalent to a fluid pressure that is 60% of the total lithostatic load.

Applying a standard Galerkin finite element discretization to 10.8 along with integration by parts on the spatial derivatives (and neglecting the resulting boundary integrals) and finite difference discretization of the time derivative results in the following discrete equation for a single element:

$$\left(\frac{\mathbf{MM}}{\Delta t} + \mathbf{KM}\right)\mathbf{p}^{n+1} = \frac{\mathbf{MM}}{\Delta t}\mathbf{p}^{n} \tag{10.11}$$

Here, the mass and stiffness element matrices are defined, respectively, as

$$\mathbf{MM} = \int\int \beta\phi\, \mathbf{N}^T\mathbf{N}\, dxdz \tag{10.12}$$

and

$$\mathbf{KM} = \int\int \frac{k}{\mu}\, \nabla\mathbf{N}^T\nabla\mathbf{N}\, dxdz \tag{10.13}$$

and Δt is the time interval between n and $n+1$. Recall that $\mathbf{p}$ is the vector of nodal excess fluid pressures and $\mathbf{N}$ are the shape functions. Discretization is performed here using linear (three-node) triangles, the shape functions for which are listed in Section 7.1. The mesh is generated using the pdetool GUI and consists of two rectangular polygons, one with the dimensions of the bounding model domain that encloses a second 200-m-wide and 22-km-long polygon tilted at an angle of 26°, which represents the fault zone (see Figure 10.1). Apart from these features, the Matlab script listed in the following text closely resembles several other programs presented earlier (e.g., see Sections 5.6, 7.1, 8.2, and 9.2).

```
%---------------------------------------
% Program: porousflow2d
% 2D FEM - diffusion equation
% 3-node triangles - Mesh generated externally with pdetool
%---------------------------------------

clear

seconds_per_yr = 60*60*24*365 ; % seconds in 1 year

% physical parameters
lx        = 50e3   ;  % width of domain (m)
lz        = 20e3   ;  % depth of domain (m)
mu        = 1.33e-4 ; % fluid viscosity (Pa s)
perm      = 1e-16   ; % ambient permeability (m^2)
perm_max = 1e-12 ; % fault permeability (m^2)
kappav    =  [perm/mu perm_max/mu];
rho       = 2700 ;  % rock density (km/m^3)
rhof      = 1000;   % water density (km/m^3)
g         = 9.8  ;  % acceleration due to gravity (m/s^2)
beta      = 1e-10 ; % bulk compresibility (1/Pa)
phi       = 0.1 ;   % porosity
lambda    = 0.6 ;   % pore pressure ratio
Pb        = lambda*rho*g*lz ;% fixed fluid pressure on base (Pa)
Ph        = rhof*g*lz ;      % hydrostatic fluid pressure on base (Pa)
Peb       = Pb - Ph ;        % excess overpressure on base (Pa)

% numerical parameters
ndim  = 2    ;    % number of spatial dimensions
nod   = 3    ;    % number of nodes in 1 element
nip   = 3   ;     % number of Gauss integration points
ntime = 100 ;     % number of time steps performed
dt    = 0.1*seconds_per_yr ; % Time step (s)

% load triangular mesh (e.g., generated using pdetool)
% the following variables are loaded:
% g_coord - mesh nodes
% edge    - mesh edge data
% g_num   - node-element connectivity
% phase   - index (1 or 2)
load mesh_flow2d
nn    = length(g_coord) ;  % total number of nodes
nels  = length(g_num)   ;  % total number of elements

figure(10) , clf % visualise the mesh
pdemesh(g_coord,edge,g_num) ; axis equal

% Find boundary nodes
bx0 = find(g_coord(1,:)==0) ;  % nodes on x=0 boundary
bxn = find(g_coord(1,:)==lx) ; % nodes on x=lx boundary
bz0 = find(g_coord(2,:)==0) ;  % nodes on z=0 boundary
bzn = find(g_coord(2,:)==-lz) ;% nodes on z=-lz boundary

% define fixed boundary conditions
bcdof = [bz0 bzn ] ;  % boundary nodes
bcval = [zeros(1,length(bz0)) Peb*ones(1,length(bzn))] ; % values

% gauss integration data
points = zeros(nip,ndim); % location of points
```

```
points(1,1)=0.5  ; points(1,2)=0.5 ;
points(2,1)=0.5  ; points(2,2)=0   ;
points(3,1)=0    ; points(3,2)=0.5 ;
c   = 0.5 ; % triangle factor
wts = c*1/3*ones(1,nip) ; % weights

% save shape functions and their derivatives in local coordinates
% evaluated at integration points
for k=1:nip
   L1=points(k,1); L2=points(k,2); L3=1.-L1-L2  ;
   fun = [L1 L2 L3] ;
   fun_s(k,:) = fun ; % shape functions
   der(1,1)=1 ; der(1,2)=0 ; der(1,3)=-1  ;
   der(2,1)=0 ; der(2,2)=1 ; der(2,3)=-1  ;
   der_s(:,:,k) = der ; % derivative of shape function
end

% initialise arrays
ff     = zeros(nn,1);      % global load vector
b      = zeros(nn,1);      % global rhs vector
lhs    = sparse(nn,nn);    % global lhs matrix
rhs    = sparse(nn,nn);    % global rhs matrix

% initial conditions
displ = -g_coord(2,:)'/lz*Peb;

%-----------------------------------------------------
% time loop
%-----------------------------------------------------
time = 0 ; % initialise time
for n=1:ntime

time = time + dt ;  % update time

%-----------------------------------------------------
% matrix integration and assembly
%-----------------------------------------------------
lhs    = sparse(nn,nn);    % reinitialise global lhs matrix
rhs    = sparse(nn,nn);    % reinitialise global rhs matrix
ff     = zeros(nn,1);      % global load vector
for iel=1:nels  % sum over elements
  num    = g_num(:,iel)      ; % element nodes
  coord  = g_coord(:,num)' ; % element coordinates
  KM     = zeros(nod,nod)    ; % initialisation
  MM     = zeros(nod,nod)    ;
  F      = zeros(nod,1)      ;
  kappa  = kappav(phase(iel));
  for k = 1:nip % integration loop
    fun    = fun_s(k,:) ; % shape functions
    der    = der_s(:,:,k) ; % der. of shape functions
    jac    = der*coord  ;    % jacobian matrix
    detjac = det(jac) ;      % det. of jacobian
    invjac = inv(jac) ;      % inv. of jacobian
    deriv  = invjac*der ;    % der. of shape fun. in physical coords.
    dwt    = detjac*wts(k) ;
    KM     = KM + deriv'*kappa*deriv*dwt ;% stiffness matrix
    MM     = MM + phi*beta*fun'*fun*dwt ; % mass matrix
  end
  % assemble global matrices
```

```
    lhs(num,num) = lhs(num,num) + MM/dt + KM  ;
    rhs(num,num) = rhs(num,num) + MM/dt       ;
  end
  %-----------------------------------------------------------------

    b = rhs*displ + ff ;              % form rhs global vector

    % impose boundary conditions
    lhs(bcdof,:) = 0 ;                % zero the boundary equations
    tmp = spdiags(lhs,0) ;            % store diagonal
    tmp(bcdof)=1 ;                    % place 1 on stored-diagonal
    lhs=spdiags(tmp,0,lhs);           % reinsert diagonal
    b(bcdof) = bcval ;                % set boundary values

    displ = lhs \ b ;                 % solve system of equations

    %-----------------------------------------------------------------
    % postprocessing to compute Darcy velocities at integration points
    for iel=1:nels  % sum over elements
      num    = g_num(:,iel)       ; % element nodes
      coord  = g_coord(:,num)' ; % element coordinates
      for k = 1:nip % integration loop
        fun      = fun_s(k,:) ;    % shape functions
        der      = der_s(:,:,k) ; % der. of shape functions
        jac      = der*coord  ;    % jacobian matrix
        detjac   = det(jac) ;      % det. of jacobian
        invjac   = inv(jac) ;      % inv. of jacobian
        deriv    = invjac*der ;    % shape fun. der in physical coords.
        u(k,iel) = -kappa*deriv(1,:)*displ(num); % hor.  velocity
        v(k,iel) = -kappa*deriv(2,:)*displ(num); % vert. velocity
      end
    end

    % plot Excess fluid pressure with Darcy flow vectors
    figure(1)
    um = mean(u); % ave. integration point velocities for each element
    vm = mean(v);
    um(find(phase==2))=0; % zero velocites in fault to focus on flow
    vm(find(phase==2))=0; % in the surrounding crust
    pdeplot(g_coord/1e3,edge,g_num,'xydata',displ/1e6,'mesh','off','contour','on');
    hold on
    pdeplot(g_coord/1e3,edge,g_num,'flowdata',[um' vm'],'mesh','off','contour','off');
    title(['Excess fluid pressure (MPa) after ',num2str(time/seconds_per_yr),' years'])
    xlabel('Distance (km)' )
    ylabel('Depth(km)' )
    colormap('jet')
    axis equal
    drawnow

    %-----------------------------------------------------------------
    % Plot fluid pressure vs depth at different positions
    figure(2)
    zp = linspace(0,-lz,100);% z vector
    pl = -rho*g*zp/1e6 ;      % total lithostatic load (MPa)
    ph = -rhof*g*zp/1e6 ;     % hydrostatic fluid pressure (MPa)
    xp = 15e3*ones(1,100);    % x vector at 15 km
    pp1=griddata(g_coord(1,:),g_coord(2,:),displ,xp,zp)/1e6; % excess p at 15 km
    xp = 35e3*ones(1,100);    % x vector at 35 km
    pp2=griddata(g_coord(1,:),g_coord(2,:),displ,xp,zp)/1e6;% excess p at 35 km
```

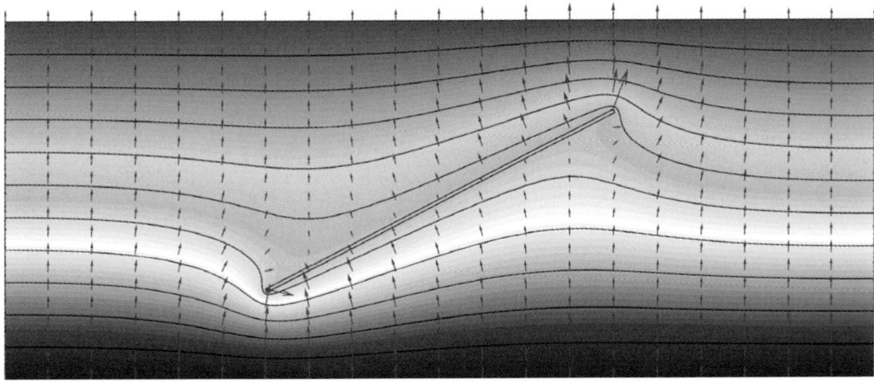

Figure 10.2 Excess fluid pressure and flow vectors in the crust (with a uniform permeability of 10^{-16} m^2) 10 years after introduction of a high permeability (10^{-12} m^2) fault zone.

```
    xp = ones(1,100);         % x vector close to x=0
    pp3=griddata(g_coord(1,:),g_coord(2,:),displ,xp,zp)/1e6;% excess p at ~0 km
    plot(pp1+ph,zp/1e3,'b',pp2+ph,zp/1e3,'r',pp3+ph,zp/1e3,'k',ph,zp/1e3, ...
         'k--',pl,zp/1e3,'k-.')
    legend('p at x=15 km','p at x=35 km','p at x=0 ...
        km','Hydrostatic p','Lithostatic p')
    xlabel('Fluid pressure (MPa)' )
    ylabel('Depth(km)' )

    %-------------------------------------------------------------
end

%-----------------------------------------------------
% end of time loop
%-----------------------------------------------------
```

An example of output generated with this script is illustrated in Figure 10.2. The problem was solved with a total of 5248 triangles, a time step of 0.1 years, and the following physical parameters: $k_{crust}=10^{-16}$ m^2, $k_{fault}=10^{-12}$ m^2, $\mu = 1.33 \times 10^{-4}$ Pa s, $\beta = 10^{-10}$ Pa^{-1}, and $\phi = 0.1$. The results, after an approximately steady situation has been achieved (after 10 years), show how the fluid overpressure and fluid flow vectors in the crust are strongly perturbed by the presence of the high permeability fault zone. Fluid is focused toward the base of the fault, while it diverges away from its upper tip.

10.2 Viscous Fingering

In this section we treat two important aspects that have either not yet been encountered or have been treated only briefly. First, we solve a coupled problem involving two rather than just one unknown (see Section 7.3). Second, we study a hyperbolic problem that is advection dominated (i.e., involving high Peclet numbers) that requires a more specialized treatment to obtain stable solutions when sharp discontinuities exist. To illustrate these features, we consider the 2D displacement of a relatively viscous fluid such as oil by a less viscous miscible solvent, a situation commonly encountered during enhanced oil recovery. The equations governing this so-called Peaceman problem are (e.g., see Peaceman and Rachford (1962)) as follows:

$$\nabla \cdot \mathbf{u} = 0 \tag{10.14}$$

$$\mathbf{u} = -\frac{k}{\mu(c)} \nabla p \tag{10.15}$$

$$\phi \frac{\partial c}{\partial t} + \nabla \cdot (\mathbf{u}c) = \nabla \cdot (\kappa \nabla c) \tag{10.16}$$

$$\mu(c) = \left(\frac{c}{\mu_s^{1/4}} + \frac{1-c}{\mu_o^{1/4}} \right)^{-4} \tag{10.17}$$

Here, p is the pore pressure, c is the concentration of the solvent, k is the permeability, $\mu(c)$ is the bulk viscosity of the mixture, μ_s and μ_o are viscosities of the solvent and oil, respectively, ϕ is the porosity, $\mathbf{u}$ is the velocity vector, and κ is the diffusivity of the solvent. These equations are to be solved with the boundary conditions and initial conditions illustrated in Figure 10.3. Note that this problem has two unknowns, the fluid pressure and the solvent concentration, and that they depend on each other (i.e., they are coupled). Note also that these equations bear a close resemblance to those introduced earlier for coupled fluid and heat transfer (i.e., Equations 10.4–10.7) and can solved using exactly the same approach. Equations 10.14 and 10.15 are mass conversation and Darcy's law, respectively. Equation 10.16 is an advection–diffusion equation describing the evolution of the solvent concentration. Equation 10.16 is linked to the behavior of the flowing fluid due to dependency of the fluid viscosity on the concentration, as described by Equation 10.17. As shown in the following text, this nonlinear dependency gives rise to a well-known fingering instability.

In Sections 7.2 and 8.1, we mentioned that hyperbolic equations involving first-order spatial derivatives (i.e., the term $\nabla \cdot (\mathbf{u}c)$ in Equation 10.16) make the equations non-self-adjoint, which are not optimally solved using the standard Galerkin FEM used until now. This is not necessarily a problem for flows that involve low Peclet (*Pe*) numbers, as, for example, was the case in the problem studied in Section 8.1. However, when $Pe \gg 1$, standard discretization of advection terms can cause numerical oscillations that completely destroy the accuracy of the solution (see Section 7.2). This problem is analogous to that related to use of central (as opposed to upwind) discretization of advection terms when using the finite difference method.

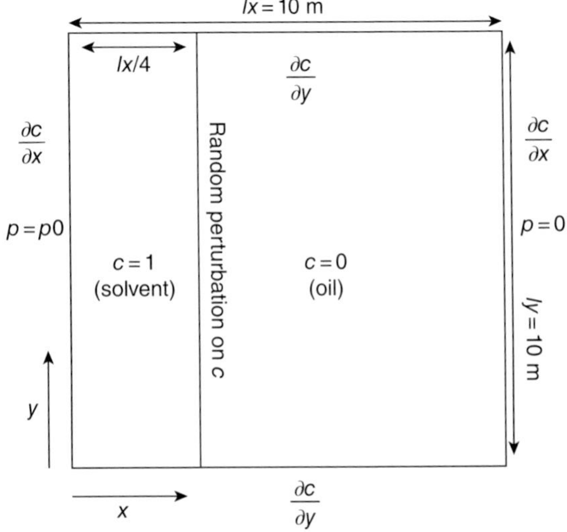

Figure 10.3 Initial and boundary conditions for displacement of a viscous fluid ($c = 0$) by a miscible solvent ($c = 1$) entering from the left-hand boundary. The concentration on the interface between the two fluids is perturbed by a random value between 0 and 1.

A simple, reasonably successful approach to deal with hyperbolic advection-dominated problems is to use the streamline (upwind) Petrov–Galerkin FEM (Zienkiewicz and Taylor, 2000). Recall that with the Galerkin FEM used until now, the weighting functions **W** are assumed to be equal to the shape functions **N**. Conversely, the Petrov–Galerkin method uses weighting functions that are different from the shape functions. Various forms of **W** have been proposed, but a satisfactory choice is (see Zienkiewicz and Taylor (2000))

$$W_k = N_k + \frac{\alpha h}{2}\frac{u_i}{|\mathbf{u}|}\frac{\partial N_k}{\partial x_i} \tag{10.18}$$

with

$$\alpha = \coth Pe - \frac{1}{Pe} \tag{10.19}$$

and

$$Pe = \frac{|\mathbf{u}|h}{2\kappa} \tag{10.20}$$

Here, Pe is the element Peclet number, h is the element size, $\mathbf{u}$ is the elemental flow velocity (with components u and v in the x- and y-directions), and κ is the solvent diffusivity. The effect of the second term on the right-hand side of 10.18 is to introduce additional anisotropic diffusion in the direction of flow, which stabilizes oscillations arising due to strong advection. Note that when there is no flow, $Pe = 0$ and $\mathbf{W} = \mathbf{N}$, so the method is equivalent to the standard Galerkin FEM. Note also that the modified weighting function should be applied to all terms of the governing equations that involve the shape functions (i.e., including any source terms if present).

To solve the governing system of partial differential equations of PDEs (i.e., Equations 10.14–10.17), we perform discretization based on nine-node Lagrangian quadrilateral elements (see Figures 7.6 and 7.12). The shape functions and the first spatial derivatives for this element are given in 7.53 and 7.54. Assuming that both p and c are approximated with the same shape functions, we can write

$$p \simeq [N_1\ N_2\ N_3\ N_4\ N_5\ N_6\ N_7\ N_8\ N_9]\begin{Bmatrix} p_1 \\ p_2 \\ p_3 \\ p_4 \\ p_5 \\ p_6 \\ p_7 \\ p_8 \\ p_9 \end{Bmatrix} = \mathbf{N}\,\mathbf{p} \tag{10.21}$$

and

$$c \simeq [N_1\ N_2\ N_3\ N_4\ N_5\ N_6\ N_7\ N_8\ N_9]\begin{Bmatrix} c_1 \\ c_2 \\ c_3 \\ c_4 \\ c_5 \\ c_6 \\ c_7 \\ c_8 \\ c_9 \end{Bmatrix} = \mathbf{N}\,\mathbf{c} \tag{10.22}$$

where p and c are the continuous variables and $\mathbf{p}$ and $\mathbf{c}$ are the vectors defined at the element nodes. Substituting 10.15 into 10.14 to eliminate $\mathbf{u}$, introducing the approximations defined by 10.21 and 10.22 into Equations 10.14 and 10.16, weighting each equation with the functions defined by 10.18, integrating over the element, integrating by parts where necessary (and neglecting the resulting surface integrals), and using an implicit finite difference approximation of the time derivative in 10.16, the following set of discrete element equations can be obtained:

$$\begin{bmatrix} \mathbf{KP}(c) & 0 \\ 0 & \frac{\mathbf{MM}}{\Delta t} + \mathbf{CM} + \mathbf{KM} \end{bmatrix} \begin{Bmatrix} \mathbf{p} \\ \mathbf{c} \end{Bmatrix}^{n+1} = \begin{bmatrix} 0 & 0 \\ 0 & \frac{\mathbf{MM}}{\Delta t} \end{bmatrix} \begin{Bmatrix} \mathbf{p} \\ \mathbf{c} \end{Bmatrix}^{n} \tag{10.23}$$

where

$$\mathbf{MM} = \int\int \phi \mathbf{W}^T \mathbf{N}\, dxdy \tag{10.24}$$

$$\mathbf{CM} = \int\int \left(u\, \mathbf{W}^T \frac{\partial \mathbf{N}}{\partial x} + v\, \mathbf{W}^T \frac{\partial \mathbf{N}}{\partial y} \right) dxdy \tag{10.25}$$

$$\mathbf{KW} = \int\int \kappa (\nabla \mathbf{N})^T \nabla \mathbf{N}\, dxdy \tag{10.26}$$

$$\mathbf{KP}(c) = \int\int \frac{k}{\mu(c)} (\nabla \mathbf{N})^T \nabla \mathbf{N}\, dxdy \tag{10.27}$$

Note that the individual element matrices (e.g., **MM**) have the dimensions [9,9], the total element stiffness matrix (i.e., the first bracketed matrix of Equation 10.23) has the dimensions [18,18], and the element nodal unknown vector (i.e., the term with the superscript $n+1$ in Equation 10.23) has the dimensions [18,1]. The matrix **MM** is seen to be a mass matrix; **KM** and **KP** are standard stiffness matrices similar to that encountered in diffusion problems (except that **W** replaces **N**), while **CM** is an advection matrix encountered with hyperbolic problems (see Sections 7.2 and 8.1).

In the following text, we list a Matlab script to solve the coupled fluid flow problem. This program is very similar to that presented in Section 7.3 for the system of coupled reaction–diffusion equations. The following points are noteworthy with respect to the program listed:

1) The various element matrices are evaluated using Gauss–Legendre quadrature in a manner identical to done in the previous chapters.
2) The problem has two degrees of freedom per node, the fluid pressure p, and the solvent concentration c. Within a single element, we assume that the unknowns are ordered as follows: $p_1, p_2, \ldots, p_9, c_1, c_2, \ldots, c_9$. The mesh numbering scheme for a small mesh is shown in Figure 7.12.
3) The fluid viscosity appearing in 10.27 depends on the solvent concentration, which must be obtained by interperpolating the concentration from the nodes of an element to the integration points. The integration point concentration and viscosity can be calculated in Matlab with the following snippet:

```
c     = fun*displ(gc);  % interp. c to current int. point
mu    = (c/mus^0.25+(1-c)/mu0^0.25)^(-4); % fluid viscosity
```

4) The Petrov–Galerkin weighting function `W` (Equation 10.18) can be computed in Matlab with the following lines:

```
vx      = -kpa*deriv(1,:)*displ(gp)  ; % x Darcy velocity
vy      = -kpa*deriv(2,:)*displ(gp)  ; % y Darcy velocity
mag     = sqrt(vx^2 + vy^2)  ; % magnitude of flow velocity
```

```
  pe      = mag*he/2/kappa    ; % element Peclet number
  if pe > 0 % compute Petrov-Galerkin weighting function
     a     = coth(pe)-1/pe ;
     W     = fun + a*he/2*(vx*deriv(1,:)+vy*deriv(2,:))/mag ;
     else  % use standard Galerkin approx.
     W     = fun ;
  end
```

5) The mass matrix is lumped (see Section 9.2), which helps further stabilize numerical oscillations related to the small amount of physical diffusivity in the problem.

```
%-------------------------------------------
% Program: porousflow2d_coupled.m
% Porous flow coupled to misicble solvent
% Streamline (upwind) Petrov-Galerkin Method
% The unknowns are p (dof=1) and c (dof=2)
% 9 node Langrangian quadrilateral elements
%-----------------------------------------
clear
secsperyr = 60*60*24*365;

% physical parameters
lx    = 10 ;       % length of x domain (m)
ly    = 10 ;       % length of y domain (m)
perm  = 1e-13 ;    % permeability (m^2)
phi   = 0.1  ;     % porosity
kappa = 1e-9 ;     % c diffusivity (m^2/s)
mus   = 1.33e-4 ; % solvent viscosity (Pa s)
mu0   = 20*mus  ; % oil viscosity (Pa s)
p0    = 0.1e6 ;    % applied pressure difference across domain (Pa)

% numerical parameters
nod    = 9    ;       % number of nodes per element
ndof   = 2    ;       % degrees of freedom in coupled problem
ntot   = ndof*nod ; % degrees of freedom per element
nip    = 9 ;          % number of Gauss integration points
nxe    = 100   ;      % number of elements in x direction
nye    = 100   ;      % number of elements in y direction
nels   = nxe*nye ;   % total number of elements
nx     = 2*nxe+1 ;   % number of nodes in x-direction
ny     = 2*nye+1 ;   % number of nodes in y-direction
nn     = nx*ny    ;  % total number of nodes in mesh
dx     = lx/nxe   ;  % element dimension in x-direction
dy     = ly/nye   ;  % element dimension in y-direction
he     = min(dx/2,dy/2) ;    % characteristic element dimension (m)
dt     = 0.0001*secsperyr ; % time step (s)
ntime = 100 ;                % number of time steps computed

%-----------------------------------------
% generate mesh coordinate, nodal and equation numbering
%-----------------------------------------

% mesh (numbering in y direction)
g_coord = zeros(2,nn) ;
n = 1 ;
for i = 1:nx
    for j=1:ny
      g_coord(1,n) = (i-1)*dx/2 ;
```

```
        g_coord(2,n) = (j-1)*dy/2 ;
        n = n + 1 ;
      end
end

% establish node numbering for each element
gnumbers = reshape(1:nn,[ny nx]) ; % grid of node numbers
g_num     = zeros(nod,nels);
iel = 1 ; % intialise element number
for i=1:2:nx-1 % loop over x-nodes
    for j=1:2:ny-1 % loop over y-nodes
      g_num(1,iel) = gnumbers(j,i)     ;    % node 1
      g_num(2,iel) = gnumbers(j+1,i) ;     % node 2
      g_num(3,iel) = gnumbers(j+2,i) ;     % node 3
      g_num(4,iel) = gnumbers(j+2,i+1) ;  % node 4
      g_num(5,iel) = gnumbers(j+2,i+2) ;  % node 5
      g_num(6,iel) = gnumbers(j+1,i+2) ;  % node 6
      g_num(7,iel) = gnumbers(j,i+2) ;     % node 7
      g_num(8,iel) = gnumbers(j,i+1) ;     % node 8
      g_num(9,iel) = gnumbers(j+1,i+1) ;  % node 9
      iel = iel + 1 ; % increment the element number
    end
end

% reshaped coordinates (used only for plotting)
 xg = reshape(g_coord(1,:),ny,nx);
 yg = reshape(g_coord(2,:),ny,nx);

% establish equation number for each node
sdof   = 0 ;                    % system degrees of freedom (dof) counter
nf     = zeros(ndof,nn) ;      % node degree of freedom array
for n = 1:nn                    % loop over all nodes in mesh
  for i=1:ndof                  % loop over each dof
     sdof = sdof + 1 ;          % increment sdof
     nf(i,n) =  sdof ;        % store eqn number on each node
  end
end

% equation number for each element
g     = zeros(ntot,1) ;     % equation numbers for 1 element
g_g = zeros(ntot,nels);    % equation numbers for all elements
for iel=1:nels ;             % loop over elements
    num = g_num(:,iel) ; % node numbers for this element
    inc=0 ;                  % initialise local eqn number
    % loop 2 times of nodes of an element, once for each dof
    for i=1:nod ;  inc=inc+1 ; g(inc)=nf(1,num(i)) ; end
    for i=1:nod ;  inc=inc+1 ; g(inc)=nf(2,num(i)) ; end
    g_g(:,iel) = g ;        % store the equation numbers
end

%----------------------------------------
% boundary condition definition
%----------------------------------------

% locate boundary nodes
bx0 = find(g_coord(1,:)==0)   ;
bxn = find(g_coord(1,:)==lx)  ;
by0 = find(g_coord(2,:)==0)   ;
byn = find(g_coord(2,:)==ly)  ;
```

```
% specify boundary conditions
bcdof = [       nf(1,bx0)              nf(1,bxn)          ] ;
bcval = [  p0*ones(1,length(bx0)) 0*ones(1,length(bxn))] ;

%-------------------------------------
%  integration data and shape functions
%-------------------------------------
% local coordinates of Gauss integration points for nip=3x3
  points(1:3:7,1) = -sqrt(0.6);
  points(2:3:8,1) = 0;
  points(3:3:9,1) = sqrt(0.6);
  points(1:3,2)   = sqrt(0.6);
  points(4:6,2)   = 0 ;
  points(7:9,2)   = -sqrt(0.6);

  % Gauss weights for nip=3x3
  w   = [ 5./9. 8./9. 5./9.] ;
  v   = [ 5./9.*w ; 8./9.*w ; 5./9.*w ] ;
  wts = v(:) ;

 % evaluate shape functions and their derivatives
 % at integration points and save the results
  for k = 1:nip
     xi  = points(k,1);
     eta = points(k,2);
     etam = eta - 1; etap = eta + 1 ;
     xim  = xi - 1 ; xip  = xi + 1  ;
     x2p1 = 2*xi+1 ;   x2m1 = 2*xi-1 ;
     e2p1 = 2*eta+1 ;  e2m1 = 2*eta-1 ;
     % shape functions
     fun= [ .25*xi*xim*eta*etam -.5*xi*xim*etap*etam ...
     .25*xi*xim*eta*etap -.5*xip*xim*eta*etap ...
     .25*xi*xip*eta*etap -.5*xi*xip*etap*etam ...
     .25*xi*xip*eta*etam -.5*xip*xim*eta*etam xip*xim*etap*etam ] ;
     % derivates of shape functions
     der(1,1) = 0.25*x2m1*eta*etam  ; %dN1dxi
     der(1,2) =-0.5*x2m1*etap*etam ;  %dN2dxi, etc
     der(1,3) = 0.25*x2m1*eta*etap  ;
     der(1,4) =      -xi*eta*etap ;
     der(1,5) = 0.25*x2p1*eta*etap  ;
     der(1,6) =-0.5*x2p1*etap*etam ;
     der(1,7) = 0.25*x2p1*eta*etam  ;
     der(1,8) =      -xi*eta*etam ;
     der(1,9) = 2*xi*etap*etam    ;

     der(2,1) = 0.25*xi*xim*e2m1 ; %dN1deta
     der(2,2) =-xi*xim*eta         ;
     der(2,3) = 0.25*xi*xim*e2p1 ;
     der(2,4) =-0.5*xip*xim*e2p1   ;
     der(2,5) = 0.25*xi*xip*e2p1 ;
     der(2,6) =-xi*xip*eta         ;
     der(2,7) = 0.25*xi*xip*e2m1 ;
     der(2,8) =-0.5*xip*xim*e2m1   ;
     der(2,9) = 2*xip*xim*eta ;
     % save
     fun_s(k,:) = fun ;
     der_s(:,:,k) = der ;
  end
```

```
%--------------------------------------
% initialisation
%--------------------------------------

lhs   = sparse(sdof,sdof) ;  % system stiffness matrix
b     = zeros(sdof,1)     ;  % system load vector
displ = zeros(sdof,1)     ;  % solution vector

% apply step in concentration (dof=2)
ii = find(g_coord(1,:)<lx/4);
displ(nf(2,ii))=1;
% random perturbation on interface
iir = find(g_coord(1,:)==lx/4);
displ(nf(2,iir))=rand(length(iir),1) ;

%-----------------------------------------------------
% time loop
%-----------------------------------------------------

time = 0 ;
for n=1:ntime

n
time = time + dt ;
b    = zeros(sdof,1)     ; % system load vector
lhs  = sparse(sdof,sdof) ; % system stiffness matrix

%-----------------------------------------------
%  element integration and assembly
%--------------------------------------

for iel=1:nels  % sum over elements

    num         = g_num(:,iel)     ; % list of nodes in element
    coord       = g_coord(:,num)'  ; % nodal coordinates
    g           = g_g(:,iel)       ; % all equations in element
    gp          = g(1:nod)         ; % p equations
    gc          = g(nod+1:ntot)    ; % c equations
    KM          = zeros(nod,nod)   ; % initialisation of KM
    KP          = zeros(nod,nod)   ; % initialisation of KP
    MM          = zeros(nod,nod)   ; % initialisation of MM
    CM          = zeros(nod,nod)   ; % initialisation of CM
    for k = 1:nip % loop over integration points
        fun    = fun_s(k,:) ;    % shape functions
        der    = der_s(:,:,k) ; % der. of shape functions
        jac    = der*coord    ;  % Jacobian matrix
        detjac = det(jac)     ;  % determinant of Jac
        invjac = inv(jac)     ;  % inverse of Jac
        deriv  = invjac*der   ;  % der. of N in physical coords
        c      = fun*displ(gc);  % interp. c to current int. point
        mu     = (c/mus^0.25+(1-c)/mu0^0.25)^(-4); % fluid viscosity
        kpa    = perm/mu; % permeability divided by fluid viscosity
        vx     = -kpa*deriv(1,:)*displ(gp) ; % x Darcy velocity
        vy     = -kpa*deriv(2,:)*displ(gp) ; % y Darcy velocity
        mag    = sqrt(vx^2 + vy^2) ; % magnitude of flow velocity
        pe     = mag*he/2/kappa    ; % element Peclet number
        if pe > 0 % compute Petrov-Galerkin weighting function
          a    = coth(pe)-1/pe ;
```

```
        W     = fun + a*he/2*(vx*deriv(1,:)+vy*deriv(2,:))/mag ;
      else  % standard Galerkin approx.
        W     = fun ;
      end
      dwt = detjac*wts(k) ; % det. of Jac x Gauss weight
      MM  = MM  + phi/dt*W'*fun*dwt ; % mass matrix for c
      KM  = KM  + kappa*deriv'*deriv*dwt ; % diffn matrix for c
      adv = (vx*W'*deriv(1,:)+vy*W'*deriv(2,:));% advection term
      CM  = CM  + adv*dwt ; % advection matrix for c
      KP  = KP  + kpa*deriv'*deriv*dwt ; % diffn matrix for p
  end % end of integration loop
    % assembly
    MM           = diag([sum(MM)],0); % lump the mass matrix
    lhs(gp,gp) = lhs(gp,gp) + KP  ; % p contribution to lhs
    lhs(gc,gc) = lhs(gc,gc) + MM + CM + KM ;% c contribution to lhs
    b(gc)        = b(gc) + MM*displ(gc)  ; % rhs global vector
end % end of element integration and assembly

%-------------------------------------------------

   % apply boundary conditions
   lhs(bcdof,:) = 0      ;
   tmp = spdiags(lhs,0)  ;
   tmp(bcdof)=1          ;
   lhs=spdiags(tmp,0,lhs) ;
   b(bcdof) = bcval ;

   displ = lhs \ b ; % solve system

 % -------------------------------
 % visualisation
 figure(1) % plot fluid pressure
 Pg = reshape(displ(nf(1,:)),ny,nx);
 pcolor(xg,yg,Pg)
 shading interp
 axis equal
 title('Pressure (Pa)')
 colorbar
 colormap(jet)

 figure(2) % plot solvent concentration
 Cg = reshape(displ(nf(2,:)),ny,nx);
 pcolor(xg,yg,Cg)
 shading interp
 axis equal
 title(['Concentration, time =',num2str(time/secsperyr),'years'])
 colorbar
 colormap(jet)

 drawnow
%--------------------------
end
%-------------------------------------------------
% end of time integration
%-------------------------------------------------
```

Figure 10.4 show an example of numerical results calculated with the given Matlab script. This simulation shows that the initially sharp interface between the two fluids develops a complex fingered

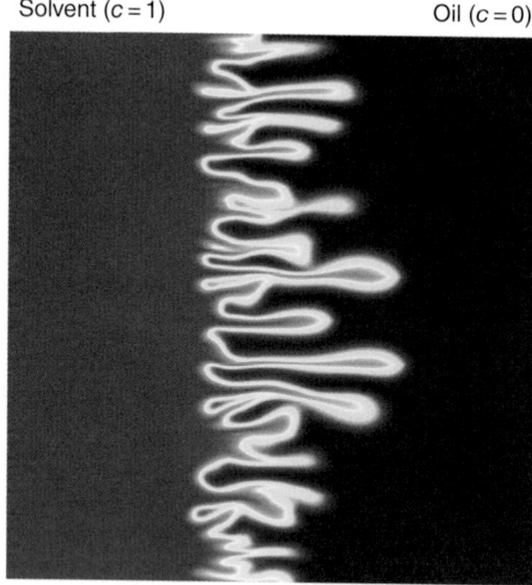

Figure 10.4 Solvent concentration showing fingering instability developed after 0.01 years when a relatively low viscosity solvent introduced from the left boundary displaces a more viscous fluid to the right (based on the numerical solution of Equations 10.14–10.17). The calculation was carried out with the following parameter values: $k=10^{-13}$ m^2, $\mu_s = 1.33 \times 10^{-4}$ Pa s, $\mu_o = 2.7 \times 10^{-3}$ Pa s, $\kappa = 10^{-9}$ m^{2}s^{-1}, $dx = dy = 0.1$ m, $\Delta t = 0.0001$ years, $\Delta p = 0.1$ MPa, and $lx = ly = 10$ m.

pattern. This occurs because dependence of the viscosity on the fraction of solvent is nonlinear; a small concentration of solvent mixed with the oil leads to a large reduction in its viscosity, which causes small perturbations on the interface between the two fluids to become amplified.

Suggested Reading

S. E. Ingebritsen, W. E. Sanford, and C. E. Neuzil, *Groundwater in Geologic Processes*, Cambridge University Press, Cambridge, 2006.

R. W. Lewis, and B. A. Schrefler, *The Finite Element Method in the Static and Dynamic Deformation and Consolidation of Porous Media*, John Wiley & Sons, Ltd/University Press, Chichester, 1998.

R. W. Lewis, P. Nithiarasu, and K. N. Seetharamu, *Fundamentals of the Finite Element Method for Heat and Fluid Flow*, John Wiley & Sons, Ltd, Chichester, 2004.

O. M. Phillips, *Flow and Reactions in Permeable Rocks*, Cambridge University Press, Cambridge, 1991.

O. M. Phillips, *Geological Fluid Dynamics*, Cambridge University Press, Cambridge, 2009.

D. L. Turcotte, and G. Schubert, *Geodynamics*, Cambridge University Press, Cambridge, 2002.

O. C. Zienkiewicz, and R. L. Taylor, *The Finite Element Method, Volume 3, Fluid Dynamics*, Butterworth-Heinemann, Oxford/Boston, MA, 2000.

11

Lithospheric Flexure

This chapter shows how the finite element method (FEM) can be applied to solve problems involving flexural deformation of the lithosphere in three-dimensions (3Ds). The presence of fourth-order spatial derivatives in the equation governing flexure requires higher order shape functions than previously considered. In addition, we show that even though flexure is governed by a single variable w, the FEM approximation of this continuous variable involves not only its nodal values but also those of its gradients. Thus, the discretized flexural model has several degrees of freedom per node, as in the coupled problem treated in Chapter 10. A single stand-alone Matlab script is listed and discussed in detail, which enables various practical problems to be investigated.

Earth's lithosphere becomes warped when subjected to vertical or horizontal loads, related, for example, to faulting, volcanism, erosion, deposition, and glaciation (Turcotte and Schubert, 1982; Watts, 2001). In situations when the vertical deflection is small in comparison to the thickness of the lithosphere itself (which is often the case), deformation of the full 3D solid can be reduced to a single quantity, w, the vertical deflection of the mid-plane of the plate. This model is quite accurate for certain situations, and it results in a large reduction in computational cost compared to solving the full displacement field in the entire 3D volume. This chapter treats this simplified "flexural" model, while Chapter 12 deals with the mechanics of "classic" solids where the simplifying assumptions involved in deriving the flexural equation are relaxed.

11.1 Governing Equations

The vertical deflection (w) of an elastic lithospheric plate underlain by a dense inviscid fluid (i.e., one with zero viscosity) is governed by the following fourth-order partial differential equation or PDE (see Timoshenko and Woinowsky-Krieger (1959) and Turcotte and Schubert (1982)):

$$(L\nabla)^T \mathbf{D}(L\nabla)\, w + \nabla^T \mathbf{P}\, \nabla\, w + \Delta\rho\, g\, w = q \tag{11.1}$$

Here,

$$(L\nabla)^T = \begin{bmatrix} \frac{\partial^2}{\partial x^2} & \frac{\partial^2}{\partial y^2} & 2\frac{\partial^2}{\partial x \partial y} \end{bmatrix} \tag{11.2}$$

$$\mathbf{D} = D \begin{bmatrix} 1 & \nu & 0 \\ \nu & 1 & 0 \\ 0 & 0 & \frac{1-\nu}{2} \end{bmatrix} \tag{11.3}$$

Practical Finite Element Modeling in Earth Science Using Matlab, First Edition. Guy Simpson.

Companion website: www.wiley.com/go/simpson

$$D = \frac{E\,T_e^3}{12(1-\nu^2)} \tag{11.4}$$

$$\mathbf{P} = \begin{bmatrix} P_x & P_{xy} \\ P_{xy} & P_y \end{bmatrix} \tag{11.5}$$

where $\Delta\rho$ (kg m^{-3}) is the density difference between the material below and above the plate, g (m s^{-2}) is the acceleration due to gravity, q (Pa) is the vertical load imposed on the surface of the plate, $\mathbf{P}$ is the matrix of imposed in-plane forces per unit length (Pa m, i.e., acting parallel to the lithosphere in the horizontal directions x or y), E is Young's modulus, ν is Poisson's ratio, and T_e is the elastic thickness of the plate. The first term of 11.1 represents the elastic resistance of the lithosphere to bending, the second term accounts for imposed in-plane forces, the third term accounts for restoring forces produced as the plate deflects vertically and replaces material of different densities above and below, and the fourth term accounts for vertical loads applied on the surface or base of the plate. In the case of isotropy and a constant flexural rigidity D, Equation 11.1 reduces to

$$D\left(\frac{\partial^4 w}{\partial x^4} + \frac{\partial^4 w}{\partial y^2} + 2\frac{\partial^4 w}{\partial x^2 \partial y^2}\right) + P_x\frac{\partial^2 w}{\partial x^2} + P_y\frac{\partial^2 w}{\partial y^2} + P_{xy}\frac{\partial^2 w}{\partial x \partial y} + \Delta\rho\, g\, w = q \tag{11.6}$$

In 1D, Equation 11.6 simplifies to

$$D\frac{\partial^4 w}{\partial x^4} + P_x\frac{\partial^2 w}{\partial x^2} + \Delta\rho\, g\, w = q \tag{11.7}$$

These various forms of the flexure equation are important because they enable one to predict in a simple manner how the lithosphere responds to different load scenarios. In addition, the equations can be used to constrain the elastic properties of the lithosphere (and in particular T_e appearing in 11.4) in cases when the loads and deflections are known. In the chapter, we demonstrate how to solve Equation 11.1 with the FEM using Matlab. The single program listed in the following text can be used to study full 3D flexure with in-plane forces, spatially variable loads, and variable flexural rigidity, as well as more restricted situations, for example, such as that represented by Equation 11.7.

The flexure model differs in several respects compared to the previously considered diffusion problems. First, the governing equations lack a time derivative, which indicates that the solution depends only on the boundary conditions and loads and can be obtained in a single step. For such a steady-state model, there is no notion of initial conditions. The flexural equation is therefore an example of an elliptic (boundary value) problem. Second, the presence of fourth-order spatial derivatives requires a more specialized consideration of shape functions. Third, the presence of fourth-order spatial derivatives indicates that four boundary conditions must be provided in each direction (two at each boundary). The most common boundary conditions are the built-in (also known as clamped) condition, where at each boundary one sets the deflection and its derivatives to zero, and the simple support condition, where one sets the deflection and its second derivatives to zero at each boundary.

11.2 FEM Discretization

Following the formulation presented by Zienkiewicz and Taylor (2000), we begin by introducing the standard finite element approximation

$$w = \mathbf{N}\,\mathbf{a}^e \tag{11.8}$$

where w is the continuous variable being approximated in the governing equation, $\mathbf{N}$ are the shape functions, and $\mathbf{a}^e$ is a vector of the element unknowns (defined later). Substituting 11.8 into 11.1,

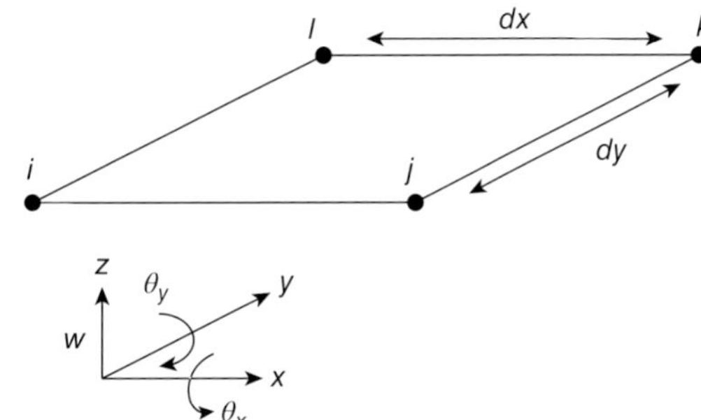

Figure 11.1 Four-node rectangular element with 12 degrees of freedom used for plate flexure. The three unknowns at each node are the vertical deflection (w), θ_x (defined as $\partial w/\partial y$), and θ_y (defined as $-\partial w/\partial x$).

weighting the residual equation with the shape functions, integrating over the element, and integrating by parts results in the discrete element equation

$$[\mathbf{KM}]\,\mathbf{a}^e + [\mathbf{KP}]\,\mathbf{a}^e + [\mathbf{MM}]\,\mathbf{a}^e = \mathbf{F} \tag{11.9}$$

where

$$\mathbf{KM} = \int\int \mathbf{B}^T\mathbf{D}\mathbf{B}\,dx\,dy \tag{11.10}$$

$$\mathbf{KP} = \int\int \mathbf{A}^T\mathbf{P}\,\mathbf{A}\,dx\,dy \tag{11.11}$$

$$\mathbf{MM} = \int\int \Delta\rho g\,\mathbf{N}^T\mathbf{N}\,dx\,dy \tag{11.12}$$

$$\mathbf{F} = \int\int \mathbf{N}^T q\,dx\,dy \tag{11.13}$$

with

$$\mathbf{B} = (\mathbf{L}\nabla)\mathbf{N} \tag{11.14}$$

$$\mathbf{A} = \nabla\,\mathbf{N} \tag{11.15}$$

$$\mathbf{L} = \begin{bmatrix} \frac{\partial}{\partial x} & 0 \\ 0 & \frac{\partial}{\partial y} \\ \frac{\partial}{\partial y} & \frac{\partial}{\partial x} \end{bmatrix} \tag{11.16}$$

$$\nabla = \begin{bmatrix} \frac{\partial}{\partial x} \\ \frac{\partial}{\partial y} \end{bmatrix} \tag{11.17}$$

Finite element discretization is performed here using four-node rectangular elements that have 12 degrees of freedom per element: w, θ_x $(= \partial w/\partial y)$ and θ_y $(= -\partial w/\partial x)$ at each node (Figure 11.1). This element ensures continuity in w and its derivatives across the interface between adjacent elements (C_1 continuity), which is an important requirement related to the presence of second-order derivatives in 11.9. For a single element, the vector of unknowns $\mathbf{a}^e$ is defined as

$$\mathbf{a}^e = \begin{Bmatrix} \mathbf{a}_i \\ \mathbf{a}_j \\ \mathbf{a}_k \\ \mathbf{a}_l \end{Bmatrix} \tag{11.18}$$

where

$$\mathbf{a}^i = \begin{Bmatrix} w_i \\ \theta_{xi} \\ \theta_{yi} \end{Bmatrix} \quad \mathbf{a}^j = \begin{Bmatrix} w_j \\ \theta_{xj} \\ \theta_{yj} \end{Bmatrix}, \quad \ldots \tag{11.19}$$

The deflection w is assumed to vary according to the following 12-term polynomial:

$$\begin{aligned} w &= \alpha_1 + \alpha_2 x + \alpha_3 y + \alpha_4 x^2 + \alpha_5 xy + \alpha_6 y^2 + \alpha_7 x^3 + \alpha_8 x^2 y \\ &\quad + \alpha_9 xy^2 + \alpha_{10} y^3 + \alpha_{11} x^3 y + \alpha_{12} xy^3 \\ &\equiv \mathbf{H}\,\boldsymbol{\alpha} \end{aligned} \tag{11.20}$$

Here,

$$\mathbf{H} = [1, x, y, x^2, xy, y^2, x^3, x^2y, xy^2, y^3, x^3y, xy^3] \tag{11.21}$$

and

$$\boldsymbol{\alpha} = [\alpha_1, \alpha_2, \alpha_3, \alpha_4, \alpha_5, \alpha_6, \alpha_7, \alpha_8, \alpha_9, \alpha_{10}, \alpha_{11}, \alpha_{12}]^T \tag{11.22}$$

The coefficients α_1 to α_{12} can be determined by writing down the 12 simultaneous equations linking the values of w and its derivatives at the nodes when the coordinates take their appropriate values. For example, for node i (Figure 11.1),

$$\begin{aligned} w_i &= \alpha_1 + \alpha_2 x_i + \alpha_3 y_i + \alpha_4 x_i^2 + \alpha_5 xy + \cdots \\ \left(\frac{\partial w}{\partial y}\right)_i &= \theta_{xi} = \alpha_3 + \alpha_5 x_i + \cdots \\ -\left(\frac{\partial w}{\partial x}\right)_i &= \theta_{yi} = -\alpha_2 - 2\alpha_4 x_i - \alpha_5 y_i - \cdots \end{aligned}$$

Listing the full set of terms for all four nodes,

$$\mathbf{a}^e = \mathbf{C}\,\boldsymbol{\alpha} \tag{11.23}$$

where $\mathbf{C}$ is a 12×12 matrix that depends on the nodal coordinates for an element and $\boldsymbol{\alpha}$ is a column vector of the 12 unknown coefficients that can be computed from

$$\boldsymbol{\alpha} = \mathbf{C}^{-1}\mathbf{a}^e \tag{11.24}$$

Although this inversion can be performed algebraically, it is also conveniently performed on a computer (e.g., using Matlab). Now that the coefficients for the approximating polynomial have been determined, one can write the shape functions in the implicit form

$$\mathbf{N} = \mathbf{H}\ \mathbf{C}^{-1} \tag{11.25}$$

with which Equation 11.8 can be rewritten as follows:

$$w = \mathbf{H}\ \mathbf{C}^{-1}\mathbf{a}^e \tag{11.26}$$

The matrix $\mathbf{B}$ (Equation 11.14) that is required to evaluate the element matrix $\mathbf{KM}$ contains second-order derivatives of the shape functions. It can be computed from the expression

$$\mathbf{B} = \mathbf{Q}\ \mathbf{C}^{-1} \tag{11.27}$$

where

$$\mathbf{Q} = \mathbf{L}\,\nabla\,\mathbf{H} = \begin{bmatrix} 0 & 0 & 0 & 2 & 0 & 0 & 6x & 2y & 0 & 0 & 6xy & 0 \\ 0 & 0 & 0 & 0 & 0 & 2 & 0 & 0 & 2x & 6y & 0 & 6xy \\ 0 & 0 & 0 & 0 & 2 & 0 & 0 & 4x & 4y & 0 & 6x^2 & 6y^2 \end{bmatrix} \tag{11.28}$$

The matrix **A** appearing in the element matrix **KP** contains the shape functions once derived and can be computed from

$$\mathbf{A} = \mathbf{R}\;\mathbf{C}^{-1} \tag{11.29}$$

where

$$\mathbf{R} = \nabla\,\mathbf{H} = \begin{bmatrix} 0 & 1 & 0 & 2x & y & 0 & 3x^2 & 2xy & y^2 & 0 & 3x^2y & y^3 \\ 0 & 0 & 1 & 0 & x & 2y & 0 & x^2 & 2xy & 3y^2 & x^3 & 3xy^2 \end{bmatrix} \tag{11.30}$$

Using these relations, the element stiffness matrices listed after Equation 11.9 can be rewritten as follows:

$$\mathbf{KM} = \int\!\!\int \mathbf{B}^T\mathbf{D}\mathbf{B}\,dx\,dy = \mathbf{C}^{-T}\int\!\!\int \mathbf{Q}^T\mathbf{D}\,\mathbf{Q}\,dx\,dy\,\mathbf{C}^{-1} \tag{11.31}$$

$$\mathbf{KP} = \int\!\!\int \mathbf{A}^T\mathbf{P}\,\mathbf{A}\,dx\,dy \;= \mathbf{C}^{-T}\int\!\!\int \mathbf{R}^T\,\mathbf{P}\,\mathbf{R}\,dx\,dy\;\mathbf{C}^{-1} \tag{11.32}$$

$$\mathbf{MM} = \int\!\!\int \Delta\rho g\;\mathbf{N}^T\mathbf{N}\;dx\,dy = \mathbf{C}^{-T}\int\!\!\int \Delta\rho g\;\mathbf{H}^T\;\mathbf{H}\;dx\,dy\;\mathbf{C}^{-1} \tag{11.33}$$

$$\mathbf{F} = \int\!\!\int \mathbf{N}^T q\;dx\,dy = \mathbf{C}^{-T}\int\!\!\int q\;\mathbf{H}^T\;dx\,dy \tag{11.34}$$

Note that because **C** depends only on the nodal coordinates, it has been removed from the integration operation.

11.3 Matlab Implementation

The various steps involved in computing a numerical solution to Equation 11.1 are best illustrated by solving a simple problem constructed to treat many features of the full flexure model. Consider, for example, a square (200 × 200 km) lithospheric plate (with $E = 10^{10}$ Pa and $\nu = 0.3$) whose thickness increases linearly from 10 to 20 km over the model domain in the x-direction (Figure 11.2). The material below the plate is assumed to have a density of 3300 kg m^{-3}, while that above the plate has a density of 1000 kg m^{-3} (e.g., water). The plate is considered to be broken (meaning that $\partial^2 w/\partial x^2 = 0$) along the boundary at $x = 0$, clamped (i.e., $w = 0$ and $\partial w/\partial x = 0$) at the boundary $x = lx$, while at the two y-boundaries the deflection gradients perpendicular to the boundaries are assumed to vanish (i.e., $\partial w/\partial y = 0$). The plate is subjected to an in-plane force of 2×10^{12} Pa m in the x-direction, while the other in-plane forces are taken to be zero. The vertical load on the upper surface of the plate is assumed to be described by the function

$$q = \rho g h_0 \exp\left(-\left(\frac{1}{2}\frac{(x - x_0)^2}{\sigma_x^2} + \frac{1}{2}\frac{(y - y_0)^2}{\sigma_y^2}\right)\right) \tag{11.35}$$

where ρ is the density of the load (assumed to be 2700 kg m^{-3}); h_0 is the maximum height of the load (taken to be 1000 m) centered at x_0 and y_0 (taken as 20 km and 100 km, respectively); and σ_x (=10 km)

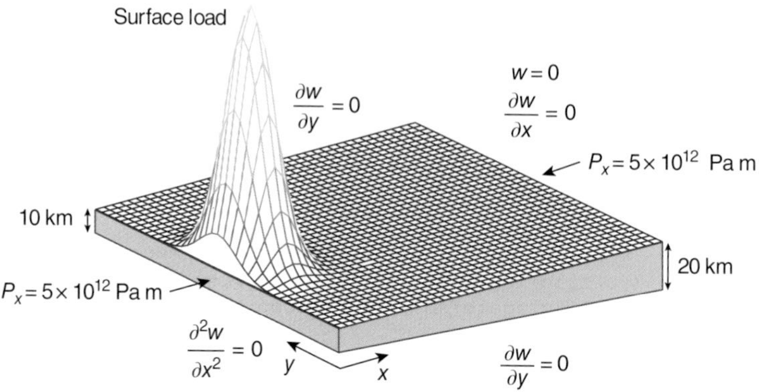

Figure 11.2 Model setup for flexure problem (not to scale). A surface load with an amplitude of 1000 m is applied to a 200×200 km elastic plate. The plate is assumed to be broken along the boundary at $x = 0$. Numerical results for this setup are shown in Figure 11.3.

and σ_y (=15 km) are load length scales in the x- and y-directions, respectively. Our task is to compute using the FEM how the plate deflects under this surface load in the presence of the in-plane force P_x.

We now turn to the practical problem of how to solve the discretized flexural problem with Matlab. A complete listing of the Matlab script is provided in the following text. Many aspects of the code share close similarities with the scripts presented in the previous chapters. The following points deserve special mention:

1) All physical parameters are converted into nondimensional form after their initial definition. This rescaling is performed here to avoid excessively large numbers that may cause the global system matrix to become singular. Nondimensionalization is performed by choosing a single characteristic scale for each fundamental physical quantity present in the problem (in this case there are only two, length, and stress) and then rescaling each parameter in such a way that it becomes dimensionless. The choice of the characteristic scales is arbitrary, but it's best to choose scales that are meaningful with respect to the problem being investigated. The following snippet shows an example of how nondimensionalization is performed:

```
% Definition of parameters as dimensional quantities
lx          = 500e3 ; % length of model domain (m)
ly          = 500e3 ; % length of model domain (m)
emod        = 1e10  ; % Young's modulus (Pa)
pois        = 0.3   ; % Poisson's ratio
ht          = 10e3  ; % plate thickness (m)

% Definition of characterisic scales
length_scale = ht   ; % m
stress_scale = emod ; % Pa

% Scale all dimensional parameters
lx         = lx / length_scale ; % dimensionless length of domain
ly         = ly / length_scale ; % dimensionless length of domain
emod       = emod/stress_scale ; % dimensionless Young's modulus
ht         = ht/length_scale   ; % dimensionless plate thickness
```

One of the important parameters in flexural problems is the flexural rigidity defined as $D = ET_e^3/(12(1 - \nu^2))$; see Equation 11.4. Using the dimensional quantities in the given snippet, the flexural rigidity has a value of 9.16×10^{20} Pa m^3. However, after rescaling the nondimensional flexural rigidity becomes 0.0916. The latter is much less prone to rounding errors than the former. It is important to remember that once scaling has been performed, the solution (i.e., w) that is computed is also dimensionless. Thus, if one wants the deflection in terms of meters, it needs to be multiplied by the characteristic length scale to retrieve its original physical dimension.

2) It should be remembered that even though the governing PDE has a single unknown (w), the discretized form of the same equation has three unknowns (also termed degrees of freedom), w, $\partial w/\partial y$, and $-\partial w/\partial x$ (in this order). Thus, as in Chapter 10 for the coupled problem, here it is important to clearly distinguish between nodes and equations, since the two aren't equivalent as they are in single degree-of-freedom problems. In the presented script, the node numbering is generated as done many times earlier in the text. For example, recalling that the element has four corner nodes (Figure 11.1), one can obtain the array that defines the nodal connectivity with the following snippet:

```
% establish nodal connectivity
gnumbers = reshape(1:nn,[nx ny]) ; % array of nodes
iel = 1 ; % element counter
for j=1:nelx % loop over x-elements
    for i=1:nely % loop over y-elements
      g_num(1,iel) = gnumbers(j,i)      ;  % node 1
      g_num(2,iel) = gnumbers(j,i+1)    ;  % node 2
      g_num(3,iel) = gnumbers(j+1,i+1)  ;  % node 3
      g_num(4,iel) = gnumbers(j+1,i)    ;  % node 4
      iel = iel + 1 ; % increment the element number
    end
end
```

Here, nx and ny are the number of nodes in the x- and y-directions, respectively, while nelx and nely are the number of elements in the x- and y-directions (nx = nelx+1, ...). The array that stores the equation numbers belonging to each node (nf) can be created with the following lines:

```
sdof  = 0 ;     % initialise system degrees of freedom
nf    = zeros(ndof,nn) ; % node degree of freedom array
for n = 1:nn              % loop over all nodes
  for i=1:ndof            % loop over each degree of freedom
     sdof = sdof + 1 ;    % increment global equation number
     nf(i,n) =  sdof ;    % save equation number on each node
  end
end
```

Here, ndof is the number of degrees of freedom on each node (=3) and sdof (after termination of the entire snippet) is the total number of degrees of freedom (i.e., the total number of equations). Finally, the equation numbers for each element can be computed and stored with the following snippet:

```
g = zeros(ntot,1) ; % equation number list for a single element
g_g = zeros(ntot,nels) ;    % equation numbers for all elements
for iel=1:nels ;            % loop over all elements
    num = g_num(:,iel) ;    % extract node numbers
    inc=0 ;                 % initialise counter
    for i=1:nod             % loop over each node
      for k=1:ndof          % loop over each degree of freedom
      inc=inc+1 ;           % local equation number
      g(inc)=nf(k,num(i)) ; % get global equation number
     end
```

```
    end
    g_g(:,iel) = g ; % store equation numbers for each element
end
```

Here, nod is the number of nodes per element element (here 4), ntot is the total number of equations per element (= $3 \times 4 = 12$), and nels is the total number of elements in the mesh.

3) The boundary conditions to be imposed are a broken plate at $x = 0$ (implying $\partial^2 w/\partial x^2 = 0$), clamped conditions at $x = lx$ (meaning $w = 0$ and $\partial w/\partial x = 0$), and zero gradient perpendicular to the $y = 0$ and $y = ly$ boundaries. Recalling that the three degrees of freedom are ordered as w (=1), $\partial w/\partial y$ (=2), and $-\partial w/\partial x$ (=3), respectively, the arrays storing the boundary equations (bcdof) and the fixed boundary values (bcval) for these conditions can be constructed with the following lines:

```
% locate boundary nodes
bx0 = find(g_coord(1,:)==0) ;
bxn = find(g_coord(1,:)==lx);
by0 = find(g_coord(2,:)==0) ;
byn = find(g_coord(2,:)==ly);

% specify boundary equations and the fixed values
% recall dof 1 = w, dof 2 = dwdy and dof 3 = -dwdx
bcdof = [ nf(1,bxn) nf(3,bxn) nf(2,byn) nf(2,byn) ] ;
bcval = zeros(1,length(bcdof)) ;
```

Note that the condition $\partial^2 w/\partial x^2 = 0$ on $x = 0$ is automatically satisfied by leaving the third degree of freedom unconstrained (since then its first derivative normal to the boundary is zero by default).

4) Most flexural problems in Earth science have loads that vary in space. Moreover, in most cases of practical interest, the loads will consist of scattered discrete values and will need to be read into a script from data rather than being governed by some continuous mathematical function (as Equation 11.35). Thus, in order to compute the load vector (Equation 11.34), one must normally interpolate q from the positions where the loads are defined to the integration points. To show how this is easily achieved, we assume here that q is defined initially at the mesh nodes, even though they could more easily be computed directly at integration points.

   A point that is less obvious is that at each node, the load will have three components, one corresponding to each degree of freedom. The load related to the vertical deflection (qw) is provided directly from Equation 11.35 (or from data), while the loads related to the second and third degrees of freedom can be computed by taking the gradient of qw in each spatial direction (refer to Figure 11.1 for definition of the three degrees of freedom).

   The following Matlab snippet performs these various tasks:

```
% Define load vector on nodes
% along with the gradients in each direction
xg = reshape(g_coord(1,:),ny,nx); % x-coords of nodes
yg = reshape(g_coord(2,:),ny,nx); % y-coords of nodes

% surface load function defined on nodes
q0 = h0*rho_load*gr/stress_scale ; % (note nondimensionalisation)
qw = q0*exp(-((xg-x0).^2/(2*sigmax^2)+(yg-y0).^2/(2*sigmay^2)));

% compute gradients of load
[FX,FY] = gradient(qw,dx,dy) ;

% storage of the three load components in vector on nodes
Qg(nf(1,:)) = qw(:) ; % Q (dof=1)
```

```
Qg(nf(2,:)) =  FY(:) ;   % dQdy (dof=2)
Qg(nf(3,:)) = -FX(:) ;   % -dQdx  (dof=3)
```

Note in this snippet that because `h0`, `rho_load`, and `gr` are still dimensional quantities, their product, which has the dimension of Pascals, is divided by the stress scale to ensure that the resulting load is dimensionless.

5) Many of the matrices that need to be constructed in the loops where element integration is performed have not been encountered previously. The first is the matrix **C** (see Equation 11.23) that can be formed within the element loop with the following lines:

```
% Form C matrix for the current element
is = 1 ; ie = 3 ; % start and end storage indices
for i=1:nod       % loop over nodes
 x = coord(i,1) ;  % x coordinates of a node
 y = coord(i,2) ;  % y coordinates of a node
 C(is:ie,:) =...   % C matrix
  [1,x,y,x^2,x*y,y^2,x^3,x^2*y,x*y^2,y^3,x^3*y,x*y^3 ; ... % w
   0,0,1,0,x,2*y,0,x^2,2*x*y,3*y^2,x^3,3*x*y^2 ; ...    % dwdy
   0,-1,0,-2*x,-y,0,-3*x^2,-2*x*y,-y^2,0,-3*x^2*y,-y^3 ]; % -dwdx
 is = ie+1   ; % increment the start storage index
 ie = is+2   ; % increment the end storage index
end
```

Note that because **C** depends only on the nodal coordinates, it can be formed before the loop over integration points. Note also that once **C** has been formed, its inverse (e.g., see Equation 11.24) is easily computed as `iC=inv(C)`.

6) The matrices **H**, **Q**, and **R** (Equations 11.21, 11.28, and 11.30, respectively) depend on the coordinates of integration points expressed in physical coordinates. Because the positions of the integration points are provided in local coordinates, they must be scaled to the physical coordinate system. The x-coordinate of a given integration point can be transformed from local (ξ) to physical (x) coordinates with the following lines:

```
xi = points(k,1);       % local x-coord. of integration point
b  = max(coord(:,1)); % max. x-coordinate in physical space
a  = min(coord(:,1)); % min. x-coordinate in physical space
m  = (b-a)/2  ;        % half element width in x direction
n  = (b+a)/2 ;         % element center in x direction
x  = xi*m+n  ;         % integration point in real x coordinates
```

The local y-coordinate (η) can be transformed in a similar manner. Once the physical coordinates of the integration points have been established, all terms appearing in the matrices **H**, **Q**, and **R** can be computed. In addition, the local plate thickness (that is assumed to depend on the x-coordinate) can now be computed, thereby enabling calculation of the flexural rigidity (D; see Equation 11.4) and the rigidity matrix (**D**; see Equation 11.3).

7) After **H**, **Q**, and **R** have been assigned, the shape functions **N** (Equation 11.25) and the matrices **B** (Equation 11.27) and **A** (Equation 11.29) can be computed with the following lines:

```
fun = H*iC ; % shape functions
B   = Q*iC ; % kinematic matrix
A   = R*iC ; % A matrix
```

The shape functions can be used to interpolate the load from the element nodes (stored in `Qg`; see point 4) to the current integration point with the following snippet:

```
q  = fun*Qg(g)' ; % interp. loads from nodes to int. pt.
```

where g as usual contains a list of the equations for the current element. All that remains to compute the element matrices **KM**, **KP**, **MM**, and **F** is to perform the operations listed in Equations 11.31–11.34, that is,

```
KM  = KM + B'*Dm*B*wts(k) ;        % elastic stiffness matrix
KP  = KP + A'*Pm*A*wts(k) ;        % in-plane force matrix
MM  = MM + fun'*fun*ks*wts(k) ; % mass matrix
f   = f  + fun'*q*wts(k) ;         % load vector
```

Here, `wts(k)` contains the Gauss–Legendre weight for the current integration point. Note that, in this formulation, there is no need to multiply the matrices by the determinant of the Jacobian matrix (as done in the previous scripts) because here, the limits of integration were converted from local to physical coordinates explicitly (see point 6). This marks the end of the integration of the element matrices. The matrices can now be assembled, boundary conditions can be imposed, and the system of discrete equations can be solved, as done in the previous chapters.

Some results computed with the script listed in the following text are illustrated in Figure 11.3. The calculation shows how the surface load causes the plate to bend down under the load near the broken boundary. However, the deflection extends considerably further than the extent of the load itself due to the rigidity of the plate. The x-directed in-plane force (of magnitude 5×10^{12} Pa m) increases the depth of the flexural depression by a factor of approximately 1.6.

```
%----------------------------------------
% Program flex2d.m
% 2-D plate flexure with 4 node
% non-conforming shape functions
% Formulated in terms if w and dwdy, -dwdx
%----------------------------------------

clear

% physical parameters
lx = 200e3 ;   % length of model in x direction
ly = 200e3 ;   % length of model in y direction

Emod   = 1e10      ;     % Young's modulus (Pa)
ht     = 10e3      ;     % elastic thickness (m) at x=0 boundary
pois   = 0.3       ;     % Poisson's ratio
h0          = 1000 ;     % height amplitude of surface load (m)
rho_load   = 2700 ;      % density of surface load (kg/m3)
rho_top    = 1000 ;      % density of material above plate(kg/m3)
rho_bot    = 3300 ;      % density of material below plate(kg/m3)
gr         = 9.8  ;      % gravity (m/s^2)
ks         = (rho_bot-rho_top)*gr ; % drho g
Px         = -500e6*ht ; % in-plane force in x direction (Pa m)
Pxy        = 0 ;           % in-plane shear force    (Pa m)
Py         = 0 ;           % in-plane force in y direction (Pa m)

% characteristic scales for non-dimensionalisation
length_scale = ht    ; % length
stress_scale = Emod ; % stress

% non-dimensionalise variables
lx    = lx/length_scale    ;
ly    = ly/length_scale    ;
Emod = Emod/stress_scale ;
ht    = ht/length_scale    ;
```

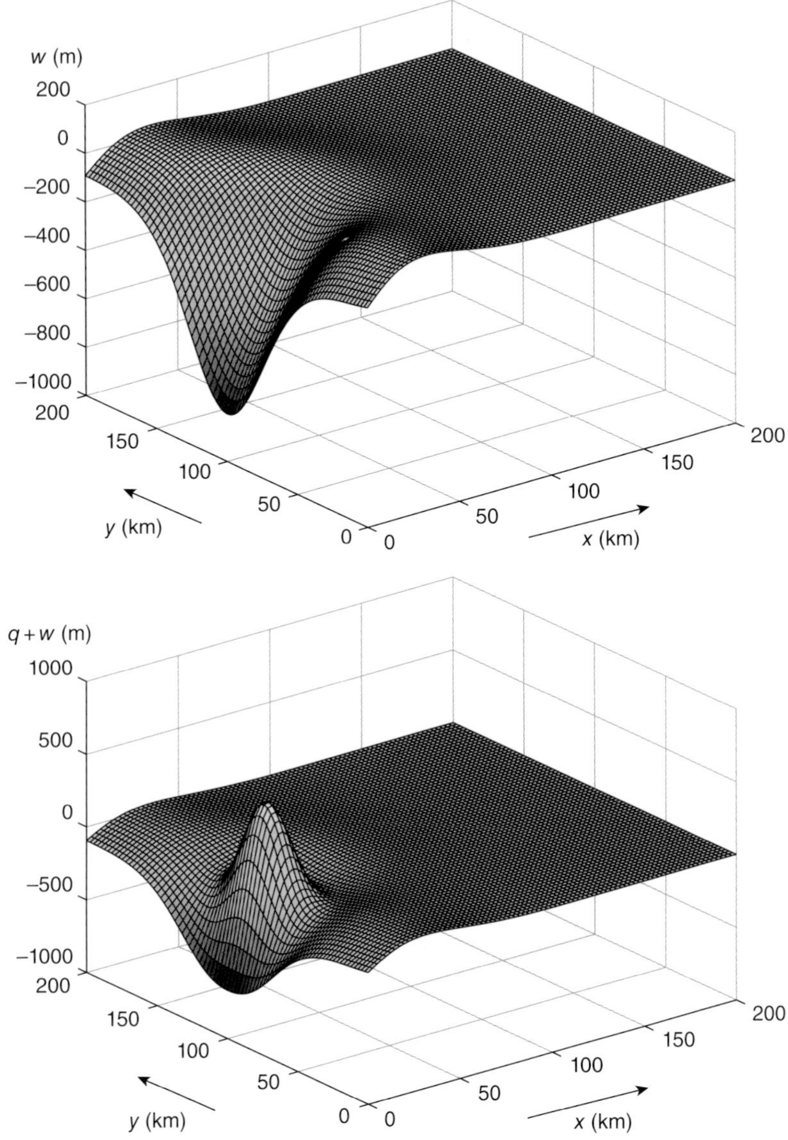

Figure 11.3 Computed deflection (above) and the total topography (i.e., the surface load deformed by the computed deflection, shown in lower panel) for the plate flexure problem defined in Figure 11.2. The calculation was performed using the program listed in the main text, using 80 × 80 finite elements.

```
ks    = ks*length_scale/stress_scale     ;
Px    = Px/(stress_scale*length_scale)   ;
Py    = Py/(stress_scale*length_scale)   ;
Pxy   = Pxy/(stress_scale*length_scale)  ;

Pm      = [ Px Pxy ; Pxy Py ] ; % in-plane force matrix
x0      = 0.1*lx ;    % x coordinate for load center
y0      = ly/2 ;      % y coordinate for load center
```

```
sigmax = 0.05*lx  ; % load length scale in x direction
sigmay = 0.075*ly ; % load length scale in y direction

% numerical parameters
nip    = 9 ;         % number of integration points
ndof   = 3 ;         % number of degrees of freedom per node
nod    = 4 ;         % number of nodes per element
nelx   = 80 ;        % number of nodes in x-direction
nely   = 80 ;        % number of nodes in y-direction
nx     = nelx+1 ;    % nodes in x direction
ny     = nely+1 ;    % nodes in y direction
nn     = nx*ny ;     % total number of nodes
nels   = nelx*nely;  % total number of elements
ntot   = ndof*nod ;  % element degrees of freedom
dx     = lx/nelx  ;  % element size in x direction
dy     = ly/nely  ;  % element size in y direction

%--------------------------------------------
% mesh specifications and equation numbering
%-------------------------------------------

% define mesh (numbering in y direction)
g_coord = zeros(2,nn) ;
n = 1 ;
for i = 1:nx
    for j=1:ny
    g_coord(1,n) = (i-1)*dx ;
    g_coord(2,n) = (j-1)*dy ;
    n = n + 1;
    end
end

% establish nodal connectivity
gnumbers = reshape(1:nn,[nx ny]) ;
iel = 1 ;
for j=1:nelx
    for i=1:nely
      g_num(1,iel) = gnumbers(j,i) ;     % node 1
      g_num(2,iel) = gnumbers(j,i+1) ;   % node 2
      g_num(3,iel) = gnumbers(j+1,i+1);  % node 3
      g_num(4,iel) = gnumbers(j+1,i) ;   % node 4
      iel = iel + 1 ;
    end
end

%-----------------------------------------
% create global-local connection arrays
%-----------------------------------------
sdof   = 0 ;                  % system degrees of freedom
nf     = zeros(ndof,nn) ;  % node degree of freedom array
for n = 1:nn
  for i=1:ndof
     sdof = sdof + 1 ;
     nf(i,n) =  sdof ;
  end
end

% equation number for elements
g = zeros(ntot,1) ;
```

```
g_g = zeros(ntot,nels);
for iel=1:nels ;
    num = g_num(:,iel) ;
    inc=0 ;
    for i=1:nod ; for k=1:ndof ; inc=inc+1 ; g(inc)=nf(k,num(i)) ; end  ;end
    g_g(:,iel) = g ;
end
%-------------------------------------------
% define boundary conditions
%-------------------------------------------
% locate boundary nodes
bx0 = find(g_coord(1,:)==0);
bxn = find(g_coord(1,:)==lx);
by0 = find(g_coord(2,:)==0);
byn = find(g_coord(2,:)==ly);

% specify boundary equations and the fixed values
% recall dof 1 = w, dof 2 = dwdy and dof 3 = -dwdx
bcdof = [ nf(1,bxn) nf(3,bxn) nf(2,byn) nf(2,byn) ] ;
bcval = zeros(1,length(bcdof)) ;

%-----------------------------------------
%  integration data
%-----------------------------------------

% local coordinates of Gauss integration points for nip=3x3
  points(1:3:7,1) = -sqrt(0.6);
  points(2:3:8,1) = 0;
  points(3:3:9,1) = sqrt(0.6);
  points(1:3,2)   = sqrt(0.6);
  points(4:6,2)   = 0 ;
  points(7:9,2)   = -sqrt(0.6);

  % Gauss weights for nip = 3x3 points
  w   = [ 5./9. 8./9. 5./9.] ;
  v   = [ 5./9.*w ; 8./9.*w ; 5./9.*w ] ;
  wts = v(:) ;

%-----------------------------------------
% initialise matrices and vectors
%-----------------------------------------

bv    = zeros(sdof,1);      % system rhs load vector
displ = zeros(sdof,1);      % system solution
lhs   = sparse(sdof,sdof) ; % system stiffness matrix

%-----------------------------------------------------------------
% define load vector on nodes
% along with the gradients in each direction
xg = reshape(g_coord(1,:),ny,nx); % x-coorindates of nodes
yg = reshape(g_coord(2,:),ny,nx); % y-coorindates of nodes

% surface load function defined on nodes
q0 = h0*rho_load*gr/stress_scale ;
qw = q0*exp(-((xg-x0).^2/(2*sigmax^2)+(yg-y0).^2/(2*sigmay^2))) ;

% compute gradients of load
[FX,FY] = gradient(qw,dx,dy) ;
```

```
% storage of the three load components in vector on nodes
Qg(nf(1,:)) =  qw(:) ;  % Q (dof=1)
Qg(nf(2,:)) =  FY(:) ;  % dQdy (dof=2)
Qg(nf(3,:)) = -FX(:) ;  % -dQdx  (dof=3)

%-----------------------------------------
%  element integration and assembly
%-----------------------------------------
for iel=1:nels  % sum over elements

    num       = g_num(:,iel)     ;
    coord     = g_coord(:,num)' ;
    g         = g_g(:,iel)        ;
    KM        = zeros(ntot,ntot) ;
    KP        = zeros(ntot,ntot) ;
    MM        = zeros(ntot,ntot) ;
    f         = zeros(ntot,1)    ;

    % form C matrix for the current element
    is = 1 ; ie = 3 ; % start and end storage indices
    for i=1:nod % loop over nodes
      x = coord(i,1) ; % x coordinates of a node
      y = coord(i,2) ; % y coordinates of a node
      C(is:ie,:) = [ 1   x   y   x^2  x*y   y^2  x^3    x^2*y ...
          x*y^2  y^3   x^3*y  x*y^3 ; ... % w
                     0   0   1   0       x   2*y   0      x^2   ...
                        2*x*y  3*y^2  x^3 3*x*y^2 ; ...% dwdy
                     0  -1   0  -2*x    -y    0    -3*x^2 -2*x*y ...
                        -y^2   0    -3*x^2*y -y^3 ];    % -dwdx
      is = ie+1   ; % increment the start storage index
      ie = is+2   ; % increment the end storage index
    end
    iC = inv(C); % invert C matrix

   for k=1:nip % integration loop
     xi  = points(k,1);      % local x-coordinate of integration point
     b   = max(coord(:,1)); % max. x-coordinate in physical space
     a   = min(coord(:,1)); % min. x-coordinate in physical space
     m   = (b-a)/2 ;         % half element width in x direction
     n   = (b+a)/2 ;         % element center in x direction
     x   = xi*m+n ;          % integration point in real x coordinates
     eta = points(k,2);      % local y-coordinate of integration point
     d   = max(coord(:,2)); % max. y-coordinate in physical space
     c   = min(coord(:,2)); % min. y-coordinate in physical space
     o   = (d-c)/2 ;         % half element width in y direction
     p   = (d+c)/2 ;         % element center in y direction
     y   = eta*o+p   ;       % integration point in real y coordinates
     h   = ht + x/lx*ht ;    % linear variation in plate thickness
     D   = Emod*h^3/12/(1-pois^2) ; % bending stiffness
     Dm  = D*[1 pois 0 ; pois 1 0 ; 0 0 (1-pois)/2] ; % bending matrix
     Q   = [0 0 0 2 0 0  6*x  2*y    0   0   6*x*y      0     ; ...
            0 0 0 0 0 2   0    0   2*x  6*y    0       6*x*y  ; ...
            0 0 0 0 2 0   0   4*x  4*y   0   6*x^2   6*y^2 ];
     H   = [1 x   y   x^2  x*y   y^2  x^3   x^2*y x*y^2 y^3  x^3*y x*y^3];
     R   = [0  1  0  2*x  y  0  3*x^2 2*x*y  y^2   0   3*x^2*y y^3 ;...
           0  0  1  0    x 2*y  0      x^2  2*x*y 3*y^2 x^3 3*x*y^2] ;
     fun = H*iC ;         % shape functions
     B   = Q*iC ;         % kinematic matrix
     A   = R*iC ;         % A matrix
```

```
    q    = fun*Qg(g)' ; % interpolate loads to integration point
    KM   = KM + B'*Dm*B*wts(k) ;     % elastic stiffness matrix
    KP   = KP + A'*Pm*A*wts(k) ;     % inplane force matrix
    MM   = MM + fun'*fun*ks*wts(k) ; % mass matrix (restoring force)
    f    = f + fun'*q*wts(k) ;       % load vector
  end

    % element storage of  matrix coefficients
  lhs(g,g) = lhs(g,g) + KM + KP + MM ;
  bv(g)    = bv(g)   - f ;
end

%----------------------------------------------------

   % apply boundary conditions
   lhs(bcdof,:) = 0 ;
   tmp = spdiags(lhs,0) ; tmp(bcdof)=1 ; lhs=spdiags(tmp,0,lhs);
   bv(bcdof) = bcval ;

   % solution
   displ = lhs \ bv ;

 %---------------------------------
   % plotting of results
 %---------------------------------
 % Extract vertical deflection of plate
 % and reshape it for plotting
 sg = reshape(displ(nf(1,:)),ny,nx);

 figure(1)
 surf(xg*length_scale/1e3,yg*length_scale/1e3,sg*length_scale)
 zlabel('Plate deflection (m)')
 xlabel('X-coordinate (km)')
 ylabel('Y-coordinate (km)')

 figure(2)
 hload = qw*stress_scale/gr/rho_load ; % load (in m)
 surf(xg*length_scale/1e3,yg*length_scale/1e3,(sg*length_scale+hload))
 zlabel('Deflected load (m)')
 xlabel('X-coordinate (km)')
 ylabel('Y-coordinate (km)')

 %---------------------------------
```

Suggested Reading

P. A. Allen, and J. R. Allen, *Basin Analysis*, Wiley Blackwell, Oxford, 2013.

A. J. M. Ferreira, *MATLAB Codes for Finite Element Analysis: Solids and Structures (Solid Mechanics and Its Applications)*, Springer, Berlin, 2009.

Y. W. Kwong, and H. C. Bang, *The Finite Element Method using Matlab*, CRC Press, New York, 2000.

S. Timoshenko, and S. Woinowsky-Krieger, *Theory of Plates and Shells*, McGraw-Hill Book Company, New York, 1959.

D. L. Turcotte, and G. Schubert, *Geodynamics*, Cambridge University Press, Cambridge, 2002.

A. B. Watts, *Isostasy and Flexure of the Lithosphere*, Cambridge University Press, Cambridge, 2001.

O. C. Zienkiewicz, and R. L. Taylor, *The Finite Element Method, Volume 2, Solid Mechanics*, Butterworth Heinemann, Oxford, 2000.

12

Deformation of Earth's Crust

This chapter shows how the finite element method (FEM) can be used to model two-dimensional deformation of viscoelastoplastic materials. In general, this is a coupled nonlinear problem that is solved here with an incremental rate formulation. We present a detailed derivation of the governing equations before listing and describing a single Matlab script that is applied to folding and strain localization. Extension of the presented formulation to 3Ds is conceptually straightforward (based on the approach outlined in Chapter 6) though it requires far greater computing resources and more specialized (efficient) treatments for integration, assembly, and solution (see Chapter 13).

In Chapter 11, we considered deformation of a thin elastic lithospheric plate that can be described by a single variable w, the vertical deflection of the plate mid-plane. This so-called "flexural" model is reasonable for relatively simple geometries and when the applied forces and the resulting deflections are not too large. When these conditions are not satisfied, it is necessary to solve the displacement or velocity field throughout the entire solid. Although the resulting model is computationally more expensive, it has numerous important applications in Earth science such as understanding surface deformation linked to earthquakes and volcanic eruptions. In this chapter, we show how the FEM can be applied to solve problems involving deformation of elastic, viscoelastic, and plastic solids in 2Ds (two dimensions).

12.1 Governing Equations

To determine the displacements, stresses, and strains in a deformable material, one needs to solve the following three governing equations: (i) force balance (i.e., the equations of motion); (ii) a kinematic relation linking stresses and displacements; and (iii) a constitutive relation linking stresses and strains. If inertia is assumed to be negligible (which is reasonable for slow tectonic deformation), the equations of motion in 2Ds can be written as (e.g., see Turcotte and Schubert (1982), Pollard and Fletcher (2005), and Jaeger et al. (2007)) follows:

$$\frac{\partial \sigma_{xx}}{\partial x} + \frac{\partial \sigma_{xz}}{\partial z} = 0$$
$$\frac{\partial \sigma_{xz}}{\partial x} + \frac{\partial \sigma_{zz}}{\partial z} + \rho g = 0 \tag{12.1}$$

Here, σ_{xx}, σ_{zz}, and σ_{xz} $(= \sigma_{zx})$ are stresses (negative in compression), ρ is the density, and g is acceleration due to gravity (Figure 12.1). The strain–displacement relationship defines the changes in geometry that a deforming body undergoes when subjected to forces. This relation has different forms

Practical Finite Element Modeling in Earth Science Using Matlab, First Edition. Guy Simpson.

Companion website: www.wiley.com/go/simpson

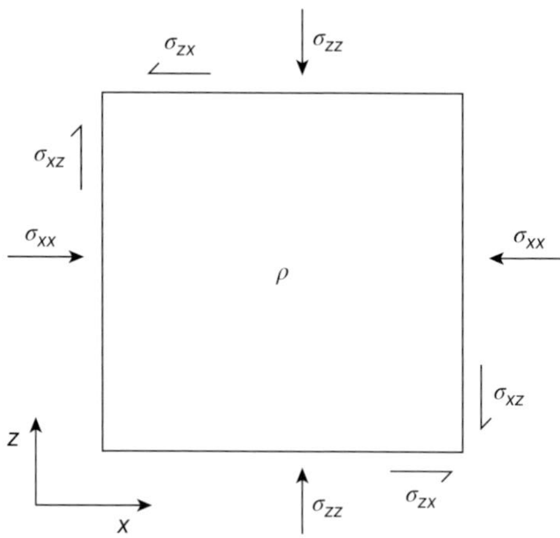

Figure 12.1 Definition of stresses in a 2D solid with density ρ.

depending on whether deformations are small (infinitesimal) or large (finite). For small strains, the kinematic relation between strains and displacements can be written as

$$\begin{Bmatrix} \epsilon_{xx} \\ \epsilon_{zz} \\ \gamma_{xz} \end{Bmatrix} = \begin{bmatrix} \frac{\partial}{\partial x} & 0 \\ 0 & \frac{\partial}{\partial z} \\ \frac{\partial}{\partial z} & \frac{\partial}{\partial x} \end{bmatrix} \begin{Bmatrix} u_x \\ u_z \end{Bmatrix} \tag{12.2}$$

where u_x and u_z are the components of displacement in the x- and z-directions, respectively, and ϵ_{xx}, ϵ_{zz}, and γ_{xz} are the strains. A constitutive relation describes different types of material behavior: for example, elastic, viscoelastic, and plastic. These relations are based on experimental observations. As an example, the stress–strain relations for an isotropic elastic material can be written as

$$\begin{Bmatrix} \sigma_{xx} \\ \sigma_{zz} \\ \sigma_{xz} \end{Bmatrix} = \begin{bmatrix} K + \frac{4}{3}G & K - \frac{2}{3}G & 0 \\ K - \frac{2}{3}G & K + \frac{4}{3}G & 0 \\ 0 & 0 & G \end{bmatrix} \begin{Bmatrix} \epsilon_{xx} \\ \epsilon_{zz} \\ \gamma_{xz} \end{Bmatrix} \tag{12.3}$$

where G is the shear modulus and K is the bulk modulus.

Using matrix notation, these three relations can be written compactly as

$$\mathbf{B}^T \sigma = -\mathbf{f} \tag{12.4}$$

$$\epsilon = \mathbf{B}\,\mathbf{u} \tag{12.5}$$

$$\sigma = \mathbf{D}\,\epsilon \tag{12.6}$$

where

$$\mathbf{u} = \begin{Bmatrix} u_x \\ u_z \end{Bmatrix} \tag{12.7}$$

$$\mathbf{f} = \begin{Bmatrix} 0 \\ \rho g \end{Bmatrix} \tag{12.8}$$

$$\mathbf{B} = \begin{bmatrix} \frac{\partial}{\partial x} & 0 \\ 0 & \frac{\partial}{\partial z} \\ \frac{\partial}{\partial z} & \frac{\partial}{\partial x} \end{bmatrix} \tag{12.9}$$

$$\mathbf{D} = \begin{bmatrix} K + \frac{4}{3}G & K - \frac{2}{3}G & 0 \\ K - \frac{2}{3}G & K + \frac{4}{3}G & 0 \\ 0 & 0 & G \end{bmatrix} \tag{12.10}$$

$$\epsilon = \begin{Bmatrix} \epsilon_{xx} \\ \epsilon_{zz} \\ \gamma_{xz} \end{Bmatrix} \tag{12.11}$$

and

$$\sigma = \begin{Bmatrix} \sigma_{xx} \\ \sigma_{zz} \\ \sigma_{xz} \end{Bmatrix} \tag{12.12}$$

In a displacement formulation, one eliminates $\boldsymbol{\sigma}$ and $\boldsymbol{\epsilon}$ in the following steps:

$$\mathbf{B}^T \sigma = -\mathbf{f} \tag{12.13}$$

$$\mathbf{B}^T \mathbf{D}\, \epsilon = -\mathbf{f} \tag{12.14}$$

$$\mathbf{B}^T \mathbf{D}\, \mathbf{B}\, \mathbf{u} = -\mathbf{f} \tag{12.15}$$

This last expression is a pair of elliptic partial differential equations (PDEs) for the two unknowns, u_x and u_z (i.e., the two components of the vector $\mathbf{u}$). Note that the governing equations do not contain any time derivative and so the displacements are always in steady state, as was the case for flexural deformation considered in Chapter 11.

12.2 Rate Formulation

If one desires, one can proceed and discretize Equation 12.15 with the FEM to obtain the nodal displacements (after applying boundary conditions), which can then be used to back-compute the stresses and strains throughout the solid. Here, we present an alternative rate formulation that expresses the unknowns in terms of velocities rather than displacements (see also Simpson (2006) and Simpson (2009)). The rate formulation is especially suitable for modeling materials that exhibit time-dependent behavior (e.g., viscous deformation). The rate formulation can be obtained from the displacement equations by differentiating the various terms with respect to time, thereby converting displacements to velocities. For example, differentiating Equation 12.5 with respect to time gives the following kinematic relationship between strain rates $\dot{\epsilon}$ and velocities $\mathbf{v}$:

$$\dot{\epsilon} = \mathbf{B}\, \mathbf{v} \tag{12.16}$$

Similarly, differentiating the constitutive relation with respect to time and discretizing the stress rate with explicit finite differences leads to

$$\dot{\sigma} = \frac{\sigma - \sigma^0}{\Delta t} = \mathbf{D}\, \dot{\epsilon} \tag{12.17}$$

$$\rightarrow \sigma = \Delta t\, \mathbf{D}\, \dot{\epsilon} + \sigma^0 \tag{12.18}$$

where Δt is the time interval between the stresses $\boldsymbol{\sigma}$ and $\boldsymbol{\sigma}^0$. In anticipation of incorporating different material behavior, we rewrite Equation 12.18 in the general form as follows:

$$\boldsymbol{\sigma} = \tilde{\mathbf{D}}\,\dot{\epsilon} + \mathbf{D}_\mathbf{s}\,\boldsymbol{\sigma}^0 \tag{12.19}$$

Here, $\tilde{\mathbf{D}}$ is the constitutive material matrix and $\mathbf{D}_\mathbf{s}$ is a matrix containing coefficients related to the old stresses $\boldsymbol{\sigma}^0$ (with components σ_{xx}, σ_{zz}, and σ_{xz}). In the case of elasticity, $\mathbf{D}_\mathbf{s}$ is simply the identity matrix, that is,

$$\mathbf{D}_\mathbf{s} = \begin{bmatrix} 1 & 0 & 0 \\ 0 & 1 & 0 \\ 0 & 0 & 1 \end{bmatrix} \tag{12.20}$$

while $\tilde{\mathbf{D}} = \Delta t\,\mathbf{D}$ (see Equation 12.10). When the constitutive relation is written as it is in Equation 12.19, different material behavior can easily be incorporated (as is done in Equation 12.21) by making appropriate choices for $\tilde{\mathbf{D}}$, and $\mathbf{D}_\mathbf{s}$. Combining Equations 12.4, 12.16, and 12.19 leads to the final governing equation

$$\mathbf{B}^T\,\tilde{\mathbf{D}}\,\mathbf{B}\,\mathbf{v} = -\mathbf{f} - \mathbf{B}^T\,\mathbf{D}_\mathbf{s}\,\boldsymbol{\sigma}^0 \tag{12.21}$$

where the only unknown is the velocity vector $\mathbf{v}$.

12.3 FEM Discretization

We now proceed and discretize Equation 12.21 with the FEM. We use nine-node quadrilaterals and the corresponding biquadratic shape functions introduced previously in Chapter 7 (see also Chapter 10). For a single element, the x- and z-components of the velocity vector can be approximated as

$$v_x \simeq [N_1\ N_2\ N_3\ N_4\ N_5\ N_6\ N_7\ N_8\ N_9] \begin{Bmatrix} v_{x_1} \\ v_{x_2} \\ v_{x_3} \\ v_{x_4} \\ v_{x_5} \\ v_{x_6} \\ v_{x_7} \\ v_{x_8} \\ v_{x_9} \end{Bmatrix} = \mathbf{N}\mathbf{v}_x \tag{12.22}$$

$$v_z \simeq [N_1\ N_2\ N_3\ N_4\ N_5\ N_6\ N_7\ N_8\ N_9] \begin{Bmatrix} v_{z_1} \\ v_{z_2} \\ v_{z_3} \\ v_{z_4} \\ v_{z_5} \\ v_{z_6} \\ v_{z_7} \\ v_{z_8} \\ v_{z_9} \end{Bmatrix} = \mathbf{N}\mathbf{v}_z \tag{12.23}$$

where v_x and v_z are the continuous velocities in the governing PDEs, $\mathbf{v}_x$ and $\mathbf{v}_z$ are the nodal velocities, and **N** are the shape functions (see Equation 7.53). Substituting these approximations into Equation 12.21, weighting each equation with the shape functions, integrating over the element, and integrating by parts where necessary, the following set of discrete element equations (after neglecting the boundary integral terms) can be obtained:

$$\mathbf{KM}\ \mathbf{v} = \mathbf{f}_s + \mathbf{f} \tag{12.24}$$

where

$$\mathbf{KM} = \int\int \mathbf{B}^T \tilde{\mathbf{D}} \mathbf{B}\ dx\, dz \tag{12.25}$$

$$\mathbf{f} = \int\int \mathbf{N}^T \rho g\ dx\, dz \tag{12.26}$$

$$\mathbf{f_s} = \int\int \mathbf{B}^T \mathbf{D}_s \sigma^0\ dx\, dz \tag{12.27}$$

$$\mathbf{B} = \begin{bmatrix} \frac{\partial N_1}{\partial x} & 0 & \frac{\partial N_2}{\partial x} & 0 & \cdots\cdots & \frac{\partial N_9}{\partial x} & 0 \\ 0 & \frac{\partial N_1}{\partial z} & 0 & \frac{\partial N_2}{\partial z} & \cdots\cdots & 0 & \frac{\partial N_9}{\partial z} \\ \frac{\partial N_1}{\partial z} & \frac{\partial N_1}{\partial x} & \frac{\partial N_2}{\partial z} & \frac{\partial N_2}{\partial x} & \cdots\cdots & \frac{\partial N_9}{\partial z} & \frac{\partial N_9}{\partial x} \end{bmatrix} \tag{12.28}$$

and

$$\mathbf{v} = \begin{Bmatrix} v_{x_1} \\ v_{z_1} \\ v_{x_2} \\ v_{z_2} \\ \vdots \\ v_{x_9} \\ v_{z_9} \end{Bmatrix} \tag{12.29}$$

In anticipation of including a nonlinear material response, it is useful to rewrite Equation 12.24 in the following incremental form:

$$\mathbf{KM}\ \Delta\mathbf{v}_{i+1} = \mathbf{R}_i \tag{12.30}$$

Here, $\Delta\mathbf{v}_{i+1}$ is the velocity increment, $\mathbf{R}_i$ is the right-hand-side load vector, and i refers to the iteration number at a given time level. The total velocity can be accumulated over each iteration as

$$\mathbf{v}_{i+1} = \mathbf{v}_i + \Delta\mathbf{v}_{i+1} \tag{12.31}$$

The first iteration is computed with $\mathbf{v}_1 = 0$ and $\mathbf{R}_1 = \mathbf{f}_s + \mathbf{f}$ (i.e., the right-hand side of Equation 12.24), while for all other iterations $\mathbf{R}_i$ is computed as

$$\mathbf{R}_i = \mathbf{B}^T \sigma_i + \mathbf{f} \tag{12.32}$$

where $\mathbf{R}_i$ is the residual (or error) of the force balance equation (see Equation 12.4). For a linear problem (e.g., with elastic material behavior), when **R** is computed after the first iteration, it equals

$\mathbf{R} = \mathbf{KM}\ \mathbf{v} - \mathbf{f}_s - \mathbf{f} = 0$ (viz. Equation 12.24). Thus, when the system is resolved on the next iteration with this $\mathbf{R}$ as a load vector, it gives $\mathbf{\Delta v} = 0$ and the solution has converged. However, if $\boldsymbol{\sigma}$ is modified in a way that is not accounted for in the relation $\boldsymbol{\sigma} = \tilde{\mathbf{D}}\,\mathbf{B}\ \mathbf{v} - \mathbf{D}_s\,\boldsymbol{\sigma}^0$ (e.g., due to plastic deformation), then $\mathbf{R}$ will not be zero after the first iteration, which will generate additional deformation ($\mathbf{\Delta v} \neq 0$) to re-establish force equilibrium (requiring iterations).

12.4 Viscoelastoplasticity

Because most rocks can be described by some combination of elastic, viscous, or plastic behavior, it is desirable to include all three material responses within a single formulation. Plasticity takes on different meanings in different disciplines. Here, plastic deformation refers to time-independent permanent deformation (e.g., Mohr–Coulomb brittle behavior). In this section, we derive the various terms appearing in Equation 12.30 for a viscoelastoplastic material. The approach taken is to incorporate the viscoelastic portion of the material behavior into the matrices $\tilde{\mathbf{D}}$, and $\mathbf{D}_s$ while plasticity in accounted for by modifying the right-hand-side load vector $\mathbf{R}$.

Before introducing the constitutive relations for a viscoelastoplastic material, we begin by decomposing the stress and strain tensors into their deviatoric and dilatational parts as follows:

$$\sigma_m = \frac{1}{3}(\sigma_{xx} + \sigma_{yy} + \sigma_{zz}) \tag{12.33}$$

$$\tilde{\boldsymbol{\sigma}} = \boldsymbol{\sigma} - \mathbf{m}\,\sigma_m \tag{12.34}$$

$$\dot{\epsilon}_v = \frac{1}{3}(\dot{\epsilon}_{xx} + 0 + \dot{\epsilon}_{zz}) \tag{12.35}$$

$$\tilde{\dot{\boldsymbol{\epsilon}}} = \dot{\boldsymbol{\epsilon}} - \mathbf{m}\,\dot{\epsilon}_v \tag{12.36}$$

Here, tildes refer to deviatoric components and

$$\mathbf{m} = \begin{Bmatrix} 1 \\ 1 \\ 0 \end{Bmatrix} \tag{12.37}$$

$$\boldsymbol{\sigma} = \begin{Bmatrix} \sigma_{xx} \\ \sigma_{zz} \\ \sigma_{xz} \end{Bmatrix} \tag{12.38}$$

and

$$\dot{\boldsymbol{\epsilon}} = \begin{Bmatrix} \dot{\epsilon}_{xx} \\ \dot{\epsilon}_{zz} \\ \dot{\epsilon}_{xz} \end{Bmatrix} \tag{12.39}$$

Stresses are assumed to be negative in compression. Note that because $\dot{\epsilon}_{yy} = 0$, plane strain conditions have been assumed.

We now consider a viscoelastoplastic material response. With respect to viscoelastic deformation, it is assumed that deviatoric components of deformation are governed by the Maxwell viscoelastic model (Figure 12.2), that is,

$$\tilde{\dot{\epsilon}} = \frac{1}{2G}\frac{\partial \tilde{\sigma}}{\partial t} + \frac{1}{2\mu}\tilde{\sigma} \tag{12.40}$$

where G is the elastic shear modulus and μ is the shear viscosity. This equation can be discretized in time to give

$$\tilde{\dot{\epsilon}} = \frac{1}{2G}\frac{(\tilde{\sigma} - \tilde{\sigma}^0)}{\Delta t} + \frac{1}{2\mu}\tilde{\sigma} \tag{12.41}$$

where $\tilde{\boldsymbol{\sigma}}$ are the new stresses and $\tilde{\boldsymbol{\sigma}}^0$ are stresses from the previous time step. Dilatational deformation is modeled with a pure elastic model:

$$\frac{1}{3K}\sigma_m = \epsilon_v \tag{12.42}$$

where K is the elastic bulk modulus. Differentiating with respect to time gives

$$\frac{1}{3K}\frac{\partial \sigma_m}{\partial t} = \dot{\epsilon}_v \tag{12.43}$$

that can be discretized in time to give

$$\frac{1}{3K}\frac{(\sigma_m - \sigma_m^0)}{\Delta t} - \dot{\epsilon}_v = 0 \tag{12.44}$$

where σ_m^0 refers to the mean stress from the previous time step. Substituting Equations 12.33–12.36 into Equations 12.41 and 12.44 and solving for the new stresses lead to a relation of the same form as Equation 12.19. Collecting coefficients in front of strain rates ($\dot{\epsilon}$) yields the viscoelastic material matrix

$$\tilde{\mathbf{D}} = \begin{bmatrix} \frac{\Delta t(4G\mu+3\mu K+3\Delta tGK)}{3(\mu+\Delta tG)} & \frac{\Delta t(3\mu K+3\Delta tGK-2G\mu)}{3(\mu+\Delta tG)} & 0 \\ \frac{\Delta t(3\mu K+3\Delta tGK-2G\mu)}{3(\mu+\Delta tG)} & \frac{\Delta t(4G\mu+3\mu K+3\Delta tGK)}{3(\mu+\Delta tG)} & 0 \\ 0 & 0 & \frac{2\mu\Delta tG}{2\mu+\Delta tG} \end{bmatrix} \tag{12.45}$$

while collecting coefficients in front of the old stresses ($\boldsymbol{\sigma}^0$) leads to

$$\mathbf{D}_s = \begin{bmatrix} \frac{3\mu+\Delta tG}{3(\mu+\Delta tG)} & \frac{\Delta tG}{3(\mu+\Delta tG)} & 0 \\ \frac{\Delta tG}{3(\mu+\Delta tG)} & \frac{3\mu+\Delta tG}{3(\mu+\Delta tG)} & 0 \\ 0 & 0 & \frac{2\mu}{(2\mu+\Delta tG)} \end{bmatrix} \tag{12.46}$$

Plastic deformation is modeled with the pressure sensitive Mohr–Coulomb law, which is governed by the following yield function (Figure 12.2):

$$F = \tau^* + \sigma^* \sin\phi - c\cos\phi \tag{12.47}$$

Here, c is the cohesive rock strength, ϕ is the angle of internal friction,

$$\sigma^* = \frac{1}{2}(\sigma_{xx} + \sigma_{zz}) \tag{12.48}$$

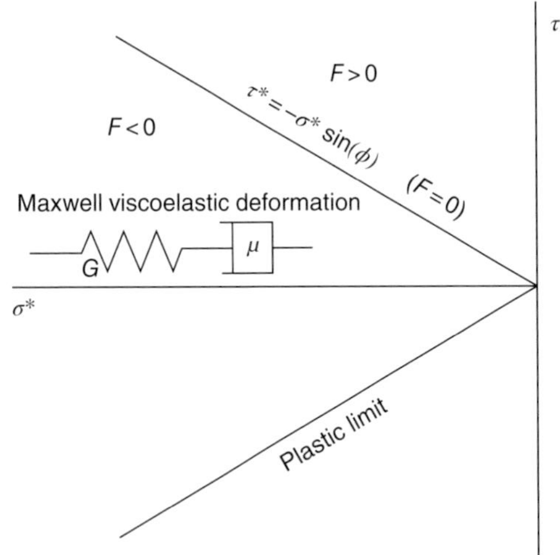

Figure 12.2 Depiction of Maxwell viscoelastic material with a stress state limited by the Mohr–Coulomb law (here with zero cohesion). Note that σ^* is taken to be negative in compression.

and

$$\tau^* = \sqrt{\frac{1}{4}(\sigma_{xx} - \sigma_{zz})^2 + \sigma_{xz}^2} \tag{12.49}$$

Deformation is viscoelastic (or elastic) when $F < 0$ and plastic when $F > 0$, requiring stresses to be reduced such that $F = 0$. Plasticity is assumed here to be nonassociated with a dilatancy angle of zero, implying no plastic volumetric strain (Vermeer and DeBorst, 1984). This condition leads to a particularly simple algorithm to return stresses of plastic points to the yield surface,

$$\sigma_{xx} = \sigma^* + \frac{1}{2}(\sigma_{xx} - \sigma_{zz})\beta \tag{12.50}$$

$$\sigma_{zz} = \sigma^* - \frac{1}{2}(\sigma_{xx} - \sigma_{zz})\beta \tag{12.51}$$

$$\sigma_{xz} = \sigma_{xz}\beta \tag{12.52}$$

where

$$\beta = \frac{|(c\cos\phi - \sigma^* \sin\phi|}{\tau^*} \tag{12.53}$$

All stresses on the right-hand side of Equations 12.50–12.52 are understood to be old stresses (i.e., $F > 0$), whereas stresses on the left satisfy $F = 0$. The difference between old (viscoelastic) stresses and new (plasticity-corrected) stresses generates out-of-balance loads (viz Equation 12.32) that are added to the right-hand side of Equation 12.30. Thus, the approach is to perform repeated viscoelastic (or elastic) solutions with stresses satisfying the plastic failure criteria and to achieve convergence by iteratively varying the load vector.

12.5 Matlab Implementation

The viscoelastoplastic model presented in Section 12.4 is straightforward to implement in a Maltab finite element code. In the following text, we present a script to show how this is performed in practice. Two problems are investigated to illustrate how the model can be applied (Figure 12.3). The

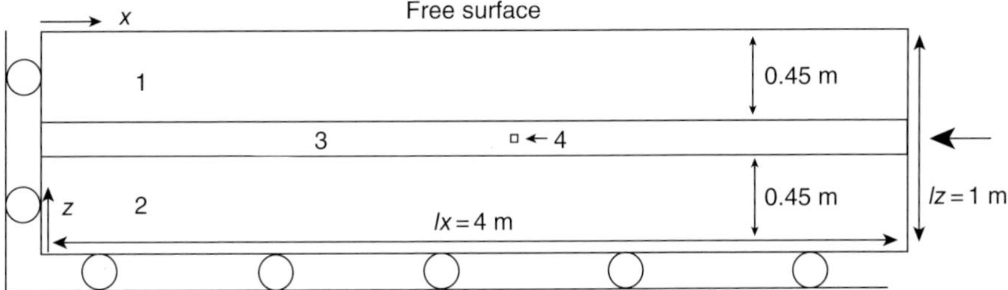

Figure 12.3 Setup for folding and shear localization numerical simulations (see results in Figures 12.4 and 12.5). Numbers indicate the lithological unit (termed the "phase" in the Matlab script). The folding experiment was performed with a viscoelastic material where the central layer (i.e., phases 3 and 4) has a shear viscosity (10^{20} Pa s) that is 100 times greater than that of the surrounding matrix (10^{18} Pa s). Other parameters are as follows: density = 2700 kg m^{-3}, gravity = 9.8 m s^{-2}, Young's modulus = 10^{11} Pa, Poisson's ratio = 0.3 (all considered uniform throughout), boundary velocity = 5 mm year^{-1}, and nxe = nze = 80. The shear localization experiment was performed with a viscoelastoplastic material with the following material properties: Young's modulus for layer = 10^{10} Pa, Young's modulus for matrix = 0.5×10^{10} Pa, Poisson's ratio = 0.3, shear viscosity = 10^{22} Pa s, cohesion = 20 MPa, cohesion for phase 4 = 18 MPa, friction angle 30°, dilatancy angle 0°, boundary velocity = 5 mm year^{-1}, and nxe = nze = 200.

first problem treats the folding of a layered viscoelastic material (neglecting plasticity), while the second case deals with strain localization in a viscoelastoplastic material. The basic steps that must be performed in both cases can be summarized as follows:

1) Assign all material properties (e.g., shear and bulk moduli, viscosity, and friction coefficient) as vectors, along with other physical parameters such as the length and depth of the model domain. This can be done, for example, with lines such as

```
lx      = 4 ;                          % Length in x-direction (m)
lz      = 1 ;                          % Length in z direction (m)
visc_v  = [ 1 1 100 100]*1e20     ;    % Viscosity (Pa s)
coh_v   =  20*[1 1 1 0.9 ]*1e6    ;    % cohesion (Pa)
phi_v   = ([1 1 1 1])*30*pi/180   ;    % friction angle (radians)
```

   In the case of the viscosity vector `visc_v`, note that the phases (or units) 1 and 2 have shear viscosities of 10^{20} Pa s, while phases 3 and 4 have viscosities of 10^{22} Pa s.

2) Nondimensionalize all physical parameters using characteristic scales for length (taken as `lz`, the initial depth extent of the model domain), stress (the maximum shear modulus), and time (inverse of the initially imposed horizontal strain rate, `edot`). Note that a similar approach was also applied and discussed in slightly more detail in Chapter 11. The following snippet shows an example of how this is performed:

```
% Define characteristic scales
length_scale = lz ;
stress_scale = max(smod_v) ;
time_scale   = 1/edot ;

% Non-dimensional scaling
lx     = lx/length_scale ;
lz     = lz/length_scale ;
visc_v = visc_v/(stress_scale*time_scale) ;
coh_v  = coh_v/stress_scale ;
```

3) Define all numerical parameters (e.g., number of elements in each direction, time step, and number of integration points).

4) Define the mesh. For example, for the nine-node mesh this can be achieved with the following Matlab snippet:

```
% Mesh coordinates
g_coord = zeros(2,nn) ; % storage for node coordinates
n = 1 ;                  % initialise node counter
for i = 1:nx             % loop over nodes in x-direction
    for j=1:nz           % loop over nodes in z-direction
      g_coord(1,n) = (i-1)*dx/2  ;
      g_coord(2,n) = (j-1)*dz/2  ;
      n = n + 1 ;        % increment node counter
    end
end
```

5) Recall that discretization of the governing equation was performed here with nine-node quadrilaterals (see Section 12.3). The array defining the nodes for each element (`g_num`) can be formed with the following snippet:

```
% establish node numbering for each element
gnumbers = reshape(1:nn,[ny nx]) ; % grid of node numbers
g_num      = zeros(nod,nels);
iel = 1 ; % intialise element number
for i=1:2:nx-1 % loop over x-nodes
    for j=1:2:nz-1 % loop over z-nodes
      g_num(1,iel) = gnumbers(j,i)     ;     % node 1
      g_num(2,iel) = gnumbers(j+1,i) ;       % node 2
      g_num(3,iel) = gnumbers(j+2,i) ;       % node 3
      g_num(4,iel) = gnumbers(j+2,i+1) ;     % node 4
      g_num(5,iel) = gnumbers(j+2,i+2) ;     % node 5
      g_num(6,iel) = gnumbers(j+1,i+2) ;     % node 6
      g_num(7,iel) = gnumbers(j,i+2) ;       % node 7
      g_num(8,iel) = gnumbers(j,i+1) ;       % node 8
      g_num(9,iel) = gnumbers(j+1,i+1) ;     % node 9
      iel = iel + 1 ; % increment the element number
    end
end
```

6) The arrays containing the equation numbers for each node (`nf`) and the equation numbers for each element (`g_g`) can be formed as done for previous coupled problems. For example, recalling that each node has two degrees of freedom (v_x and v_z) and that the degrees of freedom for each element are ordered as $v_{x_1}, v_{z_1}, v_{x_2}, v_{z_2}, \dots, v_{x_9}, v_{z_9}$, these two arrays can be constructed with the following snippet:

```
% Define equation numbering on nodes
sdof   = 0 ;                % system degrees of freedom
nf     = zeros(ndof,nn) ;   % node degree of freedom array
for n = 1:nn                % loop over all nodes
  for i=1:ndof              % loop over each degree of freedom
     sdof = sdof + 1 ;      % increment total number of equations
     nf(i,n) =  sdof ;      % save equation number for each node
  end
end

% Equation numbering for each element
g   = zeros(ntot,1) ;
g_g = zeros(ntot,nels);
for iel=1:nels ;
    num = g_num(:,iel) ; % extract nodes for the element
```

```
    inc=0 ;
    for i=1:nod
      for k=1:2
        inc=inc+1 ;
        g(inc) = nf(k,num(i)) ;
      end
    end
    g_g(:,iel) = g ;
end
```

Note that this numbering scheme differs from the nine-node quadrilateral mesh illustrated in Figure 7.12 where the two degrees of freedom were ordered as $p_1, p_2, \ldots p_9, c_1, c_2, \ldots, c_9$. The actual ordering scheme is arbitrary; but once one is chosen, it's important to maintain consistency throughout the program.

7) The boundary conditions to be imposed in both experiments are illustrated in Figure 12.3. The arrays containing the boundary equations (`bcdof`) and their fixed values (`bcval`) can be generated with the following snippet:

```
% boundary nodes
bx0 = find(g_coord(1,:)==0)    ;
bxn = find(g_coord(1,:)==lx)   ;
bz0 = find(g_coord(2,:)==0)    ;
bzn = find(g_coord(2,:)==lz) ;

% Fixed boundary equations  (vx = dof 1, vz = dof2)
%  along with their values
bcdof = [nf(1,bx0)    nf(1,bxn)    nf(2,bz0) ]  ;
bcval = [zeros(1,length(bx0)) -bvel*ones(1,length(bxn)) zeros(1,length(bz0)) ]  ;
```

Here, `bvel` is the imposed velocity at the $x = lx$ boundary. Note that the $z = 0$ boundary is located at the base of the model, not at the surface (Figure 12.3).

8) As noted already, each element is assigned a phase number to incorporate different material properties in different parts of the model. The models here include four different phases, though it's straightforward to include more. The phase numbers for the domain illustrated in Figure 12.3 can be created with the following lines:

```
% Establish phases
phase = ones(1,nels) ; % upper part of domain
phase(g_coord(2,g_num(9,:))<layer_bot)=2 ; % lower part of domain
phase(find(g_coord(2,g_num(9,:))≤layer_top & ...
    g_coord(2,g_num(9,:))≥layer_bot))=3 ; % middle layer
phase(nze*nxe/2-nze/2)=4 ;  % central inclusion
```

Here, the variables `layer_bot` and `layer_top` are the z-coordinates at the base ($z = 0.45$) and top ($z = 0.55$) of the central layer and the fourth phase is assigned to a single element located at the center of the model domain (Figure 12.3).

9) A small random perturbation is added to the mesh that serves to facilitate development of deformation (folding or shear) instabilities. This is done by randomly shifting the vertical coordinates at the boundaries of the central layer by a magnitude of up to 5% of the vertical element spacing `dz`. This can be achieved with the following lines:

```
% Perturb boundaries of central layer
ii=find(g_coord(2,:)==layer_bot | g_coord(2,:)==layer_top)    ;
g_coord(2,ii)=g_coord(2,ii) + dz*0.05*(rand(1,length(ii))-0.5) ;
```

10) The element stiffness matrix and load vectors (see Equations 12.25–12.27) are evaluated in the presented program by Gauss–Legendre Quadrature using nine integration points (three in each direction). Before the integrals are evaluated, the locations of the integration points and the weights must be provided, which can be achieved with the following lines (e.g., see Table 4.1):

```
% Local coordinates of Gauss integration points for nip=3x3
  points(1:3:7,1)  = -sqrt(0.6);
  points(2:3:8,1)  = 0;
  points(3:3:9,1)  = sqrt(0.6);
  points(1:3,2)    = sqrt(0.6);
  points(4:6,2)    = 0 ;
  points(7:9,2)    = -sqrt(0.6);

  % Gauss weights for nip=3x3
  w    = [ 5./9. 8./9. 5./9.] ;
  v    = [ 5./9.*w ; 8./9.*w ; 5./9.*w ] ;
  wts = v(:) ;
```

11) The shape functions and shape functions derivatives that correspond to the nine-node quadrilateral elements can be defined, evaluated at the integration points and saved for later use with the following snippet (see also Sections 7.3 and 10.2):

```
% Evaluate shape functions and their derivatives
% at integration points and save the results
 for k = 1:nip
    xi   = points(k,1);
    eta  = points(k,2);
    etam = eta - 1; etap = eta + 1 ;
    xim  = xi - 1 ; xip  = xi + 1  ;
    x2p1 = 2*xi+1 ;   x2m1 = 2*xi-1 ;
    e2p1 = 2*eta+1 ;  e2m1 = 2*eta-1 ;
    % shape functions
    fun= [ .25*xi*xim*eta*etam -.5*xi*xim*etap*etam ...
    .25*xi*xim*eta*etap -.5*xip*xim*eta*etap ...
    .25*xi*xip*eta*etap -.5*xi*xip*etap*etam ...
    .25*xi*xip*eta*etam -.5*xip*xim*eta*etam xip*xim*etap*etam ] ;
    % derivatives of shape functions
    der(1,1)  = 0.25*x2m1*eta*etam   ; %dN1dxi
    der(1,2)  =-0.5*x2m1*etap*etam ;  %dN2dxi, etc
    der(1,3)  = 0.25*x2m1*eta*etap   ;
    der(1,4)  =      -xi*eta*etap ;
    der(1,5)  = 0.25*x2p1*eta*etap   ;
    der(1,6)  =-0.5*x2p1*etap*etam ;
    der(1,7)  = 0.25*x2p1*eta*etam   ;
    der(1,8)  =      -xi*eta*etam ;
    der(1,9)  = 2*xi*etap*etam      ;
    der(2,1)  = 0.25*xi*xim*e2m1 ;  %dN1deta
    der(2,2)  =-xi*xim*eta         ;  %dN2deta, etc
    der(2,3)  = 0.25*xi*xim*e2p1 ;
    der(2,4)  =-0.5*xip*xim*e2p1    ;
    der(2,5)  = 0.25*xi*xip*e2p1 ;
    der(2,6)  =-xi*xip*eta           ;
    der(2,7)  = 0.25*xi*xip*e2m1 ;
    der(2,8)  =-0.5*xip*xim*e2m1    ;
    der(2,9)  = 2*xip*xim*eta ;
    fun_s(k,:)  = fun ; % save shape functions
    der_s(:,:,k)  = der ; % save derivatives
 end
```

12) Once the various system matrices and load vectors have been initialized, it is necessary to perform the element integration and assembly before solving for the nodal velocities. This must be done within a time loop (since the matrices will generally vary in time), and it must also be done within a loop involving multiple iterations at each time level, because the load vector contains loads generated by plastic deformation (see Equation 12.32) that make the problem nonlinear. The general structure of a nonlinear program is illustrated in Figure 7.13. Within these various loops, the following major tasks must be performed:

- In a loop over all elements, retrieve the material properties for the current element. This can be performed using lines such as the following:

```
smod   = smod_v(phase(iel))   ; % shear modulus
K      = bmod_v(phase(iel))   ; % bulk modulus
```

  Here, `smod_v` and `bmod_v` are vectors containing the shear and bulk moduli with values for each unit (phase) and `phase` is an array that contains the phase number (in this case, an integer between 1 and 4; see Figure 12.3). Once the material properties have been obtained, the viscoelastic material matrix $\tilde{\mathbf{D}}$ (`dee`, Equation 12.45) and the stress matrix $\mathbf{D}_s$ (`dees`, Equation 12.46) can be formed.

- In a loop over integration points, form the kinematic (**B**) matrix (Equation 12.28) with the lines

```
% kinematic matrix
  bee                = zeros(nst,ntot) ;
  bee(1,1:2:ntot-1)  = deriv(1,:) ;
  bee(2,2:2:ntot)    = deriv(2,:) ;
  bee(3,1:2:ntot-1)  = deriv(2,:) ;
  bee(3,2:2:ntot)    = deriv(1,:) ;
```

  where `deriv` contains the first spatial derivatives of the shape functions in physical coordinates. Once **B** has been constructed, the strain rates and stresses at the current integration point can be computed with the snippet

```
strain_rate   = bee*uv ; % strain rates
stress = dee*strain_rate+dees*tensor0(:,k,iel) ; % stresses
```

  which follow directly from Equations 12.16 and 12.19, respectively. The vector `uv` contains the latest velocity estimation (i.e., $\mathbf{v}_{i+1}$ in Equation 12.31) for the nodes of the current element.

- If plasticity is included, compute the plastic yield function (Equation 12.47). If $F > 0$, return stresses to the yield surface using Equations 12.50, 12.51, and 12.52. These various tasks can be performed with the following snippet:

```
if plasticity % do only if plasticity is included
  tau=(1/4*(stress(1)-stress(2))^2+stress(3)^2)^(1/2); % tau
  sigma = 1/2*(stress(1)+stress(2)); % sigma star
  F = tau + sigma*sin(phi)-coh*cos(phi); % plastic yield function
  if F>0 % return stresses to yield surface
      if (sigma≤coh/tan(phi)) % 'normal' case
         beta  =  abs(coh*cos(phi)-sin(phi)*sigma)/tau ;
         sxx_new = sigma + beta*(stress(1)-stress(2))/2 ;
         szz_new = sigma - beta*(stress(1)-stress(2))/2 ;
         sxz_new =  beta*stress(3)  ;
         else % special treatment for corners of yield surface
         sxx_new = coh/tan(phi)  ;
         szz_new = coh/tan(phi)  ;
         sxz_new = 0             ;
     end
```

```
            stress(1)  = sxx_new   ;
            stress(2)  = szz_new   ;
            stress(3)  = sxz_new   ;
        end % end of stress return algorithm
    end % end of plasticity
```

- Perform the operations required to integrate the stiffness matrix **KM** and the right-hand-side load vector **R** according to Equations 12.25 and 12.32 with the following lines:

```
    if iters==1 % normal rhs stress
        stress_rhs = dees*tensor0(:,k,iel) ;
        else % total stress
        stress_rhs = stress ;
     end
     fun    = fun_s(:,k) ;      % shape functions
     fune   = zeros(1,ntot) ; % extended shape function array
     fune(2:2:ntot) = fun   ; % shape function copied into z-dof
     dwt = detjac*wts(k)   ; % multiplier
     KM = KM + bee'*dee*bee*dwt ; % element stiffness matrix
     R  = R  + (bee'*stress_rhs+fune'*rhog)*dwt ; % load vector
```

Note here that the stress appearing in the load vector (i.e., `stress_rhs` is computed as $\mathbf{D}_s\sigma^0$ for the first iteration (see Equation 12.27), whereas it is the newly computed stresses (i.e., $\boldsymbol{\sigma}$, see Equation 12.32) for all iterations thereafter. Note also that the variable `fune` used to compute the gravity loads contains zeros in the positions linked to the x degree of freedom (since gravity doesn't act in this direction), whereas it contains the shape functions in positions linked to the z degree of freedom.

- The final task within the integration loop is to save the integration point stresses, which is performed with the following line:

```
tensor(:,k,iel) = stress ; % save stresses at int. point
```

- After the integration loop has been completed, the left-hand global matrix and the global right-hand-side load vector are assembled, as done in previous programs. This marks the end of the element loop.
- After the end of the element loop, boundary conditions must be imposed. With the incremental formulation presented, the boundary values must be set to the applied boundary velocities during the first iteration but to the change in boundary velocities (i.e., to zero) during all future iterations. In Matlab, these tasks are achieved with the following lines:

```
% apply boundary conditions
  lhs(bcdof,:) = 0 ;
  tmp = spdiags(lhs,0) ;
  tmp(bcdof)=1 ;
  lhs=spdiags(tmp,0,lhs);

  if iters==1
    b(bcdof) = bcval ;
  else
    b(bcdof) = 0 ;
  end
```

- Compute the new velocity increment by solving Equation 12.30 and update the total velocity according to Equation 12.31. These tasks can be performed with the following lines:

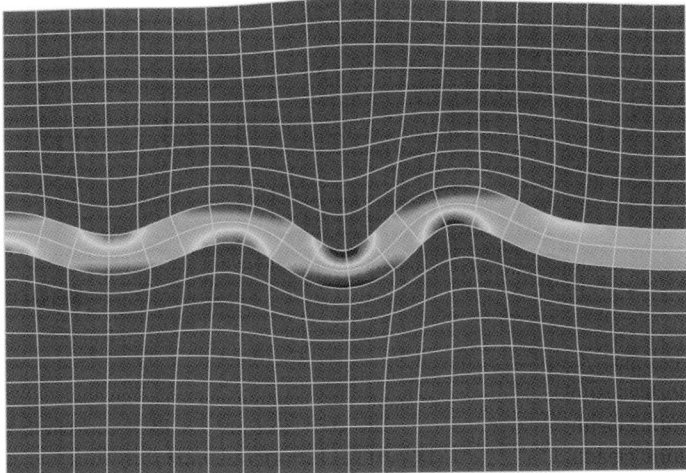

Figure 12.4 Folding in a layered viscoelastic material after 40% shortening. The central layer has a viscosity 100 higher than the surrounding matrix. The shaded colors represent mean stress, while the grid shows finite deformation (which is not the computational mesh). Parameter values are listed in the caption of Figure 12.3.

```
displ_inc = lhs \ b ; % solve for change in velocity
displ     = displ + displ_inc ; % update total velocity
```

- To evaluate whether the solution has converged, and therefore whether the iteration loop can be exited, it is necessary to estimate how much the solution changes from one iteration to the next. A simple means of checking convergence can be obtained by calculating the largest absolute velocity increment normalized to the total maximum absolute velocity, that is,

  ```
  error = max(abs(displ_inc))/max(abs(displ)) % error estimate
  ```

  If this "error" exceeds a certain tolerance, a new iteration must be computed (i.e., return to point 2); otherwise, the iteration loop can be exited.

13) The final tasks that must be performed before a new time step can be computed are to advect the mesh coordinates using the newly computed velocity field and to save a copy of the old stresses needed for the next time step. These tasks can be achieved with the following lines:

```
% Advect mesh
g_coord(1,:)=g_coord(1,:)+dt*displ(nf(1,:))';% update x-coords
g_coord(2,:)=g_coord(2,:)+dt*displ(nf(2,:))';% update z-coords
tensor0  = tensor ; % save 'old' stresses for next time step
```

Figures 12.4 and 12.5 show output produced with the script listed in the following text for the folding and shear localization experiments, respectively. Both experiments reveal features similar to observed in experiments and in deformed natural rocks (Ramsay and Huber, 1993). The folding experiment is a relatively linear problem and so can be performed at a relatively low spatial and temporal resolution. However, the calculation involving plasticity involves extreme strain localization and requires not only numerous nonlinear iterations but also a very high spatial and temporal resolution to obtain "sharp" shear bands. This strain localization also results in intense deformation of the finite element mesh, which would eventually necessitate remeshing.

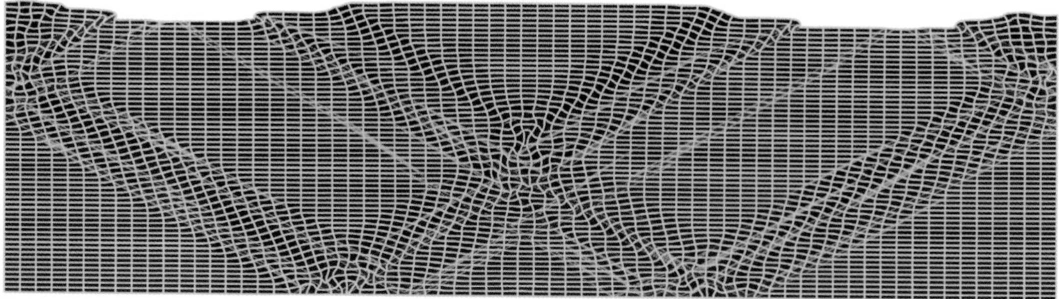

Figure 12.5 Strain localization in a viscoelastoplastic material after 14% shortening. The shaded colors represent the second invariant of the strain rate, while the grid shows finite deformation. Parameter values are listed in the caption of Figure 12.3.

```
%-----------------------------------------
% Program: deformation_vep2d.m
% 2D viscoelastoplastic plane strain deformation
% solid mechanics rate formulation
% 9-node quadrilaterals
%-----------------------------------------

clear
seconds_per_year = 60*60*24*365 ;

% physical parameters
lx          =   4  ;                            % length of x domain (m)
lz          =   1   ;                           % length of z domain (m)
emod_v      =  [1 1 1 1 ]*1e11      ;           % Young's modulus (Pa)
pois_v      =  [0.3 0.3 0.3 0.3 ]        ;      % Poisson's ratio
smod_v      =  emod_v./(2*(1+pois_v))    ;      % shear modulus (Pa)
bmod_v      =  emod_v./(3*(1-2*pois_v)) ;       % bulk modulus (Pa)
visc_v      =  [ 1 1 100 100]*1e18  ;           % shear viscosity (Pas)
coh_v       =  20*[1 1 1 0.9 ]*1e6     ;        % cohesion (Pa)
phi_v       = ([1 1 1 1])*30*pi/180  ;          % friction angle(radians)
psi_v       = [0 0 0 0]    ;                    % dilation angle (radians)
grav_v      = [ 2700 2700 2700 2700 ]*9.8 ;     % density*gravity
bvel        = 5e-3/seconds_per_year ;           % boundary velocity (m/s)
edot        = bvel/lx ;                         % initial strain rate (1/s)
plasticity= logical(0) ;    % include/ignore plasticity (true/false)
nst         = 3 ;           % number of stress/strain components

% characteristic scales
length_scale = lz ;             % length scale
stress_scale = max(smod_v) ; % stress_scale
time_scale   = 1/edot ;         % time_scale

% rescaled parameters
emod_v = emod_v/stress_scale ;
smod_v = smod_v/stress_scale ;
bmod_v = bmod_v/stress_scale ;
visc_v = visc_v/(stress_scale*time_scale) ;
coh_v  = coh_v/stress_scale ;
grav_v = grav_v*length_scale/stress_scale ;
lx     = lx/length_scale ;
lz     = lz/length_scale ;
```

```
bvel    = bvel/length_scale*time_scale ;
edot    = edot*time_scale ;

layer_top = 0.55 ; % z coord. (nondimensional) at top of layer
layer_bot = 0.45 ; % z coord. (nondimensional) at bottom of layer

% numerical parameters
ntime = 100 ;          % number of time steps to perform
nxe   = 40  ;          % n elements in x-direction
nze   = 40  ;          % n elements in z-direction
nels  = nxe*nze ;      % total number of elements
nx    = 2*nxe+1 ;      % n of nodes in x-direction
nz    = 2*nze+1 ;      % n of nodes in z-direction
nn    = nx*nz   ;      % total number of nodes
dx    = lx/nxe  ;      % element size in x-direction
dz    = lz/nze  ;      % element size in z-direction
nod   = 9   ;          % number of nodes
ndof  = 2   ;          % number of degrees of freedom
ntot  = nod + nod  ; %  total degrees of freedom in an element
nip   = 9 ;            % number of integration points in an element
eps   = 1e-3 ;         % convergence tolerance
limit = 50 ;           % maximum number of iterations
dt    = 0.01 ;         % time step (non dimensional)

%---------------------------------------
% generate coordinates and nodal numbering
%---------------------------------------

% computational mesh
% define mesh (numbering in z direction)
g_coord = zeros(2,nn) ;
n = 1 ;
for i = 1:nx
    for j=1:nz
      g_coord(1,n) = (i-1)*dx/2 ;
      g_coord(2,n) = (j-1)*dz/2 ;
      n = n + 1 ;
    end
end

% establish node numbering numbering
gnumbers = reshape(1:nn,[nz nx]) ;
iel = 1 ;
for i=1:2:nx-1
    for j=1:2:nz-1
      g_num(1,iel) = gnumbers(j,i)     ;    % node 1
      g_num(2,iel) = gnumbers(j+1,i)   ;    % node 2
      g_num(3,iel) = gnumbers(j+2,i)   ;    % node 3
      g_num(4,iel) = gnumbers(j+2,i+1) ;    % node 4
      g_num(5,iel) = gnumbers(j+2,i+2) ;    % node 5
      g_num(6,iel) = gnumbers(j+1,i+2) ;    % node 6
      g_num(7,iel) = gnumbers(j,i+2)   ;    % node 7
      g_num(8,iel) = gnumbers(j,i+1)   ;    % node 8
      g_num(9,iel) = gnumbers(j+1,i+1) ;    % node 9
      iel = iel + 1 ; % increment the element number
    end
end

%---------------------------------------
```

```
% create global-local connection arrays
%-----------------------------------------
sdof   = 0 ;                 % system degrees of freedom
nf     = zeros(ndof,nn) ; % node degree of freedom array
for n = 1:nn                 % loop over all nodes
  for i=1:ndof               % loop over each degree of freedom
     sdof = sdof + 1 ;       % increment total number of equations
     nf(i,n) =  sdof ;       % record the equation number for each node
  end
end

% equation numbering for each element
g = zeros(ntot,1) ;
g_g = zeros(ntot,nels);
for iel=1:nels ;
    num = g_num(:,iel) ; % extract nodes for the element
    inc=0 ;
    for i=1:nod ;
        for k=1:2 ;
            inc=inc+1 ;
            g(inc)=nf(k,num(i)) ;
        end
    end
    g_g(:,iel) = g ;
end

%-----------------------------------------
% define boundary conditions
%-----------------------------------------
% boundary nodes
bx0 = find(g_coord(1,:)==0)    ;
bxn = find(g_coord(1,:)==lx)   ;
bz0 = find(g_coord(2,:)==0)    ;
bzn = find(g_coord(2,:)==lz)   ;

% fixed boundary equations along with their values
bcdof = [     nf(1,bx0)   nf(1,bxn)     nf(2,bz0) ]     ;
bcval = [ zeros(1,length(bx0)) -bvel*ones(1,length(bxn))  zeros(1,length(bz0)) ]    ;

%----------------------------------------------------------
% establish phases
%----------------------------------------------------------

phase = ones(1,nels) ;                           % top
phase(g_coord(2,g_num(9,:))<layer_bot)=2 ; % bottom
phase(find(g_coord(2,g_num(9,:))≤layer_top & g_coord(2,g_num(9,:))≥layer_bot))=3 ; ...
                                                 % layer
phase(nze*nxe/2-nze/2)=4 ;                       % inclusion

%-----------------------------------------
% perturb boundaries of layer
%-----------------------------------------
iit=find(g_coord(2,:)==layer_bot);% nodes on top of layer
iib=find(g_coord(2,:)==layer_top);% nodes on bottom of layer
ii = [iit iib];                   % combined nodes
g_coord(2,ii)=g_coord(2,ii)+dz*0.05*(rand(1,length(ii))-0.5) ;
%-----------------------------------------
%  integration data and shape functions
%-----------------------------------------
```

```
% local coordinates of Gauss integration points for nip=3x3
  points(1:3:7,1) = -sqrt(0.6);
  points(2:3:8,1) = 0;
  points(3:3:9,1) = sqrt(0.6);
  points(1:3,2)   = sqrt(0.6);
  points(4:6,2)   = 0 ;
  points(7:9,2)   = -sqrt(0.6);

  % Gauss weights for nip=3x3
  w   = [ 5./9. 8./9. 5./9.] ;
  v   = [ 5./9.*w ; 8./9.*w ; 5./9.*w ] ;
  wts = v(:) ;

 % evaluate shape functions and their derivatives
 % at integration points and save the results
  for k = 1:nip
     xi  = points(k,1);
     eta = points(k,2);
     etam = eta - 1; etap = eta + 1 ;
     xim  = xi - 1 ; xip  = xi + 1  ;
     x2p1 = 2*xi+1 ;   x2m1 = 2*xi-1 ;
     e2p1 = 2*eta+1 ;  e2m1 = 2*eta-1 ;
     % shape functions
     fun= [ .25*xi*xim*eta*etam -.5*xi*xim*etap*etam ...
     .25*xi*xim*eta*etap -.5*xip*xim*eta*etap ...
     .25*xi*xip*eta*etap -.5*xi*xip*etap*etam ...
     .25*xi*xip*eta*etam -.5*xip*xim*eta*etam xip*xim*etap*etam ] ;
     % first derivatives of shape functions
     der(1,1) = 0.25*x2m1*eta*etam  ; %dN1dxi
     der(1,2) =-0.5*x2m1*etap*etam ;  %dN2dxi, etc
     der(1,3) = 0.25*x2m1*eta*etap  ;
     der(1,4) =     -xi*eta*etap ;
     der(1,5) = 0.25*x2p1*eta*etap  ;
     der(1,6) =-0.5*x2p1*etap*etam ;
     der(1,7) = 0.25*x2p1*eta*etam  ;
     der(1,8) =     -xi*eta*etam ;
     der(1,9) = 2*xi*etap*etam    ;
     der(2,1) = 0.25*xi*xim*e2m1 ; %dN1deta
     der(2,2) =-xi*xim*eta       ;%dN2deta, etc
     der(2,3) = 0.25*xi*xim*e2p1 ;
     der(2,4) =-0.5*xip*xim*e2p1   ;
     der(2,5) = 0.25*xi*xip*e2p1 ;
     der(2,6) =-xi*xip*eta        ;
     der(2,7) = 0.25*xi*xip*e2m1 ;
     der(2,8) =-0.5*xip*xim*e2m1   ;
     der(2,9) = 2*xip*xim*eta ;
     fun_s(k,:) = fun ;   % save shape functions
     der_s(:,:,k) = der ; % save derivatives
  end

%----------------------------------------
% initialisation
%----------------------------------------

  lhs     = sparse(sdof,sdof) ;     % global stiffness matrix
  b       = zeros(sdof,1)     ;     % global rhs vector
  displ   = zeros(sdof,1)     ;     % solution vector (velocities)
  tensor  = zeros(nst,nip,nels)  ; % stresses at integration points
  tensor0 = zeros(nst,nip,nels)  ; % old stresses
```

```
%---------------------------------------------------
% loading loop
%---------------------------------------------------
time = 0 ; % initialise time
for n=1:ntime

 time = time + dt ; % increment time
%-----------------------------------------
% iterations
%-----------------------------------------

error   = eps*2 ;           % initialise error
iters   = 0     ;           % initialise iteration counter
displ   = zeros(sdof,1) ; % initialise solution vector

while error > eps & iters < limit
  iters = iters + 1            % increment iteration counter
  lhs   = sparse(sdof,sdof); % initialised global stiffness matrix
  b     = zeros(sdof,1)     ; % initialised global load vector
%-----------------------------------------
%  element integration and assembly
%-----------------------------------------

for iel=1:nels  % sum over elements

    num   = g_num(:,iel)       ;  % list of element nodes
    g     = g_g(:,iel)         ;  % element equation numbers
    coord = g_coord(:,num)'   ;  % nodal coordinates
    KM    = zeros(ntot,ntot)  ;  % initialise stiffness matrix
    R     = zeros(ntot,1)      ;  % initialise stress load vector
    uv    = displ(g)           ;  % current nodal velocities

    % retieve material properties for the current element
    smod  = smod_v(phase(iel))  ; % shear modulus
    K     = bmod_v(phase(iel))  ; % bulk modulus
    mu    = visc_v(phase(iel))  ; % viscosity
    phi   = phi_v(phase(iel))   ; % friction coeff.
    psi   = psi_v(phase(iel))   ; % dilation angle
    coh   = coh_v(phase(iel))   ; % cohesion
    rhog  = grav_v(phase(iel))  ; % rho x gravity

    % viscoelastic material matrix
    di    = dt*(3*mu*K + 3*dt*smod*K + 4*smod*mu) ;
    od    = dt*(-2*smod*mu + 3*mu*K + 3*dt*smod*K);
    d     =  3*(mu + dt*smod) ;
    ed    = 2*mu*dt*smod/(2*mu + dt*smod) ;
    dee   = [di/d od/d 0 ; od/d di/d 0 ; 0 0 ed];

    % stress matrix
    di    = 3*mu + dt*smod  ;
    od    = dt*smod  ;
    ed    = 2 * mu / (2 * mu + dt * smod) ;
    dees  = [di/d od/d 0 ; od/d di/d 0 ; 0 0 ed] ;

    for k = 1:nip                    % integration loop
        fun     = fun_s(:,k) ;     % shape functions
        fune    = zeros(1,ntot) ; % extended shape function array
        fune(2:2:ntot) = fun ;    % shape function for z-dof only
```

```
        der    = der_s(:,:,k) ;   % shape functions in local coords
        jac    = der*coord   ;    % jacobian matrix
        detjac = det(jac)    ;    % det. of the Jacobian
        dwt    = detjac*wts(k) ;  % detjac x weight
        invjac = inv(jac)    ;    % inverse of the Jacobian
        deriv  = invjac*der ;     % shape functions in physical coords
        bee    = zeros(nst,ntot); % kinematic matrix
        bee(1,1:2:ntot-1) = deriv(1,:) ;
        bee(2,2:2:ntot)   = deriv(2,:) ;
        bee(3,1:2:ntot-1) = deriv(2,:) ;
        bee(3,2:2:ntot)   = deriv(1,:) ;
        strain_rate  = bee*uv   ; % strain rates
        stress =  dee*strain_rate + dees*tensor0(:,k,iel) ; % stresses
        if plasticity % do only if plasticity included
        tau   = (1/4*(stress(1)-stress(2))^2+stress(3)^2)^(1/2); % tau star
        sigma = 1/2*(stress(1)+stress(2));       % sigma star
        F = tau + sigma*sin(phi)-coh*cos(phi); % plastic yield function
           if F>0 % return stresses to yield surface
            if (sigma≤coh/tan(phi)) % 'normal' case
              beta      =   abs(coh*cos(phi)-sin(phi)*sigma)/tau ;
              sxx_new   =   sigma + beta*(stress(1)-stress(2))/2 ;
              szz_new   =   sigma - beta*(stress(1)-stress(2))/2 ;
              sxz_new   =   beta*stress(3) ;
            else % special treatment for corners of yield surface
              sxx_new   =   coh/tan(phi) ;
              szz_new   =   coh/tan(phi) ;
              sxz_new   =   0            ;
            end
            stress(1) = sxx_new  ; stress(2)=szz_new ; stress(3)=sxz_new ;
           end % end of stress return algorithm
        end % end of plasticity
        if iters==1 % normal rhs stress
          stress_rhs = dees*tensor0(:,k,iel) ;
          else % total stress
          stress_rhs = stress ;
        end
        KM   = KM + bee'*dee*bee*dwt  ; % element stiffness matrix
        R    = R  + (bee'*stress_rhs + fune'*rhog)*dwt ; % load vector
        tensor(:,k,iel)  = stress ;  % stresses at integration point
    end

   % assemble global stiffness matrix and rhs vector
    lhs(g,g) = lhs(g,g) + KM ; %  stiffness matrix
    b(g)     = b(g) - R  ;     %  load vector
end

%-------------------------------------------------
 % implement boundary conditions and solve system
 %------------------------------------------------
  % apply boundary conditions
  lhs(bcdof,:) = 0 ;
  tmp = spdiags(lhs,0) ; tmp(bcdof)=1 ; lhs=spdiags(tmp,0,lhs);
  if iters==1
    b(bcdof) = bcval ;
  else
    b(bcdof) = 0 ;
  end

   displ_inc = lhs \ b ; % solve for velocity increment
```

```
    displ = displ + displ_inc ; % update total velocity
    error = max(abs(displ_inc))/max(abs(displ)) % estimate error

end

 %-----------------------------------------
 % end of iterations
 %------------------------------------------

 % update mesh coordinates
   g_coord(1,:) = g_coord(1,:) + dt*displ(nf(1,:))' ;
   g_coord(2,:) = g_coord(2,:) + dt*displ(nf(2,:))' ;

   tensor0 = tensor ; % save stresses

   shortening_percent = (1 - (max(g_coord(1,:))-min(g_coord(1,:)))/lx)*100

 %-------------------------------
 % plotting of results
 %-------------------------------

 xgrid = reshape(g_coord(1,:),nz,nx)*length_scale ;% m
 zgrid = reshape(g_coord(2,:),nz,nx)*length_scale ;% m
 c     = length_scale/time_scale*seconds_per_year*1e3 ;
 u_solution = reshape(displ(nf(1,:)),nz,nx)*c ;% (mm/yr)
 v_solution = reshape(displ(nf(2,:)),nz,nx)*c ;% (mm/yr)
 tensor_scaled = tensor*stress_scale/1e6 ; % MPa

 figure(1) , clf  % mesh
 plot(xgrid,zgrid,'b')
 hold on
 plot(xgrid',zgrid','b')
 plot(g_coord(1,iit),g_coord(2,iit),'r')
 plot(g_coord(1,iib),g_coord(2,iib),'r')
 hold off
 axis equal
 title(['Deformed mesh after ',num2str(shortening_percent), ' % shortening'])

 figure(2)  , clf  % velocity field
 quiver(xgrid,zgrid,u_solution,v_solution)
 axis equal
 drawnow
 title('Velocity vector field')

 figure(3) , clf % x-velocity
 pcolor(xgrid,zgrid,u_solution)
 colormap(jet)
 colorbar
 shading interp
 axis equal
 title('x-velocity (mm/yr)')

 figure(4) , clf   % z-velocity
 pcolor(xgrid,zgrid,v_solution)
 colormap(jet)
 colorbar
 shading interp
 axis equal
 title('z-velocity (mm/yr)')
```

```
figure(5) , clf % plot negative mean stress
for iel=1:nels
  num        = g_num(:,iel)        ;
  coord      = g_coord(:,num(1:8))' ;
  means      = (tensor(1,:,iel)+tensor(2,:,iel))/2;
  nodevalues = fun_s\means' ;
  h = fill(coord(:,1),coord(:,2),-nodevalues(1:8)) ;
  set(h,'linestyle','none')
  hold on
end
plot(xgrid(:,1:8:end),zgrid(:,1:8:end),'Color',[0.8,0.8,0.8])
plot(xgrid(1:8:end,:)',zgrid(1:8:end,:)','Color',[0.8,0.8,0.8])
hold off
axis equal
colorbar
title('Mean stress (MPa)')

%-------------------------------------------------------

end

%-------------------------------------------------------
% end of time integration
%-------------------------------------------------------
```

Suggested Reading

A. J. M. Ferreira, *MATLAB Codes for Finite Element Analysis: Solids and Structures (Solid Mechanics and Its Applications)*, Springer, Berlin, 2009.

T. Gerya, *Introduction to Numerical Geodynamic Modelling.* Cambridge University Press, Cambridge, 2009.

Y. W. Kwong, and H. C. Bang, *The Finite Element Method Using Matlab*, CRC Press, New York, 2000.

D. D. Pollard, and R. C. Fletcher, *Fundamentals of Structural Geology*, Cambridge University Press, Cambridge, 2005.

K. Stüwer, *Geodynamics of the Lithosphere*, Springer, Berlin, 2007.

D. L. Turcotte, and G. Schubert, *Geodynamics*, Cambridge University Press, Cambridge, 2002.

O. C. Zienkiewicz, and R. L. Taylor, *The Finite Element Method, Volume 2, Solid Mechanics*, Butterworth Heinemann, Oxford, 2000.

13

Going Further

The purpose of this chapter is to provide some suggestions to Earth scientists wishing to go further in use of the finite element method (FEM). Three main directions are identified and briefly discussed: the use of more optimized computing strategies, the use of other classes of the FEM, and the use of existing FEM software.

13.1 Optimization

This text has presented a number of simple stand-alone scripts that illustrate how the FEM is easily programmed in Matlab. Many real problems can already be solved with these scripts. However, because emphasis has been placed on learnability rather than efficiency, some of these scripts may take considerable time to run on a standard personal computer (depending on the problem and spatial resolution). The main bottlenecks in the listed FEM program are related to (i) element integration, (ii) matrix assembly, and (iii) solution. In the case of the first bottleneck, poor performance is linked to matrix–matrix multiplication on small (element matrices), which is costly due to the large overhead used by the Basic Linear Algebra Subprograms (BLAS) libraries (see http://www.netlib.org/blas/faq.html and Dongarra et al. (1990)). In addition, a significant amount of time is typically spent on calculation of the determinant and the inverse of the Jacobian (i.e., when using `det(jac)` and `inv(jac)`). One approach that leads to a marked increase in efficiency during this stage of the calculation is to perform element integration simultaneously on large blocks of elements rather than on individual elements and to calculate the determinant and the inverse of the Jacobian explicitly rather than using in-built functions (see Dabrowski et al. (2008)). The second bottleneck mentioned earlier linked to matrix assembly can be improved by avoiding accessing (and manipulating) the global sparse matrices multiple times (especially one element at a time, e.g., see Chen (2016)), as is done in assembly lines such as `lhs(num,num)= lhs(num,num)+ KM`. A far better approach is to compute and save the element matrices and sparse storage indices and then construct the global sparse stiffness matrix in one step entirely after the element integration loop (e.g., see Davis, 2007). The assembly part of an FEM program can also be improved by using a faster function than `sparse` to create the sparse matrix. For example, CHOLMOD (which is part of the SuiteSparse package; Davis and Hu, 2011) includes a `sparse2` function that executes significantly faster than `sparse`. CHOLMOD can also be used to significantly speed up the last major bottleneck linked to solving the system of equations by using various functions specifically designed to solve linear systems when the coefficient matrix is sparse, symmetric, and positive definite.

To illustrate some of these possibilities, the following is a more optimized version of the program listed in Section 6.5, which uses the FEM to solve the transient heat equation in three dimensions.

Practical Finite Element Modeling in Earth Science Using Matlab, First Edition. Guy Simpson.

Companion website: www.wiley.com/go/simpson

The program uses a block procedure for calculation of the stiffness matrices, forms the sparse matrix entirely after the integration loop, and uses various CHOLMOD routines for more efficient storage, factorization, and solution of the linear system. Note that the Matlab interface to the SuiteSparse package needs to be installed (along with the possibility for "metis" ordering) before the program can be executed. The speed-up relative to the original program is nearly 20-fold for matrix calculation and assembly stages and about 70-fold for the final steps performed within the time loop. Thus, the program (which has 9261 unknowns) runs to completion (after performing 200 time steps) in less than 1 min. on a PC workstation.

```
%-----------------------------------------------------
% Program: diffusion3d_opt.m - optimised version
% FEM - Diffusion equation in 3D
% 8-node hexahedra elements
%-----------------------------------------------------

clear

seconds_per_yr = 60*60*24*365 ; % seconds in 1 year

% physical parameters
kappa = 1e-6   ;    % thermal diffusivity, m^2/s
lx    = 3e3    ;    % length of domain, m
ly    = 2e3    ;    % width of domain, m
lz    = 1e3    ;    % depth of domain, m
H     = 0*1e-9  ; % heat source, °C/s
Tb    = 0      ;    % fixed boundary temperature, °C
Ti    = 1      ;    % initial temperature, °C

% numerical parameters
ntime = 200   ;     % number of time steps
ndim  = 3     ;     % number of spatial dimensions
nod   = 8     ;     % number of nodes per element
nelx  = 20    ;     % number of elements in x-direction
nely  = 20    ;     % number of elements in y-direction
nelz  = 20    ;     % number of elements in z-direction
nels  = nelx*nely*nelz ; % total number of elements
nx    = nelx+1 ;      % number of nodes in x-direction
ny    = nely+1 ;      % number of nodes in y-direction
nz    = nelz+1 ;      % number of nodes in z-direction
nn    = nx*ny*nz ;    % total number of nodes
nip   = 2^ndim  ;   % number of integration points
dx    = lx/nelx ;     % element length in x-direction
dy    = ly/nely ;     % element length in y-direction
dz    = lz/nelz ;     % element length in z-direction
dt    = 50*seconds_per_yr ; % time step (s)

nelblo0 = 400 ;   % number of elements in a block
nelblo  = min(nels,nelblo0); % must be < nels
nblo    = ceil(nels/nelblo);

% define mesh (numbering in z direction)
g_coord = zeros(ndim,nn) ;
n = 1 ;
for i=1:ny
  for j=1:nx
      for k=1:nz
```

```
        g_coord(1,n) = (j-1)*dx ;
        g_coord(2,n) = (i-1)*dy ;
        g_coord(3,n) = (k-1)*dz ;
        n = n + 1 ;
    end
  end
end

% establish elem-node connectivity
gnumbers = reshape(1:nn,[nz nx ny]) ;
iel = 1 ;
for i=1:nelx
  for j=1:nely
    for k=1:nelz
      g_num(1,iel) = gnumbers(k,i,j) ;         % node 1
      g_num(2,iel) = gnumbers(k+1,i,j) ;       % node 2
      g_num(3,iel) = gnumbers(k+1,i+1,j);      % node 3
      g_num(4,iel) = gnumbers(k,i+1,j);        % node 4
      g_num(5,iel) = gnumbers(k,i,j+1) ;       % node 5
      g_num(6,iel) = gnumbers(k+1,i,j+1) ;     % node 6
      g_num(7,iel) = gnumbers(k+1,i+1,j+1);    % node 7
      g_num(8,iel) = gnumbers(k,i+1,j+1);      % node 8
      iel          = iel + 1 ;
    end
  end
end

% find boundary nodes
bx0 = find(g_coord(1,:)==0)  ;
bxn = find(g_coord(1,:)==lx) ;
by0 = find(g_coord(2,:)==0)  ;
byn = find(g_coord(2,:)==ly) ;
bz0 = find(g_coord(3,:)==0)  ;
bzn = find(g_coord(3,:)==lz) ;

% define boundary conditions
bcdof = unique([bx0 bxn by0 byn bz0 bzn]) ;
bcval = Tb*ones(1,length(bcdof))    ;

% gauss integration data
points = zeros(nip,ndim); % location of points
root3  = 1./sqrt(3);
points(1,1)= root3;points(1,2)= root3;points(1,3)= root3;
points(2,1)= root3;points(2,2)= root3;points(2,3)=-root3;
points(3,1)= root3;points(3,2)=-root3;points(3,3)= root3;
points(4,1)= root3;points(4,2)=-root3;points(4,3)=-root3;
points(5,1)=-root3;points(5,2)= root3;points(5,3)= root3;
points(6,1)=-root3;points(6,2)=-root3;points(6,3)= root3;
points(7,1)=-root3;points(7,2)= root3;points(7,3)=-root3;
points(8,1)=-root3;points(8,2)=-root3;points(8,3)=-root3;
wts = ones(1,nip) ; % weights

% save shape functions and their derivatives in local coordinates
% evaluated at integration points
for k=1:nip
  xi   = points(k,1) ;
  eta  = points(k,2) ;
  zeta = points(k,3) ;
  etam = 1-eta ;  xim=1-xi  ;  zetam=1-zeta ;
```

```
  etap = eta+1 ;  xip=xi+1  ;  zetap=zeta+1 ;
  fun=[0.125*xim*etam*zetam 0.125*xim*etam*zetap 0.125*xip*etam*zetap ...
       0.125*xip*etam*zetam 0.125*xim*etap*zetam 0.125*xim*etap*zetap ...
       0.125*xip*etap*zetap 0.125*xip*etap*zetam ] ;
  fun_s{k} = fun ; % shape functions
  der(1,1)=-.125*etam*zetam ; der(1,2)=-.125*etam*zetap ;
  der(1,3)=.125*etam*zetap  ; der(1,4)=.125*etam*zetam ;
  der(1,5)=-.125*etap*zetam ; der(1,6)=-.125*etap*zetap ;
  der(1,7)=.125*etap*zetap  ; der(1,8)=.125*etap*zetam ;
  der(2,1)=-.125*xim*zetam  ; der(2,2)=-.125*xim*zetap ;
  der(2,3)=-.125*xip*zetap  ; der(2,4)=-.125*xip*zetam ;
  der(2,5)=.125*xim*zetam   ; der(2,6)=.125*xim*zetap ;
  der(2,7)=.125*xip*zetap   ; der(2,8)=.125*xip*zetam ;
  der(3,1)=-.125*xim*etam   ; der(3,2)=.125*xim*etam ;
  der(3,3)=.125*xip*etam    ; der(3,4)=-.125*xip*etam ;
  der(3,5)=-.125*xim*etap   ; der(3,6)=.125*xim*etap ;
  der(3,7)=.125*xip*etap    ; der(3,8)=-.125*xip*etap  ;
  der_s{k} = der' ; % derivative of shape function
end

% initialise arrays
b        = zeros(nn,1);          % global rhs vector
displ    = zeros(nn,1);          % global solution vector
lhs_save = zeros(nod*nod,nels);  % storage for lhs element matrix
rhs_save = zeros(nod*nod,nels);  % storage for rhs element matrix
f_save   = zeros(nod,nels);      % storage for rhs load vector
kki      = zeros(nod*nod,nels) ; % storage for row equation numbers
kkj      = zeros(nod*nod,nels) ; % storage for col. equation numbers

% sparse storage indices
for iel = 1 : nels
      num        = g_num(:,iel)     ;
      rows       = num(:)*ones(1,nod);
      cols       = ones(nod,1)*num(:)';
      kki(:,iel) = rows(:);
      kkj(:,iel) = cols(:);
end

% grids for plotting
xv = linspace(0,lx,nx);
yv = linspace(0,ly,ny);
zv = linspace(0,lz,nz);
[xg,zg,yg] = meshgrid(xv,yv,zv);

%-----------------------------------------------
% matrix integration and assembly
%-----------------------------------------------

K_block   = zeros(nelblo,nod*nod);
M_block   = zeros(nelblo,nod*nod);
F_block   = zeros(nelblo,nod);
invJx     = zeros(nelblo,ndim);
invJy     = zeros(nelblo,ndim);
invJz     = zeros(nelblo,ndim);
ns        = 1 ;      % element at start of block
ne        = nelblo ; % element at end of block

for ib=1:nblo  % loop over blocks of elements
```

```
% node coordinates for elements of a block
coord_x = reshape(g_coord(1,g_num(:,ns:ne)),nod,nelblo);
coord_y = reshape(g_coord(2,g_num(:,ns:ne)),nod,nelblo);
coord_z = reshape(g_coord(3,g_num(:,ns:ne)),nod,nelblo);

% initialisation
K_block(:) = 0 ;
M_block(:) = 0 ;
F_block(:) = 0 ;

for k = 1:nip % integration loop

  fun  = fun_s{k} ; % shape function
  der  = der_s{k} ;   % shape fun. derivatives in local coords
  %  jacobian
  Jx         = coord_x'*der;
  Jy         = coord_y'*der;
  Jz         = coord_z'*der;
  %  determinant of jacobian
  detJ       = Jx(:,1).*(Jy(:,2).*Jz(:,3)-Jy(:,3).*Jz(:,2));
  detJ       = detJ-Jy(:,1).*(Jx(:,2).*Jz(:,3)-Jx(:,3).*Jz(:,2));
  detJ       = detJ+Jz(:,1).*(Jx(:,2).*Jy(:,3)-Jx(:,3).*Jy(:,2));
  % invert jacobian
  invdetJ    = 1.0./detJ;
  invJx(:,1)  = (Jy(:,2).*Jz(:,3)-Jz(:,2).*Jy(:,3)).*invdetJ;
  invJx(:,2)  = (Jz(:,2).*Jx(:,3)-Jx(:,2).*Jz(:,3)).*invdetJ;
  invJx(:,3)  = (Jx(:,2).*Jy(:,2)-Jx(:,3).*Jy(:,2)).*invdetJ;
  invJy(:,1)  = (Jy(:,3).*Jz(:,1)-Jy(:,1).*Jz(:,3)).*invdetJ;
  invJy(:,2)  = (Jz(:,3).*Jx(:,1)-Jx(:,3).*Jz(:,1)).*invdetJ;
  invJy(:,3)  = (Jx(:,3).*Jy(:,1)-Jx(:,1).*Jy(:,3)).*invdetJ;
  invJz(:,1)  = (Jy(:,1).*Jz(:,2)-Jz(:,1).*Jy(:,2)).*invdetJ;
  invJz(:,2)  = (Jz(:,1).*Jx(:,2)-Jz(:,2).*Jx(:,1)).*invdetJ;
  invJz(:,3)  = (Jy(:,2).*Jx(:,1)-Jy(:,1).*Jx(:,2)).*invdetJ;
  % derivatives of shape functions in physical coordinates
  derivx       = invJx*der' ;
  derivy       = invJy*der' ;
  derivz       = invJz*der' ;

  dwt        = wts(k)*detJ ; % multiplier

  % compute element matrices
  indx = 0;
  for i = 1:nod
    for j = 1:nod
      indx = indx + 1 ;
      KM = derivx(:,i).*derivx(:,j) + derivy(:,i).*derivy(:,j) + ...
          derivz(:,i).*derivz(:,j) ;
      K_block(:,indx)=K_block(:,indx)+KM.*dwt*kappa ; % stiffness
      M_block(:,indx)=M_block(:,indx)+fun(i)*fun(j)*dwt/dt; % mass
    end
  end

  F_block = F_block + H*dwt*fun ; % load terms

end % end of integration loop

 % save current block
lhs_save(:,ns:ne) = K_block'+M_block';
rhs_save(:,ns:ne) = M_block';
```

```
    f_save(:,ns:ne)    = F_block';

    % readjust start, end and size of block
    ns   = ns+nelblo ;
      if(ib==nblo-1)
        nelblo  = nels-ne ;
        K_block = zeros(nelblo,nod*nod);
        M_block = zeros(nelblo,nod*nod);
        invJx   = zeros(nelblo,ndim);
        invJy   = zeros(nelblo,ndim);
        invJz   = zeros(nelblo,ndim);
        F_block = zeros(nelblo,nod);
      end
      ne   = ne+nelblo ;

  end % end of element assembly loop

  % form global sparse matrices (requires CHOLMOD)
  lhs   = sparse2(kki,kkj,lhs_save) ;
  rhs   = sparse2(kki,kkj,rhs_save) ;

  % impose boundary conditions
  lhs(bcdof,:) = 0 ;                        % zero the boundary equations
  tmp          = spdiags(lhs,0) ;           % store diagonal
  tmp(bcdof)   = 1 ;                        % place 1 on stored-diagonal
  lhs          = spdiags(tmp,0,lhs);        % reinsert diagonal

  % factorise matrix (requires CHOLMOD)
  perm    = metis(lhs);              % compute reording
  lhs     = cs_transpose(lhs);       % transpose
  lhs     = cs_symperm(lhs,perm);    % reorder matrix
  lhs     = cs_transpose(lhs);       % transpose
  L       = lchol(lhs);              % factorise

  ff = accumarray(g_num(:),f_save(:)) ; % form load vector

  %-------------------------------------------------
  % time loop
  %-------------------------------------------------

  displ(1:nn) = Ti ;  % initial conditions
  time        = 0  ;  % intialise time

  for n=1:ntime

    n
    time          = time + dt ;       % update time
    b             = ff + rhs*displ ;  % form rhs global vector
    b(bcdof)      = bcval ;           % set boundary values in rhs vector
    displ(perm)   = cs_ltsolve(L,cs_lsolve(L,b(perm))) ; % solution

  %-------------------------------------------------
    % evaluate exact solution
    % Carslaw and Jaeger (1959) eqs 4 & 5, page 184
    x = 0 ; y = 0 ; z = 0 ; % point where solution is evaluated
    nterms = 100 ;
    a = lx/2; b = ly/2; c = lz/2;
    psix = 0 ; psiy = 0 ; psiz = 0 ;
    for m=0:nterms
```

```
    nt = 2*m+1;
    t0 = (-1)^m/nt;
    cx = cos(nt*pi*x/(2*a)) ;
    ex = exp(-kappa*nt^2*pi^2*time/(4*a^2)) ;
    psix = psix + 4/pi*t0*ex*cx ;
    cy = cos(nt*pi*y/(2*b)) ;
    ey = exp(-kappa*nt^2*pi^2*time/(4*b^2)) ;
    psiy = psiy + 4/pi*t0*ey*cy ;
    cz = cos(nt*pi*z/(2*c)) ;
    ez = exp(-kappa*nt^2*pi^2*time/(4*c^2)) ;
    psiz = psiz + 4/pi*t0*ez*cz ;
  end
  exact = psix.*psiy.*psiz ;
%-----------------------------------------------------

   % plot solution
  figure(1)
  solution = reshape(displ,ny,nx,nz) ;
  xslice = lx/2 ;
  yslice = ly/2 ;
  zslice = lz/2 ;
  slice(xg,zg,yg,solution,xslice,yslice,zslice)
  xlabel('x-distance (m)')
  ylabel('y-distance (m)')
  zlabel('z-distance (m)')
  colorbar
  view([45,30])
  axis equal

  figure(2)
  ix = find(xv==x+a) ;
  iy = find(yv==y+b) ;
  iz = find(zv==z+c) ;
  numerical_s(n) = solution(ix,iz,iy) ;
  exact_s(n)     = exact ;
  time_s(n)      = time   ;
  plot(time_s/seconds_per_yr,exact_s,'r',time_s/seconds_per_yr,numerical_s,'bo')
  xlabel('Time (years)')
  ylabel('Temperature (°C)')

  drawnow

end

%-----------------------------------------------------
% end of time loop
%-----------------------------------------------------
```

13.2 Using Other FEMs

This text is based almost exclusively on the Galerkin FEM, which is a weighted residual method (WRM) (Finlayson, 1972) where the weighting functions are chosen to be from the same family as the approximating (i.e., the shape or trial) functions. For self-adjoint problems (e.g., if the order of spatial differentiation in the governing equation is even), the Galerkin FEM is known to be optimal (Zienkiewicz and Taylor, 2000a), which partially explains its popularity. However, other problems, for example, those that are not self-adjoint, may be best solved using a different choice for the weighting function, or a different method altogether. Some other common choices for weighting functions are the following:

1) **Petrov–Galerkin**. This method uses weighting functions that are similar to the shape functions, but they contain additional (stabilizing) terms that depend on the derivatives of the shape functions and the flow velocity (Zienkiewicz and Taylor, 2000c). This method is quite successful in the solution of convection-dominated (hyperbolic) problems (see Section 10.2).
2) **Point Collocation**. In this case, the weighting functions are taken to be the Dirac delta function (i.e., $\mathbf{W}(x) = \delta(x - x_n)$), defined as being zero everywhere except at x_n, with an integral of one over the entire domain (Finlayson, 1972). This approximation is equivalent to setting the residual to zero at certain points x_n and so is closely related to the finite difference method without the use of an approximate solution. The main advantage of point collocation compared to Galerkin methods is a significant reduction in the computational cost associated with computing and assembling the stiffness matrix (e.g., see Houstis et al. (1978)), which can be beneficial especially for very nonlinear time-dependent problems.
3) **Subdomain Collocation**. This method assumes $\mathbf{W}(x) = 1$ within an element while it is zero elsewhere, which is equivalent to making the integral of the residual zero over the specified subdomain. Subdomain collocation is closely related to the finite volume method and is especially suitable for fluid flow problems where mass conservation at the scale of individual subdomains is important.
4) **Least Squares**. This method takes weighting functions given by $\partial R/\partial a_i$ where R is the residual and a_i are the nodal unknowns. This is equivalent to assuming that $\iiint R^2 dxdydz$ is minimized, which is the motivation for the method's name. Although this method results in larger systems of equations, it may offer certain advantages over other FEM methods for some problems. For example, when modeling deformation of incompressible materials, this method naturally satisfies the Babuška–Brezzi condition, and so it allows flexible choice of element types while avoiding "locking" (Bochev and Gunzburger, 2009; Hughes et al., 1989; Yang and Liu, 1997).

In addition to these various methods, which are all related to the choice of the weighting function, there are a number of other FEMs that differ in other respects. The following is a list of some of the more widely used submethods:

- **Mesh-Free FEMs**. As the name suggests, mesh-free (or element-free) FEMs (of which there are many varieties) are those that do not require connectivity between the various nodes, which is normally a central component of classical FEMs (Belytschko et al., 1994; Liu, 2003). An advantage of mesh-free methods is that because the nodes can be nonuniformly distributed in space and are unconnected, they are well suited to problems involving large deformation, which normally require remeshing. The downside is that mesh-free methods are normally less stable and less accurate than methods with elements.
- **Extended FEMs (XFEMs)**. These methods extend the classical FEM by enriching the standard approximating functions with discontinuous functions in small portions of the computational domain, typically near sharp discontinuities or singularities (e.g., cracks) (Khoei, 2015; Moës et al., 1999). A key advantage of these methods is that they avoid the remeshing normally required to obtain accurate solutions when sharp discontinuities are present.
- **Discontinuous Galerkin FEMs**. These methods differ from most other finite element techniques in that there are no explicit continuity requirements for the solution at element boundaries (Li, 2006). This feature is especially desirable for hyperbolic problems with sharp internal discontinuities, which often cause numerical oscillations with naive schemes.

13.3 Use of Existing Finite Element Software

How to proceed if you made it this far and you know you want to continue using numerical models but you have no intention of developing your own finite element programs? One option is to work with existing numerical modeling software (based on the FEM or other methods). If this is your desired path, the following points might be worth considering:

1) There are many options available.[1] However, because much of the software was originally designed and is currently used by civil and structural engineers, many of the packages may not be easily adapted and applied to real (and often complicated) Earth science problems. Some of the well-known modeling packages that are widely used by Earth scientists include FLAC, ANSYS, ABAQUS, COMSOL (all commercial) along with GALE, PyLith, Escript, Ellipsis3D, and ADELI.
2) Because existing (especially commercial) software are usually very sophisticated and powerful, there is a real risk of using them incorrectly and/or of misinterpreting the output. Users are therefore advised to spend the time saved in programming and developing code to understand how the solution is computed and how the results should be correctly interpreted and applied.
3) Whether you are using your own programs or existing software, you should always analyze numerical results critically. Are the results mathematically correct? Are you interpreting them properly? Are they explainable in terms of the governing physics? Try to validate numerical solutions with exact solutions or other published numerical/experimental results, which often must be done for simplified cases. Use your intuition to test whether the numerical results are expected, based on your knowledge of the problem being solved, and always be suspicious of unexpected results, no matter how exciting they may appear to be.

1 For example, see https://en.wikipedia.org/wiki/List_of_finite_element_software_packages

Appendix A

Derivation of the Diffusion Equation

Whether one is interested in heat conduction, chemical diffusion, or fluid flow in porous media, the diffusion equations governing these diverse physical processes are derived by essentially the same principle: that is, by combining a conservation statement (e.g., conservation of mass or energy) with a constitutive law relating mass flux to gradients in the conserved potential. To demonstrate this, here we consider the example of heat conduction.

A.1 Conservation Equation

An equation for the conservation of energy states that energy flowing into a control volume must be compensated either by energy flowing out of the volume, by a change in total energy, or a combination of both. We consider here only thermal energy (neglecting potential and kinetic energy). Consider a small rectangular volume with edges having lengths dx, dy, and dz embedded in a Cartesian coordinate system (Figure A.1). The total quantity of heat (with units Joules, denoted by J) flowing into a control volume through the face $dy\,dz$ during a small interval of time dt is $q_x\,dy\,dz\,dt$ where q_x is the heat flux with units J m^{-2} s^{-1}. The quantity of heat flowing out of the control volume in the direction of positive x during dt must be determined using a Taylor expansion. Expanding q_x from the entrance of the control volume gives

$$(q_x)_{out} = q_x(x + dx) = q_x + \frac{\partial q_x}{\partial x} dx + \frac{\partial^2 q_x}{\partial x^2}\frac{dx^2}{2!} + \cdots . \tag{A.1}$$

Using this, the heat out can be approximated, to first order, as

$$h_{out} = \left(q_x + \frac{\partial q_x}{\partial x} dx \right) dy\,dz\,dt \tag{A.2}$$

The net flow of heat N_x out of the control volume in the direction of positive x is $N_x = h_{out} - h_{in}$. Thus,

$$N_x = \left(q_x + \frac{\partial q_x}{\partial x} dx \right) dy\,dz\,dt - q_x\,dy\,dz\,dt \tag{A.3}$$

$$N_x = \frac{\partial q_x}{\partial x}\,dx\,dy\,dz\,dt \tag{A.4}$$

Making similar use of Taylor expansions, the net flows N_y and N_z in the y- and z-directions are

$$N_y = \frac{\partial q_y}{\partial y}\,dx\,dy\,dz\,dt \tag{A.5}$$

Practical Finite Element Modeling in Earth Science Using Matlab, First Edition. Guy Simpson.

Companion website: www.wiley.com/go/simpson

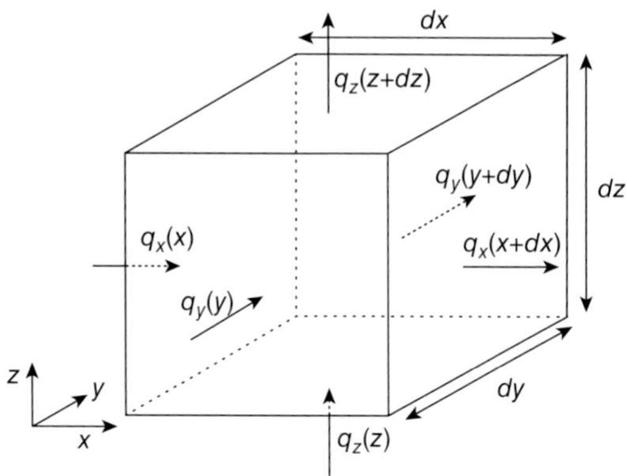

Figure A.1 Heat flow into and out of a small rectangular element.

$$N_z = \frac{\partial q_z}{\partial z} \, dx \, dy \, dz \, dt \tag{A.6}$$

The total net flow of heat N out of the control volume is $N = N_x + N_y + N_z$ which gives

$$N = \left(\frac{\partial q_x}{\partial x} + \frac{\partial q_y}{\partial y} + \frac{\partial q_z}{\partial z} \right) dx \, dy \, dz \, dt \tag{A.7}$$

According to the sign convention chosen, a positive N implies that heat is lost from the control volume. Due to energy conservation, this must be compensated by an equivalent decrease of energy within the volume. Since the volume $dx\,dy\,dz$ remains constant over the interval dt, the only way to decrease the energy is to decrease the temperature (T), specific heat (c), density (ρ), or some combination of all the three. The quantity of heat $h(t)$ within the control volume at time t is $\rho cT \, dx\,dy\,dz$. Using a Taylor expansion with time as the independent variable, the quantity of heat $h(t + dt)$ within the volume after a small interval of time dt is to first order

$$h(t + dt) = \left(\rho cT + \frac{\partial}{\partial t}(\rho cT)dt \right) dx \, dy \, dz \tag{A.8}$$

The change in energy $dh = h(t + dt) - h(t)$. Thus,

$$dh = \left(\rho cT + \frac{\partial}{\partial t}(\rho cT)dt \right) dx \, dy \, dz - \rho cT \, dx \, dy \, dz \tag{A.9}$$

$$dh = \frac{\partial}{\partial t}(\rho cT) \, dt \, dx \, dy \, dz \tag{A.10}$$

By conservation of energy, $N + dh = 0$. Therefore, summing Equations A.7 and A.10 and dividing by $dx\,dy\,dz\,dt$, the following equation can be obtained:

$$\frac{\partial(\rho cT)}{\partial t} + \frac{\partial q_x}{\partial x} + \frac{\partial q_y}{\partial y} + \frac{\partial q_z}{\partial z} = 0 \tag{A.11}$$

It is normally assumed that the density and specific heat do not change with time in which case A.11 simplifies to

$$\rho c \frac{\partial T}{\partial t} = -\frac{\partial q_x}{\partial x} - \frac{\partial q_y}{\partial y} - \frac{\partial q_z}{\partial z} = -\nabla \cdot \mathbf{q} \tag{A.12}$$

This equation is known as the energy conservation equation (in this case for thermal energy). Conservation of mass leads to a very similar equation.

A.2 Constitutive Equation

The derivation presented so far is incomplete in that we have one equation (i.e., Equation A.12) but two unknown functions T and $\mathbf{q}$. In order to complete the formulation, we must provide one additional equation. This additional equation is phenomenological (derived from experiments) and is generally called a "constitutive relation". For heat flow, the equation is called Fourier's law

$$\mathbf{q} = -k\nabla T \tag{A.13}$$

which is a relationship between the heat fluxes and the temperature gradients. The parameter k is called the "thermal conductivity", with units J m^{-1}. By inserting A.13 into Equation A.12, one obtains

$$\frac{\partial T}{\partial t} = \nabla(\boldsymbol{\kappa}\nabla T) = \frac{\partial}{\partial x}\left(\kappa\frac{\partial T}{\partial x}\right) + \frac{\partial}{\partial y}\left(\kappa\frac{\partial T}{\partial y}\right) + \frac{\partial}{\partial z}\left(\kappa\frac{\partial T}{\partial z}\right) \tag{A.14}$$

where $\kappa = k/(\rho c)$ is the thermal diffusivity with units m^2 s^{-1}. If the thermal diffusivity does not vary in space this simplifies to

$$\frac{\partial T}{\partial t} = \kappa\left(\frac{\partial^2 T}{\partial x^2} + \frac{\partial^2 T}{\partial y^2} + \frac{\partial^2 T}{\partial z^2}\right) \tag{A.15}$$

Both these are diffusion equations. In one dimension, Equation A.15 simplifies to

$$\frac{\partial T}{\partial t} = \kappa\frac{\partial^2 T}{\partial x^2} \tag{A.16}$$

Appendix B

Basics of Linear Algebra with Matlab

This appendix recalls some basic aspects of linear algebra, demonstrated with Matlab, which are necessary to understand the main text.

B.1 Matrices, Vectors, and Scalars

A matrix is an array of numbers that contains two or more columns. For example,

$$\begin{bmatrix} 2 & 5 \\ 3 & 1 \end{bmatrix}, \quad \begin{bmatrix} 1 & 5 & 6 \\ 8 & 2 & 2 \end{bmatrix} \text{ and } \begin{bmatrix} 1 & 5 \\ 8 & 2 \\ 3 & 4 \end{bmatrix} \tag{B.1}$$

are all matrices. If a matrix A has m rows and n columns, we say that A is a $m \times n$ matrix. Here, m and n refer to the dimensions of A. The terms or coefficients in a matrix can be referred to using the following notation:

$$A = \begin{bmatrix} a_{11} & a_{12} & a_{13} & \cdots & a_{1n} \\ a_{21} & a_{22} & a_{23} & \cdots & a_{2n} \\ \vdots & \vdots & \vdots & & \vdots \\ a_{m1} & a_{m2} & a_{m3} & \cdots & a_{mn} \end{bmatrix} \tag{B.2}$$

A matrix is easily constructed and manipulated with Matlab. For example, the last matrix appearing in B.1 can be formed with the snippet

```
matrix = [ 1  5 ; 8  2 ; 3  4]   ; % construct a matrix in matlab
```

The dimensions of this matrix are 3×2, which can be computed using Matlab with the command

```
dimen = size(matrix)  ; % compute dimensions of a matrix
```

For this matrix, matrix(2,2) = 2, matrix(3,2) = 4, and matrix(1,:) returns the values 1 and 5.

A vector is a matrix that contains only one column or row. For example,

$$\begin{bmatrix} 2 & 5 & 3 \end{bmatrix} \tag{B.3}$$

is a row vector, whereas

$$\begin{bmatrix} 2 \\ 3 \\ 1 \end{bmatrix} \tag{B.4}$$

Practical Finite Element Modeling in Earth Science Using Matlab, First Edition. Guy Simpson.

Companion website: www.wiley.com/go/simpson

is a column vector. These two vectors can be constructed in Matlab with the lines

```
RowVector = [ 2  5  3]     ;  % construct a row vector
ColVector = [ 2 ; 3 ; 1]   ;  % construct a column vector
```

A scalar can be regarded as a matrix with only one column and one row, so it's a single number, that is,

```
scalar = 4 ; % define a scalar
```

B.2 Matrix and Vector Operations

As with scalars, there are a number of basic operations that can be applied to modify matrices, notably matrix addition, scalar multiplication, transposition, and matrix multiplication.

B.2.1 Matrix Addition

Matrix addition is defined only for matrices of the same dimension, when they are said to be compatible. Matrix addition is performed term by term. Thus, if

$$A = \begin{bmatrix} 1 & 5 & 3 \\ 7 & 9 & 1 \end{bmatrix} \quad \text{and} \quad B = \begin{bmatrix} 2 & 6 & 2 \\ 8 & 2 & 3 \end{bmatrix} \tag{B.5}$$

then

$$C = A + B = \begin{bmatrix} 3 & 11 & 5 \\ 15 & 11 & 4 \end{bmatrix} \tag{B.6}$$

This calculation is easily performed in Matlab with the script

```
% matrix addition in matlab
A = [ 1  5  3 ; 7  9  1]  ;
B = [ 2  6  2 ; 8  2  3]  ;
C = A + B ;
```

B.2.2 Multiplication by a Scalar

Multiplication of matrix A with a scalar k is performed by multiplying each term in A by the scalar. For example, given

$$A = \begin{bmatrix} 1 & 3 \\ 7 & 2 \end{bmatrix} \quad \text{and} \quad k = 3 \tag{B.7}$$

then

$$B = kA = \begin{bmatrix} 3 & 9 \\ 21 & 6 \end{bmatrix} \tag{B.8}$$

which can be achieved in Matlab with the following snippet:

```
% scalar multiplication in matlab
A = [ 1  3 ; 7  2 ]  ; % matrix
k = 3 ;                % scalar
B = k*A ;
```

B.2.3 Transpose

The transpose of a matrix

$$A = \begin{bmatrix} 1 & 3 & 5 \\ 7 & 2 & 4 \end{bmatrix} \tag{B.9}$$

is found by exchanging its rows and columns, that is,

$$B = A^T = \begin{bmatrix} 1 & 7 \\ 3 & 2 \\ 5 & 4 \end{bmatrix} \tag{B.10}$$

In Matlab, this can be achieved with the lines

```
% transpose of a matrix
A = [ 1  3  5 ; 7  2  4]  ; % matrix A
B = A' ;                    % transpose of A
```

B.2.4 Scalar Product

The scalar (or dot) product of the two vectors

$$a = \begin{bmatrix} 1 & 5 & 2 \end{bmatrix} \quad \text{and} \quad b = \begin{bmatrix} 3 \\ 8 \\ 7 \end{bmatrix} \tag{B.11}$$

is given by

$$c = a \cdot b = (1 \times 3) + (5 \times 8) + (2 \times 7) = 52 \tag{B.12}$$

where it should be noted that the result is a scalar. This calculation can be reproduced in Matlab with the lines

```
% scalar (dot) product in matlab
a = [ 1   5  2 ]    ; % row vector
b = [ 3 ; 8  ; 7 ]  ; % column vector
c = a * b ;
```

Here, it is important that the number of columns in a equals the number of rows in b. If a and b have the same dimensions, then it is necessary to use the transpose function to ensure the dimensions are compatible. This is illustrated in the following snippet:

```
% scalar (dot) product in matlab
a = [ 1  5  2 ]   ; % row vector
b = [ 3  8  7 ]   ; % row vector
c = a * b' ; % scalar product using transpose of b
```

B.2.5 Matrix Multiplication

Matrix multiplication between two matrices A and B is defined only if the number of columns in A equals the number of rows in B. Matrix multiplication between A and B is performed by taking the scalar product between row i of A and column j of B. As an example, consider the following matrices:

$$A = \begin{bmatrix} 1 & 5 & 3 \\ 7 & 9 & 1 \end{bmatrix} \quad \text{and} \quad B = \begin{bmatrix} 2 & 6 \\ 8 & 2 \\ 4 & 8 \end{bmatrix} \tag{B.13}$$

First, note that because A contains three columns and B contains three rows, the multiplication AB is defined, and it will lead to a new matrix with dimensions 2×2 (i.e., the number of rows in $A \times$ the number of columns in B). The different terms of this new matrix C can be computed as follows:

$$c_{11} = \begin{bmatrix} 1 & 5 & 3 \end{bmatrix} \begin{bmatrix} 2 \\ 8 \\ 4 \end{bmatrix} = (1 \times 2) + (5 \times 8) + (3 \times 4) = 54 \tag{B.14}$$

$$c_{12} = \begin{bmatrix} 1 & 5 & 3 \end{bmatrix} \begin{bmatrix} 6 \\ 2 \\ 8 \end{bmatrix} = (1 \times 6) + (5 \times 2) + (3 \times 8) = 40 \tag{B.15}$$

$$c_{21} = \begin{bmatrix} 7 & 9 & 1 \end{bmatrix} \begin{bmatrix} 2 \\ 8 \\ 4 \end{bmatrix} = (7 \times 2) + (9 \times 8) + (1 \times 4) = 90 \tag{B.16}$$

$$c_{22} = \begin{bmatrix} 7 & 9 & 1 \end{bmatrix} \begin{bmatrix} 6 \\ 2 \\ 8 \end{bmatrix} = (7 \times 6) + (9 \times 2) + (1 \times 8) = 68 \tag{B.17}$$

Thus, the final result is the matrix

$$C = AB = \begin{bmatrix} 54 & 40 \\ 90 & 68 \end{bmatrix} \tag{B.18}$$

These steps are easily performed in Matlab with the following snippet:

```
% matrix multiplication in matlab
A = [ 1   5   3 ; 7   9   1 ]     ; % matrix
B = [ 2   6 ; 8   2 ; 4   8 ]     ; % matrix
C = A * B   ;
```

B.3 Matrix Operations versus Term-by-Term Operations

It is important not to confuse matrix operations with normal term-by-term operations performed between two matrices or vectors. For example, say one has the two vectors

$$a = \begin{bmatrix} 1 & 5 & 2 \end{bmatrix} \quad \text{and} \quad b = \begin{bmatrix} 3 & 8 & 7 \end{bmatrix} \tag{B.19}$$

and one wants to multiply each term in a with each term in b, which results in

$$c = ab = \begin{bmatrix} (1 \times 3) & (5 \times 8) & (2 \times 7) \end{bmatrix} = \begin{bmatrix} 3 & 40 & 14 \end{bmatrix} \tag{B.20}$$

This can be performed in Matlab with the snippet

```
% term-by-term multiplication of two vectors
a = [ 1    5   2 ]    ; % vector
b = [ 3    8   7 ]    ; % vector
c = a . * b   ;         % note use of the 'dot'
```

Here, it is important to note the dot that precedes the operator, which informs Matlab that the operation must be performed term by term. If one forgets this dot, Matlab will attempt to compute the scalar product of a and b, which will result in an error because the dimensions of a and b are incompatible (unless b is first transposed; see Section B.2.4). In general, to perform term-by-term operations, the dimensions of the matrices must be identical, whereas to perform matrix operations, the dimensions of the matrices must be compatible for the specific operation.

B.4 Simultaneous Equations

Finite element discretization of partial differential equations leads to systems of equations that must be solved simultaneously for the nodal unknowns. These equations take the form

$$\begin{aligned} k_{11}a_1 + k_{12}a_2 &= f_1 \\ k_{21}a_1 + k_{22}a_2 &= f_2 \end{aligned} \tag{B.21}$$

that can be rewritten as

$$Ka = f \tag{B.22}$$

with

$$K = \begin{bmatrix} k_{11} & k_{12} \\ k_{21} & k_{22} \end{bmatrix} \quad a = \begin{bmatrix} a_1 \\ a_2 \end{bmatrix} \quad f = \begin{bmatrix} f_1 \\ f_2 \end{bmatrix} \tag{B.23}$$

In the context of the finite element method, K is a matrix of known coefficients (commonly referred to as a stiffness matrix), f is a known column vector of loads, and a is the unknown column vector of nodal displacements. The solution of B.22 can be written as

$$a = K^{-1}f \tag{B.24}$$

where K^{-1} is the inverse of K.

Matlab provides a built-in function that computes the solution to a system of simultaneous equations that will be used extensively in the text. As an example of how to solve a small system of equations simultaneously using Matlab, consider the following equations:

$$\begin{aligned} 2a_1 + 2a_2 + a_3 &= 9 \\ 2a_1 - a_2 + 2a_3 &= 6 \\ a_1 - a_2 + 2a_3 &= 5 \end{aligned} \tag{B.25}$$

Using matrix form these equations can be rewritten as

$$K = \begin{bmatrix} 2 & 2 & 1 \\ 2 & -1 & 2 \\ 1 & -1 & 2 \end{bmatrix} \quad a = \begin{bmatrix} a_1 \\ a_2 \\ a_3 \end{bmatrix} \quad f = \begin{bmatrix} 9 \\ 6 \\ 5 \end{bmatrix} \tag{B.26}$$

which are in the form shown in Equation B.22. The solution of this system is

$$a = K^{-1}f = \begin{bmatrix} 1 \\ 2 \\ 3 \end{bmatrix} \tag{B.27}$$

which can be computed in Matlab with the following snippet:

```
% solution of a system of equations
K = [ 2  2  1  ; 2  -1  2 ;  1  -1  2 ]  ; % coefficient matrix
f = [  9  ; 6  ;  5 ]  ; % right hand side vector
a = K \ f   ; % solve system using 'backslash'
```

Note that the result is roughly the same as using `a = inv(K)* f` (see Equation B.24) although by using the backslash command (\), the solution is computed in a different (far more efficient) manner.

Appendix C

Comparison between Different Numerical Methods

The purpose of this section is to compare the solution of a single partial differential equation (PDE) using three different numerical methods: the finite element method (FEM), the finite difference method (FDM), and a spectral method. The intention is to show some of the differences and similarities in approach between the different methods rather than to compare their accuracy and speed. To illustrate these points, we will solve the parabolic equation

$$\frac{\partial u}{\partial t} = \frac{\partial^2 u}{\partial x^2} \tag{C.1}$$

with the initial conditions

$$u(x, t = 0) = 2 + \sin(x) + \sin(2x) \qquad \forall\, x \in [0; 2\pi] \tag{C.2}$$

and periodic boundary conditions, that is,

$$u(x = 0, t) = u(x = 2\pi, t) \tag{C.3}$$

The exact solution is

$$u = 2 + \sin(x)\, e^{-t} + \sin(2x)\, e^{-4t} \tag{C.4}$$

The FEM, FDM, and spectral method are different techniques for replacing continuous problems (i.e., Equation C.1) with discrete ones. Each technique leads to a system of algebraic equations that can then be solved using other numerical methods such as Gaussian elimination. In the following text, we list three Matlab scripts to solve the problem by each method.

There are two main approaches to make a continuous problem discrete (Strang, 1986). One way is to choose a finite number of *points* and to replace derivatives with differences, which is the idea behind the FDM. The other way is to approximate the exact solution using a finite number of *functions*. If these functions consist of a series of sines and cosines, then we are dealing with a spectral method. If the functions are piecewise polynomials defined over local regions, then we have a FEM. Thus, although the FDM is an approximation to the governing differential equation, both the FEM and spectral method are approximations to its solution.

C.1 Finite Difference Method

The FDM discretizes Equation C.1 by replacing derivatives (both temporal and spatial) with discrete (difference) approximations. These approximations come from the Taylor series, which can be written using finite difference notation as follows:

Practical Finite Element Modeling in Earth Science Using Matlab, First Edition. Guy Simpson.

Companion website: www.wiley.com/go/simpson

$$u_i^{n+1} \approx u_i^n + \Delta t \left[\frac{\partial u}{\partial t}\right]_i^n + \frac{\Delta t^2}{2}\left[\frac{\partial^2 u}{\partial t^2}\right]_i^n + O(\Delta t)^3 \tag{C.5}$$

The notation $O(\Delta t)^3$ means that as the distance Δt is made smaller, the error decreases as Δt^3. Equation C.5 suggests that the derivative $\partial u/\partial t$ in Equation C.1 can be estimated to first order (i.e., $O(\Delta t)$) as

$$\frac{\partial u}{\partial t} \approx \frac{u_i^{n+1} - u_i^n}{\Delta t} \tag{C.6}$$

where the superscripts n and $n+1$ refer to the time level (not powers of n). The Taylor series (in space) can also be used to approximate the second-order spatial derivative in Equation C.1, leading to

$$\begin{aligned}\frac{\partial^2 u}{\partial x^2} &= \frac{\partial}{\partial x}\left(\frac{\partial u}{\partial x}\right) \approx \frac{\frac{u_{i+1}^{n+1}-u_i^{n+1}}{\Delta x} - \frac{u_i^{n+1}-u_{i-1}^{n+1}}{\Delta x}}{\Delta x} \\ &= \frac{u_{i-1}^{n+1} - 2u_i^{n+1} + u_{i+1}^{n+1}}{\Delta x^2}\end{aligned} \tag{C.7}$$

which is accurate to $O(\Delta x^2)$. Note that we have chosen to discretize the spatial derivative at the $n+1$ time level, which implies an implicit time-stepping scheme (see Appendix E). Substituting these expressions into Equation C.1, we obtain the discrete equation

$$\frac{u_i^{n+1} - u_i^n}{\Delta t} = \frac{u_{i-1}^{n+1} - 2u_i^{n+1} + u_{i+1}^{n+1}}{\Delta x^2} \tag{C.8}$$

or using $s = \Delta t/\Delta x^2$,

$$u_i^{n+1} - u_i^n = s\,u_{i-1}^{n+1} - 2\,s\,u_i^{n+1} + s\,u_{i+1}^{n+1} \tag{C.9}$$

Rearranging such that the known quantities are on the right-hand side and the unknowns are on the left-hand side gives

$$-\,s\,u_{i-1}^{n+1} + (1+2s)u_i^{n+1} - s\,u_{i+1}^{n+1} = u_i^n \tag{C.10}$$

This equation applies to all internal points of the model domain (i.e., $i = 2, 3, \ldots, N-2, N-1$). At the boundaries, we need to apply the periodic conditions $u_1 = u_N$ where u_1 is the first node on the left side of the model domain and u_N is the last node on the right side. Thus, the first and last equations of C.10 need to be replaced, respectively, with

$$-\,su_N^{n+1} + (1+2s)u_1^{n+1} - su_2^{n+1} = u_1^n \tag{C.11}$$

and

$$-\,su_{N-1}^{n+1} + (1+2s)u_N^{n+1} - su_1^{n+1} = u_N^n \tag{C.12}$$

Rewriting Equations C.10, C.11, and C.12 in matrix form for a small domain consisting of $N = 5$ grid points gives

$$\mathbf{A}\mathrm{u} = \mathrm{b} \tag{C.13}$$

where **A** is the coefficient matrix

$$\mathbf{A} = \begin{pmatrix} (1+2s) & -s & 0 & 0 & -s \\ -s & (1+2s) & -s & 0 & 0 \\ 0 & -s & (1+2s) & -s & 0 \\ 0 & 0 & -s & (1+2s) & -s \\ -s & 0 & 0 & -s & (1+2s) \end{pmatrix} \tag{C.14}$$

u is a vector of the unknowns defined at discrete points

$$\mathbf{u} = \begin{pmatrix} u_1 \\ u_2 \\ u_3 \\ u_4 \\ u_5 \end{pmatrix}^{n+1} \tag{C.15}$$

and **b** is the vector containing the solution at the n time level

$$\mathbf{b} = \begin{pmatrix} u_1 \\ u_2 \\ u_3 \\ u_4 \\ u_5 \end{pmatrix}^{n} \tag{C.16}$$

Note that **A** is tridiagonal, except for the terms introduced by the periodic boundary conditions. The system of equations represented by Equation C.13 can now be solved using some standard direct or iterative method. The following script shows how the solution can be computed using Matlab:

```
%---------------------------------------------------------
% Program diffn1d_fdm.m
% Implicit solution of diffusion equation in 1d
% with periodic boundary conditions using
% the Finite Difference Method
%---------------------------------------------------------
clear
nx      = 100 ;                % number of space nodes
dx      = 2*pi/nx ;            % grid spacing
x       = [0:dx:2*pi-dx]' ;    % x-grid
dt      = 0.01 ;               % time increment
ntime   = 100  ;               % number of time steps
s       = dt/dx^2  ;           % multiplier
u       = 2+sin(x)+sin(2*x) ;  % initial condition

% Assemble coefficient matrix A
coeffs       = zeros(nx,3)   ;
coeffs(:,1) = -s ; coeffs(:,2) = 1+2*s ;  coeffs(:,3) = -s ;
A = diag(coeffs(1:nx-1,1),-1) + diag(coeffs(:,2),0) + diag(coeffs(1:nx-1,3),1)  ;
A(1,nx) = -s ; A(nx,1) = -s ; % Add terms for period boundaries

for n=1:ntime % time loop
  t      = dt*n ;  % time
  u      = A\u  ;  % solve linear system
  uexact = 2+exp(-t)*sin(x)+exp(-4*t)*sin(2*x); % exact solution
```

```
    plot(x,u,'o',x,uexact,'r') % plot
    xlabel('x'); ylabel('u(x,t)')
    title('Comparison between Finite Difference and Exact solutions')
    pause
end
%------------------------------------------------------------------
```

C.2 Finite Element Method

As mentioned before, the FEM is completely different from the FDM in that it assumes that the solution $u(x, t)$ can be represented by simple analytical functions. However, as we will see, both methods may lead to very similar systems of discrete equations. With the FEM, one normally begins by writing

$$u \simeq \sum_{i=1}^{n} N_i u_i = \mathbf{Nu} \tag{C.17}$$

where u is the exact solution, N_i are known analytical functions, and u_i are unknown values of the solution defined at certain positions in space and the summation is carried out of a certain number of nodes n in a local region (the finite element). Equation C.17 can be interpreted as an interpolation of the local nodal values u_i throughout the element. The functions N_i (called shape functions) are normally chosen to be low-order piecewise polynomials restricted to contiguous elements. Substitution of this approximation into the governing equation produces the residual equation

$$R = \frac{\partial}{\partial t}\mathbf{Nu} - \frac{\partial^2}{\partial x^2}\mathbf{Nu} \tag{C.18}$$

where R is the residual or error introduced due to the fact that the trial functions ($\mathbf{N}$) do not exactly describe the true variation in u. With the Galerkin FEM, the unknown nodal values $\mathbf{u}$ are determined by requiring that the integral of the weighted residual is zero over small elements (of width Δx), that is,

$$\int_0^{\Delta x} \mathbf{N}^T R\, dx = \int_0^{\Delta x} \mathbf{N}^T \left(\frac{\partial}{\partial t}\mathbf{Nu} - \frac{\partial^2}{\partial x^2}\mathbf{Nu} \right) dx = 0 \tag{C.19}$$

Using piecewise linear functions for N, applying integration by parts, and ignoring the resulting surface integral (see Appendix D) lead to the following system of equations for a single element:

$$\mathbf{MM}\frac{\partial}{\partial t}\mathbf{u} + \mathbf{KM}\ \mathbf{u} = \mathbf{0} \tag{C.20}$$

Here,

$$\mathbf{MM} = \int_0^{\Delta x} \mathbf{N}^T\mathbf{N}\, dx = \begin{bmatrix} \frac{\Delta x}{3} & \frac{\Delta x}{6} \\ \frac{\Delta x}{6} & \frac{\Delta x}{3} \end{bmatrix} \tag{C.21}$$

$$\mathbf{KM} = \int_0^{\Delta x} \frac{\partial \mathbf{N}^T}{\partial x}\frac{\partial \mathbf{N}}{\partial x} dx = \begin{bmatrix} \frac{1}{\Delta x} & -\frac{1}{\Delta x} \\ -\frac{1}{\Delta x} & \frac{1}{\Delta x} \end{bmatrix} \tag{C.22}$$

and

$$\mathbf{u} = \begin{Bmatrix} u_1 \\ u_2 \end{Bmatrix} \tag{C.23}$$

Normally the FEM is applied only in space, which still leaves the time derivative in Equation C.20. Replacing the time derivative with an implicit finite difference approximation and rearranging lead to

$$\left(\frac{\mathbf{MM}}{\Delta t}+\mathbf{KM}\right)\mathbf{u}^{n+1}=\frac{\mathbf{MM}}{\Delta t}\mathbf{u}^{n} \tag{C.24}$$

where Δt is the time interval between n and $n + 1$. To facilitate comparison with the FDM, we rewrite this last equation after dividing by Δx and multiplying by Δt as

$$\mathbf{A}^{e}\mathbf{u}^{n+1}=\mathbf{B}^{e}\,\mathbf{u}^{n}=\mathbf{b} \tag{C.25}$$

where

$$\mathbf{A}^{e}=\frac{\mathbf{MM}}{\Delta x}+\Delta t\,\frac{\mathbf{KM}}{\Delta x}=\begin{bmatrix}\frac{1}{3}+s & \frac{1}{6}-s\\ \frac{1}{6}-s & \frac{1}{3}+s\end{bmatrix} \tag{C.26}$$

$$\mathbf{B}^{e}=\frac{\mathbf{MM}}{\Delta x}=\begin{bmatrix}\frac{1}{3} & \frac{1}{6}\\ \frac{1}{6} & \frac{1}{3}\end{bmatrix} \tag{C.27}$$

and $s = \Delta t/\Delta x^2$, as also defined before. Here, the e superscripts are used to remind readers that the matrices refer to a single element. Consider now that we have four identical elements, each of which has two nodes. Thus, the entire mesh has five nodes. We require periodic boundary conditions, so the two end nodes (i.e., 1 and 5) are required to have the same value, which is itself unknown. This is accomplished by setting these equation numbers to the same value, and discarding the last one, leaving only four unknowns. The system of equations for this mesh can be obtained by summing the stiffness matrices for each element, node by node. This procedure (which is outlined in detail in Chapter 2) results in the system

$$\begin{bmatrix}\frac{2}{3}+2s & \frac{1}{6}-s & 0 & \frac{1}{6}-s\\ \frac{1}{6}-s & \frac{2}{3}+2s & \frac{1}{6}-s & 0\\ 0 & \frac{1}{6}-s & \frac{2}{3}+2s & \frac{1}{6}-s\\ \frac{1}{6}-s & 0 & \frac{1}{6}-s & \frac{2}{3}+2s\end{bmatrix}\begin{Bmatrix}u_1\\u_2\\u_3\\u_4\end{Bmatrix}^{n+1}=\begin{bmatrix}\frac{1}{3}+\frac{1}{3} & \frac{1}{6} & 0 & \frac{1}{6}\\ \frac{1}{6} & \frac{1}{3}+\frac{1}{3} & \frac{1}{6} & 0\\ 0 & \frac{1}{6} & \frac{1}{3}+\frac{1}{3} & \frac{1}{6}\\ \frac{1}{6} & 0 & \frac{1}{6} & \frac{1}{3}+\frac{1}{3}\end{bmatrix}\begin{Bmatrix}u_1\\u_2\\u_3\\u_4\end{Bmatrix}^{n}$$

which can be written compactly as

$$\mathbf{A}\mathbf{u}^{n+1}=\mathbf{B}\mathbf{u}^{n}=\mathbf{b} \tag{C.28}$$

Note that the equation for any internal node (i.e., lines two or three) can be written as

$$\left(\frac{1}{6}-s\right)u_{i-1}^{n+1}+\left(\frac{2}{3}+2s\right)u_{i}^{n+1}+\left(\frac{1}{6}-s\right)u_{i+1}^{n+1}=\frac{1}{6}u_{i-1}^{n}+\frac{2}{3}u_{i}^{n}+\frac{1}{6}u_{i+1}^{n} \tag{C.29}$$

which can be rearranged to give

$$\frac{1}{6}\left(u_{i-1}^{n+1}-u_{i-1}^{n}\right)+\frac{2}{3}\left(u_{i}^{n+1}-u_{i}^{n}\right)+\frac{1}{6}\left(u_{i+1}^{n+1}-u_{i+1}^{n}\right)=s\,u_{i-1}^{n+1}-2\,s\,u_{i}^{n+1}+s\,u_{i+1}^{n+1} \tag{C.30}$$

It is remarkable that this equation bears a strong resemblance to Equation C.10 derived using the FDM. In fact, the only difference is the spreading of the time derivative over adjacent nodes in the case of the FEM. The two formulations can be made identical if the mass matrix is lumped, as is sometimes done. This can be seen by summing each row of **MM** separately, placing each result on the diagonal and zeroing off-diagonal entries, resulting in

$$\mathbf{MM} = \begin{bmatrix} \frac{\Delta x}{2} & 0 \\ 0 & \frac{\Delta x}{2} \end{bmatrix} \tag{C.31}$$

Using this modified matrix for **MM** in Equation C.26 and repeating the steps leading to Equation C.30 lead to an equation which is the same as C.10.

The following is a script showing how the finite element solution (without mass lumping) is computed with Matlab:

```
%-------------------------------------------------------
% Program diffn1d_fem.m
% Implicit solution of diffusion equation in 1d
% with periodic boundary conditions using
% the Finite Element Method
%-------------------------------------------------------

clear

nx      = 100  ;                 % number of space nodes
sdof    = nx-1 ;                 % number of unknowns
nels    = nx-1 ;                 % number of finite elements
dx      = 2*pi/nx ;              % grid spacing
x       = [0:dx:2*pi-dx]' ;      % x-grid for nodes
dt      = 0.01 ;                 % time increment
ntime   = 100  ;                 % number of time steps
nod     = 2    ;                 % number of nodes per element
u       = 2+sin(x)+sin(2*x)  ;  % initial condition

% Assemble coefficient matrices
lhs = zeros(sdof,sdof) ;
rhs = zeros(sdof,sdof) ;
  for iel=1:nels
    num = [iel ; iel+1]  ; % equation numbers
    if iel == nels % periodic boundary condition
      num = [iel ; 1] ;
    end
    MM   = dx*[1/3  1/6 ; 1/6 1/3 ] ;   % mass matrix
    KM   = 1/dx*[1 -1 ; -1 1 ]   ;       % diffn matrix
    lhs(num,num) = lhs(num,num) + MM/dt + KM ; % add coefficients
    rhs(num,num) = rhs(num,num) + MM/dt    ;    % add coefficients
  end

um = u(1:sdof) ;
for n=1:ntime % time loop
  t       = dt*n   ;  % time
  b       = rhs*um ; % compute rhs
  um      = lhs\b ;  % solve linear system
  u       = [um ; um(1) ] ; % reconstruct solution vector
  uexact  = 2+exp(-t)*sin(x)+exp(-4*t)*sin(2*x);  % exact solution
  plot(x,u,'o',x,uexact,'r') % plot
  xlabel('x'); ylabel('u(x,t)')
  title('Comparison between Finite Element and Exact solutions')
  pause
end

%-------------------------------------------------------
```

C.3 Spectral Methods

Spectral methods are closely related to the FEM in that the approach is to approximate the solution to the governing differential equation using analytical functions. However, although the FEM normally uses low-order piecewise functions that are nonzero only within local regions (elements), spectral methods use finite series such as

$$u(x,t) = \sum_{k=0}^{N} c_k(t)\phi_k(x) \tag{C.32}$$

where $\phi_k(x)$ is a known analytical (basis) function and $c_k(t)$ are coefficients to be determined. Note the close similarity between this and Equation C.17. However, in the case of a spectral method, the approximate solution can be regarded as global because it "touches" the entire spatial domain, rather than being defined only locally in small elements. The choice of the basis function depends on the problem being investigated. For our problem that is periodic, a good choice is to take trigonometric functions (e.g., Fourier series)

$$u(x,t) = \frac{1}{2}a_o + \sum_{k=0}^{N-1} a_k(t)\cos(kx) + \sum_{k=0}^{N-1} b_k(t)\sin(kx) \tag{C.33}$$

or in complex form

$$u(x,t) = \sum_{k=0}^{N-1} c_k(t)e^{ikx} \tag{C.34}$$

where $i = \sqrt{-1}$, k is the wave number, a_k, b_k, and c_k are unknown wave amplitudes (Fourier coefficients), and N is the number of terms (waves or harmonics) used in the approximation. Substituting our approximation for $u(x, t)$ (Equation C.34) into Equation C.1 leads to

$$\sum_{k=0}^{N-1} \frac{\partial c_k(t)}{\partial t} e^{ikx} = \sum_{k=0}^{N-1} -|k|^2 c_k(t)e^{ikx} \tag{C.35}$$

which can be simplified (because the wave frequencies are uncoupled) to

$$\frac{\partial c_k(t)}{\partial t} = -|k|^2 c_k(t) \tag{C.36}$$

An important point here is that the Fourier series has been differentiated in space exactly, so that the original (PDE) in space and time has been converted to an ordinary differential equation in time with the Fourier coefficients as unknowns. Taking an implicit finite difference approximation for the time derivative in Equation C.36 leads to

$$\frac{c_k^{n+1} - c_k^n}{\Delta t} = -|k|^2 c_k^{n+1} \tag{C.37}$$

that can be rearranged to give

$$c_k^{n+1} = \frac{|k|^2 c_k^n}{1 + \Delta t|k|^2} \tag{C.38}$$

The coefficients c_k are a representation of $u(x, t)$ in the frequency domain. They can be computed from

$$c_k = \frac{1}{N} \sum_{k=0}^{N-1} u(x, t)\, e^{ikx} \tag{C.39}$$

which is known as the discrete Fourier transform and can be computed efficiently using the fast Fourier transform (FFT). Equation C.39 is closely related to the original approximation (i.e., Equation C.34), which is the inverse discrete Fourier transform that can be computed using the inverse FFT (IFFT).

To summarize, the following steps must be performed to compute a spectral solution to Equation C.1:

1) Transform the initial conditions (Equation C.2) to the frequency domain using the discrete Fourier transform (i.e., Equation C.39 using the FFT) to obtain the initial Fourier coefficients ($c(0)_k$), that is,

$$c(0)_k = \frac{1}{N} \sum_{k=0}^{N-1} u(x, t = 0)\, e^{ikx} \tag{C.40}$$

2) In a loop, advance the Fourier coefficients c_k in time using Equation C.38.
3) If one wants to plot the solution in the physical (as opposed to the frequency) domain, apply the inverse discrete Fourier transform given by Equation C.34 using the now known Fourier coefficients (using the IFFT).

The following is a script showing how the solution can be computed with Matlab:

```
%------------------------------------------------------------
% Program diffn1d_spectral.m
% Implicit solution of diffusion equation in 1d
% with periodic boundary conditions using
% a Fourier Spectral Method
%------------------------------------------------------------
clear
nk      = 100 ;                % number of waves
dx      = 2*pi/nk ;            % grid spacing
x       = 0:dx:2*pi-dx ;       % x-grid
dt      = 0.01 ;               % time increment
ntime   = 100 ;                % number of time steps
nv      = fftshift(-nk/2:1:nk/2-1); % vector of wave numbers
mu      = 1./(1+nv.^2*dt)   ;       % multiplier
u       = 2+sin(x)+sin(2*x) ;       % initial condition
ck      = fft(u) ; % transform initial condition to freq. domain

for n=1:ntime       %time loop
  t   = dt*n    ;   % time
  ck = mu.*ck ;     % update Fourier coefficients
  u   = ifft(ck); % transform back to physical domain (for plotting only)
  uexact = 2+exp(-t)*sin(x)+exp(-4*t)*sin(2*x); % exact solution
  plot(x,u,'o',x,uexact,'r') % plot
  xlabel('x'); ylabel('u(x,t)')
  title('Comparison between Spectral and Exact solutions')
  pause
end
%------------------------------------------------------------
```

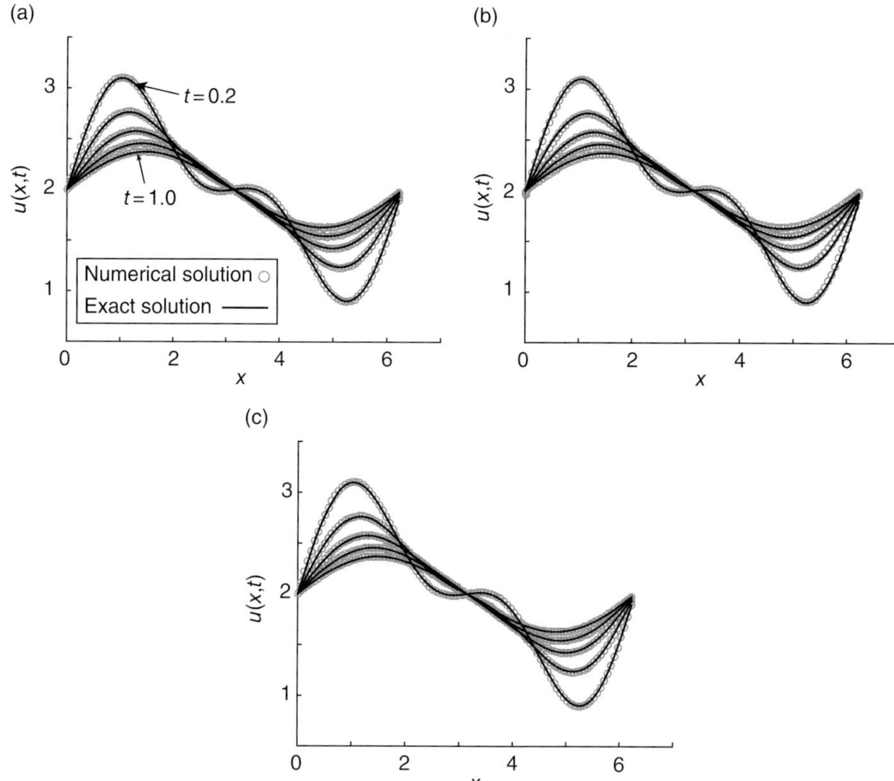

Figure C.1 Comparison between the exact solution to Equation C.1 (with C.2 and C.3) along with the numerical solutions computed at $t = 0.2, 0.4, \ldots, 1.0$ using the FDM (a), FEM (b), and a fourier spectral method (c). All results were calculated using the Matlab scripts listed.

Figure C.1 shows output computed using the three different numerical methods along with the exact solution. Despite the distinctly different approaches of the various methods, each produces results in good agreement with the exact solution.

Appendix D

Integration by Parts

Suppose $u(x)$ and $v(x)$ are two continuously differentiable functions. Then using the product rule, one can write

$$\frac{\partial}{\partial x}u(x)v(x) = u(x)\frac{\partial}{\partial x}v(x) + v(x)\frac{\partial}{\partial x}u(x) = u\,v' + v\,u' \tag{D.1}$$

Integrating both sides with respect to x gives

$$\int \frac{\partial}{\partial x}u\,v\,dx = \int u\,v'dx + \int v\,u'dx \tag{D.2}$$

which simplifies to

$$u\,v = \int u\,v'\,dx + \int v\,u'\,dx \tag{D.3}$$

Introducing $du = u'dx$ and $dv = v'dx$ and rearranging leads to the following formula that forms the basis for integration by parts:

$$\int u\,dv = u\,v - \int v\,du \tag{D.4}$$

For definite integrals, the rule can be written as follows:

$$\int_a^b u\,dv = [u\,v]_a^b - \int_a^b v\,du \tag{D.5}$$

Here, the notation $[u\,v]_a^b$ implies $(u\,v)_{x=b} - (u\,v)_{x=a}$.

To show how these formulas can be used, say we want to evaluate a definite integral that is the product of two functions $f(x)$ and $g(x)$, that is,

$$\int_a^b f(x)g(x)dx \tag{D.6}$$

To compute this integral using integration by parts, begin by introducing two intermediary functions $u(x)$ and $v(x)$ as

$$\begin{cases} u = f(x) & \text{(D.7)} \\ dv = g(x)dx & \text{(D.8)} \end{cases}$$

so that

$$\int_a^b f(x)g(x)dx = \int_a^b u\,dv \tag{D.9}$$

Practical Finite Element Modeling in Earth Science Using Matlab, First Edition. Guy Simpson.

Companion website: www.wiley.com/go/simpson

Now take the derivative of D.7 and the integral of D.8 to obtain the following:

$$\begin{cases} du = f'(x)dx & \text{(D.10)} \\ v = \int g(x)dx & \text{(D.11)} \end{cases}$$

Finally, evaluate the integral using the integration by parts formula, that is,

$$\int_a^b u(x)dv = [u(x)v(x)]_a^b - \int_a^b v(x)du \tag{D.12}$$

Integration by parts is commonly used during the evaluation of certain element matrices resulting from discretization with the finite element method (FEM). As an example, consider the integral

$$\int_0^L N(x)\frac{\partial^2 N(x)}{\partial x^2}dx \tag{D.13}$$

which arises when the governing partial differential equation contains a second-order spatial derivative. In most cases, the FEM used low-order functions for $N(x)$. However, if one assumes that $N(x)$ is linear (as often the case), then the term involving double differentiation in D.13 would give zero, making the integral also zero. This would be unrealistic since the influence of the second-order derivative would disappear completely. This problem can be avoided by applying integration by parts. First, note that D.13 has the same form as Equation D.6. We will evaluate this integral using integration by parts by choosing

$$\begin{cases} u = N(x) & \text{(D.14)} \\ dv = \dfrac{\partial^2 N(x)}{\partial x^2}dx & \text{(D.15)} \end{cases}$$

Taking the derivative of the former and the integral of the latter gives

$$\begin{cases} \dfrac{du}{dx} = \dfrac{\partial N(x)}{\partial x} \qquad \Rightarrow du = \dfrac{\partial N(x)}{\partial x}dx & \text{(D.16)} \\ v = \dfrac{\partial N(x)}{\partial x} & \text{(D.17)} \end{cases}$$

Introducing these definitions into the integration by parts formula (Equation D.5) gives the final result

$$\int_0^L N(x)\frac{\partial^2 N(x)}{\partial x^2}dx = \left[N(x)\frac{\partial N(x)}{\partial x}\right]_0^L - \int_0^L \frac{\partial N(x)}{\partial x}\frac{\partial N(x)}{\partial x}dx \tag{D.18}$$

Note that integration by parts converts the integral involving a second-order derivative into a boundary term and an integral involving only first-order derivatives. This latter term will be nonzero even when $N(x)$ is linear, thus avoiding the aforementioned problem.

Appendix E

Time Discretization

When the finite element method (FEM) is used to discretize partial differential equations such as

$$\frac{\partial T}{\partial t} = \kappa \frac{\partial^2 T}{\partial x^2} + H \tag{E.1}$$

it results in ordinary differential equations of the form

$$[\mathbf{MM}]\frac{\partial}{\partial t}\mathbf{T} + [\mathbf{KM}]\ \mathbf{T} = \mathbf{F} \tag{E.2}$$

Here, **MM**, **KM**, and **F** are known element matrices and **T** is the vector of nodal unknowns that approximates $T(x, t)$ in the original equation. Before Equation E.2 can be solved, the time derivative needs to be replaced with some discrete approximation. One common choice is to take

$$[\mathbf{MM}]\frac{\mathbf{T}^{n+1} - \mathbf{T}^{n}}{\Delta t} + [\mathbf{KM}]\ \mathbf{T}^{n} = \mathbf{F} \tag{E.3}$$

which is known as the forward Euler method. Rearranging Equation E.3 gives

$$[\mathbf{MM}]\ \mathbf{T}^{n+1} = [\mathbf{MM} - \Delta t\mathbf{KM}]\ \mathbf{T}^{n} + \Delta t\mathbf{F} \tag{E.4}$$

Note that although the forward Euler method can be regarded as an explicit scheme, the left-hand side of E.3 still contains a matrix (i.e., **MM**), rather than a single unknown, as would be the case if spatial and temporal discretization was carried out entirely with the finite difference method (FDM). Thus, in order to obtain $\mathbf{T}^{n+1}$, the matrix **MM** needs to be inverted, that is,

$$\mathbf{T}^{n+1} = [\mathbf{MM}]^{-1}\left([\mathbf{MM} - \Delta t\mathbf{KM}]\ \mathbf{T}^{n} + \Delta t\mathbf{F}\right) \tag{E.5}$$

In other words, a system of equations needs to be solved simultaneously. Explicit FEM schemes are therefore less advantageous (more expensive) than explicit schemes based on the FDM. An approach that retrieves the benefits of FDM explicit schemes is to replace **MM** with some simplified representation ($\mathbf{MM}^L$) that contains only diagonal entries, known as mass lumping. For example, one could sum the columns of **MM** separately, put each result on the diagonal, and zero-out all off-diagonal entries (known as the row sum method). In this case, the matrix inversion appearing in E.4 can be replaced with standard division, which is far more efficient. This scheme might be advantageous for very nonlinear problems, though it comes with loss of accuracy.

A drawback of the forward Euler method (with or without mass lumping) is that the scheme is only stable for certain conditions, notably, if Δt is sufficiently small. An alternative time-stepping scheme that is vastly superior in terms of stability is the backward Euler method, which approximates E.2 as

$$[\mathbf{MM}]\frac{\mathbf{T}^{n+1} - \mathbf{T}^{n}}{\Delta t} + [\mathbf{KM}]\ \mathbf{T}^{n+1} = \mathbf{F} \tag{E.6}$$

Practical Finite Element Modeling in Earth Science Using Matlab, First Edition. Guy Simpson.

Companion website: www.wiley.com/go/simpson

In this case, rearranging leads to

$$[\mathbf{MM} + \Delta t\mathbf{KM}]\,\mathbf{T}^{n+1} = [\mathbf{MM}]\,\mathbf{T}^{n} + \Delta t\mathbf{F} \tag{E.7}$$

This system of equations is fully implicit, requiring the inversion of the matrix (bracketed term) appearing on the left-hand side. The backward Euler method is also unconditionally stable, meaning that there is no stability limit to the magnitude of Δt. Remember, however, that the accuracy of the numerical approximation deteriorates as Δt is increased, even if the method is stable.

The forward and backward Euler schemes can be combined by writing

$$[\mathbf{MM}]\,\frac{\mathbf{T}^{n+1} - \mathbf{T}^{n}}{\Delta t} + \theta[\mathbf{KM}]\,\mathbf{T}^{n+1} + (1-\theta)[\mathbf{KM}]\,\mathbf{T}^{n} = \theta\mathbf{F}^{n+1} + (1-\theta)\mathbf{F}^{n} \tag{E.8}$$

where θ is a numerical parameter ($0 \geq \theta \leq 1$) that described the degree of implicitness. Note that in E.8, we have also considered the possibility that **F** varies in time. Using this approach, the scheme is fully explicit when $\theta = 0$, fully implicit $\theta = 1$, while the choice $\theta = 0.5$ is known as the Crank–Nicholson method. Rearranging E.8 leads to

$$[\mathbf{MM} + \theta\Delta t\mathbf{KM}]\,\mathbf{T}^{n+1} = [\mathbf{MM} - (1-\theta)\Delta t\mathbf{KM}]\,\mathbf{T}^{n} + \theta\Delta t\mathbf{F}^{n+1} + (1-\theta)\Delta t\mathbf{F}^{n} \tag{E.9}$$

which is the system that must be solved to advance the solution between the n and $n+1$ time levels. This time-stepping scheme, along with the possibility for mass-lumping, is implemented in the Matlab program listed in Section 1.6.

References

M. Abramowitz, and I. A. Stegun, *Handbook of Mathematical Functions with Formulas, Graphs, and Mathematical Tables*, Applied Mathematics Series 55, United States Department of Commerce, National Bureau of Standards (NBS), New York, 1983.

T. Belytschko, Y. Y. Lu, and L. Gu, Element-free Galerkin methods, *International Journal for Numerical Methods in Engineering,* 37, 229–256, 1994.

P. B. Bochev, and M. D. Gunzburger, *Least Squares Finite Element Methods,* Springer, Berlin, 2009.

H. S. Carslaw, and J. C. Jaeger, *Conduction of Heat in Solids*, Oxford University Press, Oxford, 1959.

L. Chen, Programming of Finite Element Methods in Matlab, http://www.math.uci.edu/~chenlong/226/Ch3FEMCode.pdf, 2016 (accessed October 7, 2016).

M. Dabrowski, M. Krotkiewski, and D. Schmid, MILAMIN: MATLAB-based finite element method solver for large problems, *Geochemistry, Geophysics, Geosystems,* 9, 1–24, 2008.

T. A. Davis, Creating Sparse Finite-Element Matrices in MATLAB, http://blogs.mathworks.com/loren/2007/03/01/creating-sparse-finite-element-matrices-in-matlab/, 2007 (accessed October 7, 2016).

T. A. Davis, and Y. Hu, The University of Florida Sparse Matrix Collection, *ACM Transactions on Mathematical Software,* 38, 1–25, 2011.

J. J. Dongarra, J. Du Cros, S. Hammarling, and I. Duff, A set of level 3 basic linear algebra subprograms, *ACM Transations on Mathematical Software*, 16, 1–17, 1990.

B. A. Finlayson, *The Method of Weighted Residuals and Variational Principles,* Mathematics in Science and Engineering, 87, Edited by R. Bellman. Academic Press, New York, 1972, 412 pp.

P. Garabedian, *Partial Differential Equations*, John Wiley & Sons, Inc., New York, 1964.

A. Gierer, and H. Meinhard, A theory of biological pattern formation, *Biological Cybernetics*, 12, 30–39, 1972.

E. N. Houstis, R. E. Lynch, J. R. Rice, and T.S. Papatheodorou, Evaluation of numerical methods for elliptic partial differential equations, *Journal of Computational Physics,* 27, 323–350, 1978.

T. J. R. Hughes, *The Finite Element Method*, Dover Publications, Inc., New York, 2000.

T. J. R. Hughes, L. P. Franca, and G. M. Hilbert, A new finite element formulation for computational fluid dynamics: VIII. The Galerkin/least squares method for advective-diffusive equations, *Computer Methods in Applied Mechanics and Engineering*, 73, 173–189, 1989.

S. E. Ingebritsen, W. E. Sanford, and C. E. Neuzil, *Groundwater in Geologic Processes*, Cambridge University Press, Cambridge, 2006.

J. C. Jaeger, N. G. W. Cook, and R. W. Zimmerman, *Fundamentals of Rock Mechanics*, Blackwell Publishing, Oxford, 2007.

A. R. Khoei, *Extended Finite Element Method*, John Wiley & Sons, Ltd, Chichester, 2015.

Y. W. Kwong, and H. C. Bang, *The Finite Element Method Using Matlab*, CRC Press, New York, 2000.

Practical Finite Element Modeling in Earth Science Using Matlab, First Edition. Guy Simpson.

Companion website: www.wiley.com/go/simpson

R. W. Lewis, and B. A. Schrefler, *The Finite Element Method in the Static and Dynamic Deformation and Consolidation of Porous Media*, John Wiley & Sons, Ltd, Chichester, 1998.

R. W. Lewis, P. Nithiarasu, and K. N. Seetharamu, *Fundamentals of the Finite Element Method for Heat and Fluid Flow*, John Wiley & Sons, Ltd, Chichester, 2004.

B. Q. Li, *Discontinuous Finite Element Methods in Fluid Dynamics and Heat Transfer*, Springer, London, 2006.

G.-R. Liu, *Mesh Free Methods: Moving Beyond the Finite Element Method*, CRC Press, London, 2003.

P. K. Maini, T. E. Woolley, R. E. Baker, E. A. Gaffney, and S. S. Lee, Turing's model for biological pattern formation and the robustness problem, *Interface Focus*, 2, 487–496, 2012.

D. McKenzie, and J. N. Brune, Melting on fault planes during large earthquakes, *Geophysical Journal of the Royal Astronomical Society*, 29, 65–78, 1972.

N. Moës, J. Dolbow, and T. Belytschenko, A Finite Element Method for crack growth without remeshing, *International Journal for Numerical Methods in Engineering*, 46, 131–150, 1999.

P. M. Morse, and H. Feshbach, *Methods of Theoretical Physics*, McGraw-Hill, New York, 1953.

D. W. Peaceman, and H. H. Rachford, Numerical calculation of multidimensional miscible displacement, *Society of Petroleum Engineers Journal*, 2, 327–339, 1962.

O. M. Phillips, *Flow and Reactions in Permeable Rocks*, Cambridge University Press, Cambridge, 1991.

A. J. Piwinskii, and P. J. Wyllie, Experimental studies of igneous rock series—a zoned pluton in Wallowa batholith Oregon, *Journal of Geology*, 76, 205–234, 1968.

D. D. Pollard, and R. C. Fletcher, *Fundamentals of Structural Geology*, Cambridge University Press, Cambridge, 2005.

J. G. Ramsay, and M. I. Huber, *The Techniques of Modern Structural Geology*, Academic Press, London, 1993.

J. Schnakenberg, Simple chemical reaction systems with limit cycle behaviour, *Journal of Theoretical Biology*, 81, 389–400, 1979.

T. R. Smith, and F. P. Bretherton, Stability and the conservation of mass in drainage basin evolution, *Water Resources Research*, 8, 1506–1529, 1972.

G. Strang, *Introduction to Applied Mathematics*, Wellesley-Cambridge Press, Cambridge, MA, 1986.

G. D. H. Simpson, and F. Schlunegger, Topographic evolution and morphology of surfaces evolving in response to coupled fluvial and hillslope sediment transport, *Journal of Geophysical Research*, 108, 205–234, 2003.

G. D. H. Simpson, Modelling interactions between fold-thrust belt deformation, foreland flexure and surface mass transport, *Basin Research*, 18, 125–143, 2006, doi: 10.1111/j.1365-2117.2006.00287.x.

G. D. H. Simpson, Mechanical modelling of folding versus faulting in brittle-ductile wedges, *Journal of Structural Geology*, 31, 369–381, 2009.

G. D. H. Simpson, and S. Castelltort, Coupled model of surface water flow, sediment transport and morphological evolution, *Computers and Geosciences*, 32, 1600–1614, 2006.

I. M. Smith, and D. V. Griffiths *Programming the Finite Element Method*, John Wiley & Sons, Inc., New York, 1998.

I. C. Taig, Structural analysis by the matrix displacement method, *English Electric Aviation Report*, SO17, The English Electric Company Limited, Strand, London, 1961.

S. Timoshenko, and S. Woinowsky-Krieger, *Theory of Plates and Shells*, McGraw-Hill Book Company, New York, 1959.

D. L. Turcotte, and G. Schubert, *Geodynamics*, John Wiley & Sons, Inc., New York, 1982.

P. A. Vermeer, and R. DeBorst, Non-associated plasticity for soils, concrete and rock, *HERON*, 29, 1984.

J. Walder, and A. Nur, Porosity reduction and crustal pore pressure development, *Journal of Geophysical Research*, 89, 11539–11548, 1984.

A. B. Watts, *Isostasy and Flexure of the Lithosphere*, Cambridge University Press, Cambridge, 2001.

S.-Y. Yang, and J.-L. Liu, Least squares finite element methods for the elasticity problem, *Journal of Computational and Applied Mathematics*, 87, 39–60, 1997.

O. C. Zienkiewicz, and R. L. Taylor, *The Finite Element Method, Volume 1, The Basis*, Butterworth-Heinemann, Oxford, 2000a.

O. C. Zienkiewicz, and R. L. Taylor, *The Finite Element Method, Volume 2, Solid Mechanics*, Butterworth-Heinemann, Oxford, 2000b.

O. C. Zienkiewicz, and R. L. Taylor, *The Finite Element Method, Volume 3, Fluid Mechanics*, Butterworth-Heinemann, Oxford, 2000c.

Index

Practical Finite Element Modeling in Earth Science Using Matlab, First Edition. Guy Simpson.

Companion website: www.wiley.com/go/simpson

n

o

p

r

s